FROM GSM TO
LTE-ADVANCED

FROM GSM TO LTE-ADVANCED

AN INTRODUCTION TO MOBILE NETWORKS AND MOBILE BROADBAND

Revised 2nd Edition

Martin Sauter
WirelessMoves, Germany

This edition first published 2014
© 2014 John Wiley & Sons, Ltd

Registered office
John Wiley & Sons Ltd, The Atrium, Southern Gate, Chichester, West Sussex, PO19 8SQ, United Kingdom

For details of our global editorial offices, for customer services and for information about how to apply for permission to reuse the copyright material in this book please see our website at www.wiley.com.

Library of Congress Cataloging-in-Publication Data

Sauter, Martin.
  [From GSM to LTE]
  From GSM to LTE-advanced : an introduction to mobile networks and mobile broadband / Martin Sauter. – Revised 2nd edition.
      pages cm
  Includes bibliographical references and index.
  ISBN 978-1-118-86195-0 (cloth)
  1. Mobile communication systems. 2. Wireless metropolitan area networks. 3. Wireless LANs. I. Title.
  TK5103.2.S28 2014
  621.3845'6 – dc23

                                                                                    2014016545

A catalogue record for this book is available from the British Library.
Print ISBN: 9781118861950

Set in 10/12pt, TimeLTStd by Laserwords Private Limited, Chennai, India

Contents

Preface

Wireless technologies like GSM, UMTS, LTE, Wireless LAN and Bluetooth have revolutionized the way we communicate and exchange data by making services like telephony and Internet access available anytime and from almost anywhere. Today, a great variety of technical publications offer background information about these technologies but they all fall short in one way or another. Books covering these technologies usually describe only one of the systems in detail and are generally too complex as a first introduction. The Internet is also a good source, but the articles one finds are usually too short and superficial or only deal with a specific mechanism of one of the systems. For this reason, it was difficult for me to recommend a single publication to students in my telecommunication classes, which I have been teaching in addition to my work in the wireless telecommunication industry. This book aims to change this.

Each of the six chapters in this book gives a detailed introduction and overview of one of the wireless systems mentioned above. Special emphasis has also been put into explaining the thoughts and reasoning behind the development of each system. Not only the 'how' but also the 'why' is of central importance in each chapter. Furthermore, comparisons are made to show the differences and commonalities between the technologies. For some applications, several technologies compete directly with each other, while in other cases only a combination of different wireless technologies creates a practical application for the end user. For readers who want to test their understanding of a system, each chapter concludes with a list of questions. For further investigation, all chapters contain references to the relevant standards and other documents. These provide an ideal additional source to find out more about a specific system or topic. Please note that there is a companion website for this book. Please go to http://www.wirelessmoves.com.

While working on the book, I have gained tremendous benefit from wireless technologies that are already available today. Whether at home or while traveling, Wireless LAN, Bluetooth, UMTS and LTE have provided reliable connectivity for my research and have allowed me to communicate with friends and loved ones at anytime, from anywhere. In a way, the book is a child of the technologies it describes.

Many people have been involved in revising the different chapters and have given invaluable suggestions on content, style and grammar. I would therefore like to thank Prashant John, Timothy Longman, Tim Smith, Peter van den Broek, Prem Jayaraj, Kevin Wriston, Greg Beyer, Ed Illidge, Debby Maxwell and John Edwards for their kind help and good advice.

Furthermore, my sincere thanks go to Berenike, who has stood by me during this project with her love, friendship and good advice.

Readers familiar with previous editions of this book will find many updates in this revision. In Chapter 1, additional information has been included on the 3GPP Release 4 Mobile Switching Center architecture that is now used in most networks. In Chapter 2, only few updates were necessary because the deployed feature sets of GPRS and EDGE networks have remained stable in recent years. Chapter 3 was significantly enhanced as High-Speed Packet Access (HSPA) features such as higher order modulation, dual carrier operation and enhanced mobility management states are now in widespread use. While only a few LTE networks were in operation at the publication of the previous edition, the technology has since spread and significantly matured. Chapter 4 was therefore extended to describe Circuit-Switched Fallback (CSFB) for voice telephony in more detail. In addition, a section on Voice over LTE (VoLTE) was added to give a solid introduction to standardized voice over IP telephony in LTE networks. Furthermore, a description of LTE-Advanced features was added at the end of the chapter. As the global success of LTE has significantly reduced the importance of WiMAX, the chapter on this technology was removed from this revised edition. In Chapter 5, on Wi-Fi, a new section was added on the new 802.11ac air interface. Also, a new section was added to describe the Wi-Fi-Protected Setup (WPS) mechanism that is part of commercial products today. Finally, the chapter on Bluetooth has also seen some changes as some applications such as dial-up networking have been replaced by other technologies such as Wi-Fi tethering. Bluetooth has become popular for other uses, however, such as for connecting keyboards to smartphones and tablets. This chapter has therefore been extended to cover these developments.

Martin Sauter
Cologne
January 2014

1

Global System for Mobile Communications (GSM)

At the beginning of the 1990s, GSM, the Global System for Mobile Communications triggered an unprecedented change in the way people communicate with each other. While earlier analog wireless systems were used by only a few people, GSM is used by over 5 billion subscribers worldwide in 2014. This has mostly been achieved by the steady improvements in all areas of telecommunication technology and the resulting steady price reductions for both infrastructure equipment and mobile devices. This chapter discusses the architecture of this system, which also forms the basis for the packet-switched extension called General Packet Radio Service (GPRS), discussed in Chapter 2, for the Universal Mobile Telecommunications System (UMTS), which is described in Chapter 3 and Long-Term Evolution (LTE), which is discussed in Chapter 4. While the first designs of GSM date back to the middle of the 1980s, GSM is still the most widely used wireless technology worldwide and it is not expected to change any time soon. Despite its age and the evolution toward UMTS and LTE, GSM itself continues to be developed. As shown in this chapter, GSM has been enhanced with many new features in recent years. Therefore, many operators continue to invest in their GSM networks in addition to their UMTS and LTE activities to introduce new functionality and to lower their operational cost.

In addition, it should be mentioned at this point that the industry has standardized on a new solution for voice telephony for LTE that has only little in common with GSM anymore. Although standardization is complete, efforts to roll out the new system are significant, and at the time of writing, there were only few voice-over LTE systems to be found in practice. Current LTE-capable devices thus continue using GSM and UMTS networks for voice telephony with a fallback mechanism.

1.1 Circuit-Switched Data Transmission

Initially, GSM was designed as a circuit-switched system that establishes a direct and exclusive connection between two users on every interface between all network nodes of the system. Section 1.1.1 gives a first overview of this traditional architecture. Over time, this physical

From GSM to LTE-Advanced: An Introduction to Mobile Networks and Mobile Broadband,
Revised Second Edition. Martin Sauter.
© 2014 John Wiley & Sons, Ltd. Published 2014 by John Wiley & Sons, Ltd.

circuit switching has been virtualized and many network nodes are connected over IP-based broadband connections today. The reasons for this and further details on virtual circuit switching can be found in Section 1.1.2.

1.1.1 Classic Circuit Switching

The GSM mobile telecommunication network has been designed as a circuit-switched network in a similar way to fixed-line phone networks. At the beginning of a call, the network establishes a direct connection between two parties, which is then used exclusively for this conversation. As shown in Figure 1.1, the switching center uses a switching matrix to connect any originating party to any destination party. Once the connection has been established, the conversation is then transparently transmitted via the switching matrix between the two parties. The switching center only becomes active again to clear the connection in the switching matrix if one of the parties wants to end the call. This approach is identical in both mobile and fixed-line networks. Early fixed-line telecommunication networks were designed only for voice communication, for which an analog connection between the parties was established. In the mid-1980s, analog technology was superseded by digital technology in the switching center. This meant that calls were no longer sent over an analog line from the originator to the terminator. Instead, the switching center digitized the analog signal that it received from the subscribers, which were directly attached to it, and forwarded the digitized signal to the terminating switching center. There, the digital signal was again converted back to an analog

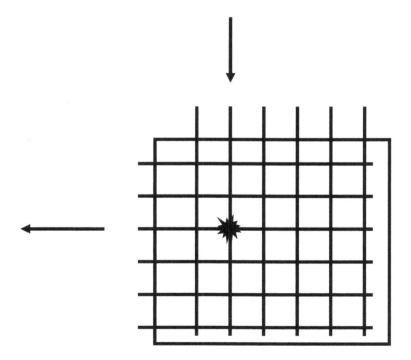

Figure 1.1 Switching matrix in a switching center

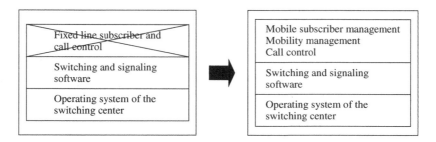

Figure 1.2   Necessary software changes to adapt a fixed-line switching center for a wireless network

signal, which was then sent over the copper cable to the terminating party. In some countries, ISDN (Integrated Services Digital Network) lines were quite popular. With this system, the transmission became fully digital and the conversion back into an analog audio signal was done directly in the phone.

GSM reuses much of the fixed-line technology that was already available at the time the standards were created. Thus, existing technologies such as switching centers and long-distance communication equipment were used. The main development for GSM, as shown in Figure 1.2, was the means to wirelessly connect the subscribers to the network. In fixed-line networks, subscriber connectivity is very simple as only two dedicated wires are necessary per user. In a GSM network, however, the subscribers are mobile and can change their location at any time. Thus, it is not possible to use the same input and output in the switching matrix for a user for each call as is the case in fixed-line networks.

As a mobile network consists of many switching centers, with each covering a certain geographical area, it is not even possible to predict in advance which switching center a call should be forwarded to for a certain subscriber. This means that the software for subscriber management and routing of calls of fixed-line networks cannot be used for GSM. Instead of a static call-routing mechanism, a flexible mobility management architecture became necessary in the core network, which needed to be aware of the current location of the subscriber and was thus able to route calls to the subscriber at any time.

It was also necessary to be able to flexibly change the routing of an ongoing call as a subscriber can roam freely and thus might leave the coverage area of the radio transmitter of the network over which the call was established. While there was a big difference in the software of a fixed and a Mobile Switching Center (MSC), the hardware as well as the lower layers of the software which are responsible, for example, for the handling of the switching matrix were mostly identical. Therefore, most telecommunication equipment vendors like Ericsson, Nokia Solutions and Networks, Huawei and Alcatel-Lucent offered their switching center hardware both for fixed-line and mobile networks. Only the software in the switching center decided if the hardware was used in a fixed or mobile network (see Figure 1.2).

1.1.2   Virtual Circuit Switching over IP

While in the 1990s voice calls were the dominating form of communication, this has significantly changed today with the rise of the Internet. While voice calls still remain important, other forms of communication such as e-mail, instant messaging (IM), social networks (e.g. Facebook), blogs, wikis and many more play an even bigger role. All these services share the Internet Protocol (IP) as a transport protocol and globally connect people via the Internet.

While circuit switching establishes an exclusive channel between two parties, the Internet is based on transferring individual data packets. A link with a high bandwidth is used to transfer the packets of many users. By using the destination address contained in each packet, each network node that the packet traverses decides over which outgoing link to forward the packet. Further details can be found in Chapter 2.

Owing to the rise of the Internet and IP-based applications, network operators thus had to maintain two separate networks: a circuit-switched network for voice calls and a packet-switched network for Internet-based services.

As the simultaneous operation of two different networks is very inefficient and costly, most network operators have, in the meantime, replaced the switching matrix in the MSC with a device referred to as media gateway. This allows them to virtualize circuit switching and to transfer voice calls over IP packets. The physical presence of a circuit-switched infrastructure is thus no longer necessary and the network operator can concentrate on maintaining and expanding a single IP-based network. This approach has been standardized under the name 'Bearer-Independent Core Network' (BICN).

The basic operation of GSM is not changed by this virtualization. The main differences can be found in the lower protocol levels for call signaling and voice call transmission. This will be looked at in more detail in the remainder of this chapter.

The trend toward IP-based communication can also be observed in the GSM radio network, even though it is still dominated today by classic circuit-switched technology. This is due to the wide distribution of the network that makes it difficult to change transport technology quickly and because the datarates required for GSM are low.

The air interface between the mobile devices and the network is not affected by the transition from circuit to packet switching. For mobile devices, it is therefore completely transparent if the network uses classic or virtual circuit switching.

1.2 Standards

As many telecom companies compete globally for orders of telecommunication network operators, standardization of interfaces and procedures is necessary. Without standards, which are defined by the International Telecommunication Union (ITU), it would not be possible to make phone calls internationally and network operators would be bound to the supplier they initially select for the delivery of their network components. One of the most important ITU standards discussed in Section 1.4 is the Signaling System Number 7 (SS-7), which is used for call routing. Many ITU standards, however, only represent the smallest common denominator as most countries have specified their own national extensions. In practice, this incurs a high cost for software development for each country as a different set of extensions needs to be implemented in order for a vendor to be able to sell its equipment. Furthermore, the interconnection of networks of different countries is complicated by this.

GSM, for the first time, set a common standard for Europe for wireless networks, which has also been adopted by many countries outside Europe. This is the main reason why subscribers can roam in GSM networks across the world that have roaming agreements with each other. The common standard also substantially reduces research and development costs as hardware and software can now be sold worldwide with only minor adaptations for the local market. The European Telecommunication Standards Institute (ETSI), which is also responsible for a number of other standards, was the main body responsible for the creation of the GSM standard. The ETSI GSM standards are composed of a substantial number of standards documents, each of which is called a technical specification (TS), which describe a particular part of the system. In the following chapters, many of these specifications are referenced and can thus be used for further information about a specific topic. All standards are freely available on the Internet at http://www.etsi.org [1] or at http://www.3gpp.org [2]. 3GPP is the organization that took over the standards maintenance and enhancement at the beginning of the UMTS standardization, as described in Chapter 3.

1.3   Transmission Speeds

The smallest transmission speed unit in a classic circuit-switched telecommunication network is the digital signal level 0 (DS0) channel. It has a fixed transmission speed of 64 kbit/s. Such a channel can be used to transfer voice or data, and thus it is usually not called a speech channel but simply referred to as a user data channel.

The reference unit of a telecommunication network is an E-1 connection in Europe and a T-1 connection in the United States, which use either a twisted pair or coaxial copper cable. The gross datarate is 2.048 Mbit/s for an E-1 connection and 1.544 Mbit/s for a T-1. An E-1 is divided into 32 timeslots of 64 kbit/s each, as shown in Figure 1.3 while a T-1 is divided into 24 timeslots of 64 kbit/s each. One of the timeslots is used for synchronization, which means that 31 timeslots for an E-1 or 23 timeslots for a T-1, respectively, can be used to transfer data. In practice, only 29 or 30 timeslots are used for user data transmission while the rest (usually one or two) are used for SS-7 signaling data (see Figure 1.3). More about SS-7 can be found in Section 1.4.

A single E-1 connection with 31 DS0s is not enough to connect two switching centers with each other. An alternative is an E-3 connection over twisted pair or coaxial cables. An E-3 connection is defined at a speed of 34.368 Mbit/s, which corresponds to 512 DS0s.

For higher transmission speeds and for long distances, optical systems that use the synchronous transfer mode (STM) standard are used. Table 1.1 shows some datarates and the number of 64- kbit/s DS0 channels that are transmitted per pair of fibers.

For virtual circuit switching over IP, optical Ethernet links are often used between network nodes at the same location. Transmission speeds of 1 Gbit/s or more are used on these links.

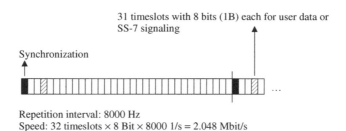

31 timeslots with 8 bits (1B) each for user data or SS-7 signaling

Synchronization

Repetition interval: 8000 Hz
Speed: 32 timeslots × 8 Bit × 8000 1/s = 2.048 Mbit/s

Figure 1.3   Timeslot architecture of an E-1 connection

Table 1.1 STM transmission speeds and number of DS0s

STM level	Speed (Mbit/s)	Approximate number of DS0 connections
STM-1	155.52	2300
STM-4	622.08	9500
STM-16	2488.32	37,000
STM-64	9953.28	148,279

Unlike the circuit-switched technology described above, Ethernet is the de facto standard for IP-based communication over fiber and copper cables and is widely used. As a consequence, network equipment can be built much cheaper.

1.4   The Signaling System Number 7

For establishing, maintaining and clearing a connection, signaling information needs to be exchanged between the end user and network devices. In the fixed-line network, analog phones signal their connection request when the receiver is lifted off the hook and by dialing a phone number that is sent to the network either via pulses (pulse dialing) or via tone dialing, which is called dual tone multifrequency (DTMF) dialing. With fixed-line ISDN phones and GSM mobile phones, the signaling is done via a separate dedicated signaling channel, and information such as the destination phone number is sent as messages.

If several components in the network are involved in the call establishment, for example, if originating and terminating parties are not connected to the same switching center, it is also necessary that the different nodes in the network exchange information with each other. This signaling is transparent for the user, and a protocol called the SS-7 is used for this purpose. SS-7 is also used in GSM networks and the standard has been enhanced by ETSI to fulfill the special requirements of mobile networks, for example, subscriber mobility management.

The SS-7 standard defines three basic types of network nodes:

- Service Switching Points (SSPs) are switching centers that are more generally referred to as network elements that are able to establish, transport or forward voice and data connections.
- Service Control Points (SCPs) are databases and application software that can influence the establishment of a connection. In a GSM network, SCPs can be used, for example, for storing the current location of a subscriber. During call establishment to a mobile subscriber, the switching centers query the database for the current location of the subscriber to be able to forward the call. More about this procedure can be found in Section 1.6.3 about the Home Location Register (HLR).
- Signaling Transfer Points (STPs) are responsible for the forwarding of signaling messages between SSPs and SCPs as not all network nodes have a dedicated link to all other nodes of the network. The principal functionality of an STP can be compared to an IP router in the Internet, which also forwards packets to different branches of the network. Unlike IP routers, however, STPs only forward signaling messages that are necessary for establishing,

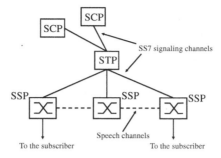

Figure 1.4   An SS-7 network with an STP, two SCP databases and three switching centers

maintaining and clearing a call. The calls themselves are directly carried on dedicated links between the SSPs.

Figure 1.4 shows the general structure of an SS-7 circuit-switched telecommunication network and the way the nodes described above are interconnected with each other.

The SS-7 protocol stack is also used in virtual circuit-switched networks for communication between the network nodes. Instead of dedicated signaling timeslots on an E-1 link, signaling messages are transported in IP packets. The following section describes the classic SS-7 protocol stack and afterward, the way SS-7 messages are transported over IP networks.

1.4.1   The Classic SS-7 Protocol Stack

SS-7 comprises a number of protocols and layers. A well-known model for describing telecommunication protocols and different layers is the Open System Interconnection (OSI) 7 layer model, which is used in Figure 1.5 to show the layers on which the different SS-7 protocols reside.

The Message Transfer Part 1 (MTP-1) protocol describes the physical properties of the transmission medium on layer 1 of the OSI model. Thus, this layer is also called the physical layer.

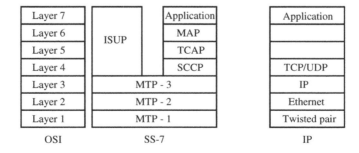

Figure 1.5   Comparison of the SS-7, OSI and TCP/IP protocol stacks

Properties that are standardized in MTP-1 are, for example, the definition of the different kinds of cables that can be used to carry the signal, signal levels and transmission speeds.

On layer 2, the data link layer, messages are framed into packets and a start and stop identification at the beginning and end of each packet are inserted into the data stream so that the receiver is able to detect where a message ends and where a new message begins.

Layer 3 of the OSI model, which is called the network layer, is responsible for packet routing. To enable network nodes to forward incoming packets to other nodes, each packet gets a source and destination address on this layer. This is done by the MTP-3 protocol of the SS-7 stack. For readers who are already familiar with the Transmission Control Protocol (TCP)/IP protocol stack, it may be noted at this point that the MTP-3 protocol fulfills the same tasks as the IP protocol. Instead of IP addresses, however, the MTP-3 protocol uses the so-called point codes to identify the source and the destination of a message.

A number of different protocols are used on layers 4–7 depending on the application. If a message needs to be sent for the establishment or clearing of a call, the Integrated Services Digital Network User Part (ISUP) protocol is used. Figure 1.6 shows how a call is established between two parties by using ISUP messages. In the example, party A is a mobile subscriber while party B is a fixed-line subscriber. Thus, A is connected to the network via an MSC, while B is connected via a fixed-line switching center.

To call B, the phone number of B is sent by A to the MSC. The MSC then analyzes the national destination code (NDC) of the phone number, which usually comprises the first two to four digits of the number, and detects that the number belongs to a subscriber in the fixed-line network. In the example shown in Figure 1.6, the MSC and the fixed-line switching center are directly connected with each other. Therefore, the call can be directly forwarded to the terminating switching center. This is quite a realistic scenario as direct connections are often used if, for example, a mobile subscriber calls a fixed-line phone in the same city.

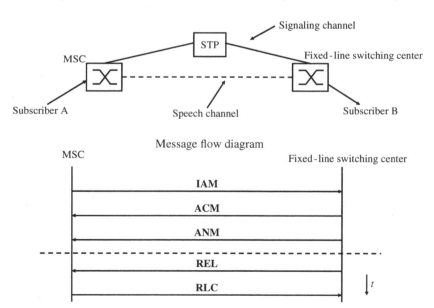

Figure 1.6 Establishment of a voice call between two switching centers

As B is a fixed-line subscriber, the next step for the MSC is to establish a voice channel to the fixed-line switching center. This is done by sending an ISUP Initial Address Message (IAM). The message contains, among other data, the phone number of B and informs the fixed-line switching center and the channel that the MSC would like to use for the voice path. In the example, the IAM message is not sent directly to the fixed-line switching center. Instead, an STP is used to forward the message.

At the other end, the fixed-line switching center receives the message, analyzes the phone number and establishes a connection via its switching matrix to subscriber B. Once the connection is established via the switching matrix, the switch applies a periodic current to the line of the fixed-line subscriber so that the fixed-line phone can generate an alerting tone. To indicate to the originating subscriber that the phone number is complete and the destination party has been found, the fixed-line switch sends back an Address Complete Message (ACM). The MSC then knows that the number is complete and that the terminating party is being alerted about the incoming call.

If B answers the call, the fixed-line switching center sends an Answer Message (ANM) to the MSC and conversation can start.

When B ends the call, the fixed-line switching center resets the connection in the switching matrix and sends a release (REL) message to the MSC. The MSC confirms the termination of the connection by sending back a Release Complete (RLC) message. If A had terminated the call, the messages would have been identical, with only the direction of the REL and RLC reversed.

For the communication between the switching centers (SSPs) and the databases (SCPs), the Signaling Connection and Control Part (SCCP) is used on layer 4. SCCP is very similar to TCP and User Datagram Protocol (UDP) in the IP world. Protocols on layer 4 of the protocol stack enable the distinction of different applications on a single system. TCP and UDP use ports to do this. If a personal computer (PC), for example, is used as both a web server and a File Transfer Protocol (FTP) server at the same time, both applications would be accessed over the network via the same IP address. However, while the web server can be reached via port 80, the FTP server waits for the incoming data on port 21. Therefore, it is quite easy for the network protocol stack to decide the application to which incoming data packets should be forwarded. In the SS-7 world, the task of forwarding incoming messages to the right application is done by SCCP. Instead of port numbers, SCCP uses Subsystem Numbers (SSNs).

For database access, the Transaction Capability Application Part (TCAP) protocol has been designed as part of the SS-7 family of protocols. TCAP defines a number of different modules and messages that can be used to query all kinds of different databases in a uniform way.

1.4.2   SS-7 Protocols for GSM

Apart from the fixed-line network SS-7 protocols, the following additional protocols were defined to address the special needs of a GSM network.

- **The Mobile Application Part (MAP).** This protocol has been standardized in 3GPP TS 29.002 [3] and is used for the communication between an MSC and the HLR, which maintains subscriber information. The HLR is queried, for example, if the MSC wants to establish a connection to a mobile subscriber. In this case, the HLR returns the information about the

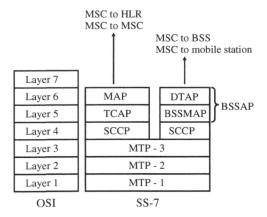

Figure 1.7   Enhancement of the SS-7 protocol stack for GSM

current location of the subscriber. The MSC is then able to forward the call to the mobile subscriber's switching center establishing a voice channel between itself and the next hop by using the ISUP message flow that has been shown in Figure 1.6. MAP is also used between two MSCs if the subscriber moves into the coverage area of a different MSC while a call is ongoing. As shown in Figure 1.7, the MAP protocol uses the TCAP, SCCP and MTP protocols on lower layers.

- **The Base Station Subsystem Mobile Application Part (BSSMAP)**. This protocol is used for communication between the MSC and the radio network. Here, the additional protocol is necessary, for example to establish a dedicated radio channel for a new connection to a mobile subscriber. As BSSMAP is not a database query language like the MAP protocol, it is based on SCCP directly instead of using TCAP in between.
- **The Direct Transfer Application Part (DTAP)**. This protocol is used between the user's mobile device, which is also called mobile station (MS), and the MSC to communicate transparently. To establish a voice call, the MS sends a setup message to the MSC. As in the example in Section 1.4.1, this message contains among other things the phone number of the called subscriber. As it is only the MSC's task to forward calls, all network nodes between the MS and the MSC forward the message transparently and thus need not understand the DTAP protocol.

1.4.3   IP-Based SS-7 Protocol Stack

When using an IP network for the transmission of SS-7 signaling messages, the MTP-1 and MTP-2 protocols are replaced by the IP and the transport medium-dependent lower layer protocols (e.g. Ethernet). Figure 1.8 shows the difference between the IP and the classic stack presented in the previous section.

In the IP stack, layer-4 protocols are either UDP or TCP for most services. For the transmission of SS-7 messages, however, a new protocol has been specified, which is referred to as Stream Control Transmission Protocol (SCTP). When compared to TCP and UDP, it offers advantages when many signaling connections between two network nodes are active at the same time.

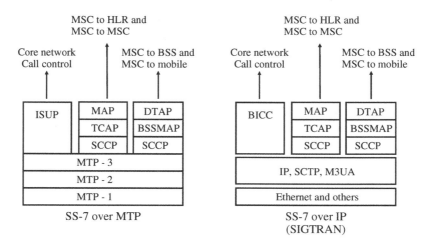

Figure 1.8 Comparison of the classic and IP-based SS-7 protocol stacks

On the next protocol layer, SCTP is followed by the M3UA (MTP-3 User Adaptation Layer) protocol. As the name implies, the protocol is used to transfer information that is contained in the classic MTP-3 protocol. For higher protocol layers such as SCCP, M3UA simulates all functionalities of MTP-3. As a consequence, the use of an IP protocol stack is transparent to all higher layer SS-7 protocols.

In the industry, the IP-based SS-7 protocol stack or the IP-based transmission of SS-7 messages is often referred to as SIGTRAN (signaling transmission). The abbreviation originated from the name of the IETF (Internet Engineering Task Force) working group that was created for the definition of these protocols.

As described in Section 1.1.1, the ISUP protocol is used for the establishment of voice calls between switching centers and the assignment of a 64 kbit/s timeslot. In an IP-based network, voice calls are transmitted in IP packets. As a consequence, the ISUP protocol has to be adapted as well. The resulting protocol is referred to as Bearer-Independent Call Control (BICC) protocol, which largely resembles ISUP.

As IP links cannot be introduced on all interfaces in live networks at once, Signaling Gateways (SGWs) have been defined to bridge E-1 based and IP-based SS-7 communication. The SGWs adapt the lower layers of the protocol stack and thus make the differences transparent for both sides. This is necessary, for example, if the subscriber database has already been converted for IP interfaces while other components such as the switching centers are still using traditional signaling links.

To bridge voice calls between E-1 based and IP-based networks, Media Gateways (MGWs) are used. Connected to an MSC Server, an MGW handles both IP-based and E-1-based voice calls transparently as it implements both the classic and IP-based signaling protocol stacks.

1.5 The GSM Subsystems

A GSM network is split into three subsystems which are described in more detail below:

- The Base Station Subsystem (BSS), which is also called 'radio network', contains all nodes and functionalities that are necessary to wirelessly connect mobile subscribers over the radio interface to the network. The radio interface is usually also referred to as the 'air interface'.

- The Network Subsystem (NSS), which is also called 'core network', contains all nodes and functionalities that are necessary for switching of calls, for subscriber management and mobility management.
- The Intelligent Network Subsystem (IN) comprises SCP databases that add optional functionality to the network. One of the most important optional IN functionalities of a mobile network is the prepaid service, which allows subscribers to first fund an account with a certain amount of money which can then be used for network services like phone calls, Short Messaging Service (SMS) messages and, of course, data services via GPRS and UMTS as described in Chapters 2 and 3. When a prepaid subscriber uses a service of the network, the responsible IN node is contacted and the amount the network operator charges for a service is deducted from the account in real time.

1.6 The Network Subsystem

The most important responsibilities of the NSS are call establishment, call control and routing of calls between different fixed and mobile switching centers and other networks. Other networks are, for example, the national fixed-line network, which is also called the Public Switched Telephone Network (PSTN), international fixed-line networks, other national and international mobile networks and Voice over Internet Protocol (VoIP) networks. Furthermore, the NSS is responsible for subscriber management. The nodes necessary for these tasks in a classic network architecture are shown in Figure 1.9. Figure 1.10 shows the nodes required in IP-based core networks. Both designs are further described in the following sections.

1.6.1 The Mobile Switching Center (MSC), Server and Gateway

The MSC is the central element of a mobile telecommunication network, which is also called a Public Land Mobile Network (PLMN) in the standards. In a classic circuit-switched network,

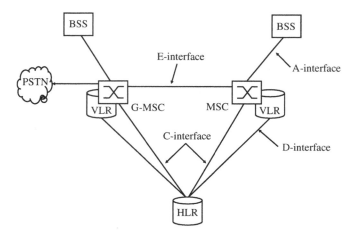

Figure 1.9 Interfaces and nodes in a classic NSS architecture

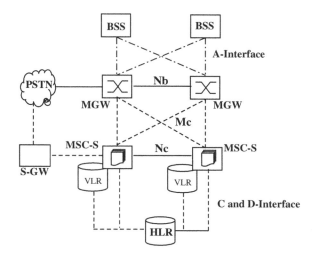

Figure 1.10   Interfaces and nodes in an IP-based NSS architecture

all connections between subscribers are managed by the MSC and are always routed over the switching matrix even if two subscribers that have established a connection communicate over the same radio cell.

The management activities to establish and maintain a connection are part of the Call Control (CC) Protocol, which is generally responsible for the following tasks:

- Registration of mobile subscribers: When the mobile device, also referred to as MS, is switched on, it registers to the network and is then reachable by all other subscribers of the network.
- Call establishment and call routing between two subscribers.
- Forwarding of SMS messages.

As subscribers can roam freely in the network, the MSC is also responsible for the Mobility Management (MM) of subscribers. This activity comprises the following tasks:

- Authentication of subscribers at connection establishment is necessary because a subscriber cannot be identified as in the fixed network by the pair of copper cables over which the signal arrives. Authentication of subscribers and the authentication center (AuC) are further discussed in Section 1.6.4.
- If no active connection exists between the network and the mobile device, the MSC has to report a change of location to the network to be reachable for incoming calls and SMS messages. This procedure is called location update and is further described in Section 1.8.1.
- If the subscriber changes its location while a connection is established with the network, the MSC is part of the process that ensures that the connection is not interrupted and is rerouted to the next cell. This procedure is called handover and is described in more detail in Section 1.8.3.

To enable the MSC to communicate with other nodes of the network, it is connected to them via standardized interfaces as shown in Figure 1.9. This allows network operators to acquire different components for the network from different network equipment vendors. The interfaces discussed below are either transmitted over timeslots in circuit-switched E-1 lines or over an IP-based network. As described earlier, only the lower protocol layers are affected by this. On the application layer, both variants are identical.

The BSS, which connects all subscribers to the core network, is connected to the MSCs via a number of 2-Mbit/s E-1 connections. This interface is called the A-interface. As has been shown in Section 1.4, the BSSMAP and DTAP protocols are used over the A-interface for communication between the MSC, the BSS and the mobile devices. As an E-1 connection can only carry 31 channels, many E-1 connections are necessary to connect an MSC to the BSS. In practice, this means that many E-1s are bundled and sent over optical connections such as STM-1 to the BSS. Another reason to use an optical connection is that electrical signals can only be carried over long distances with great effort and it is not unusual that an MSC is over 100 km away from the next BSS node.

As an MSC only has a limited switching capacity and processing power, a PLMN is usually composed of dozens of independent MSCs. Each MSC thus covers only a certain area of the network. To ensure connectivity beyond the immediate coverage area of an MSC, E-1s, which are again bundled into optical connections, are used to interconnect the different MSCs of a network. As a subscriber can roam into the area that is controlled by a different MSC while a connection is active, it is necessary to change the route of an active connection to the new MSC (handover). The necessary signaling connection is called the E-interface. ISUP is used for the establishment of the speech path between different MSCs and the MAP protocol is used for the handover signaling between the MSCs. Further information on the handover process can be found in Section 1.8.3.

The C-interface is used to connect the MSCs of a network with the HLR of the mobile network. While the A-and E-interfaces that were described previously always consist of signaling and speech path links, the C-interface is a pure signaling link. Speech channels are not necessary for the C-interface as the HLR is a pure database that cannot accept or forward calls. Despite being only a signaling interface, E-1 connections are used for this interface. All timeslots are used for signaling purposes or are unused.

As has been shown in Section 1.3, a voice connection is carried over a 64 kbit/s E-1 timeslot in a classic circuit-switched fixed-line or mobile network. Before the voice signal can be forwarded, it needs to be digitized. For an analog fixed-line connection, this is done in the switching center, while an ISDN fixed-line phone or a GSM mobile phone digitizes the voice signal by itself.

An analog voice signal is digitized in three steps, as shown in Figure 1.11: in the first step, the bandwidth of the input signal is limited to 300−3400 Hz to be able to carry the signal with the limited bandwidth of a 64- kbit/s timeslot. Afterward, the signal is sampled at a rate of 8000 times per second. The next processing step is the quantization of the samples, which means that the analog samples are converted into 8-bit digital values that can each have a value from 0 to 255.

The higher the volume of the input signal, the higher the amplitude of the sampled value and its digital representation. To be able to also transmit low-volume conversations, the quantization is not linear over the whole input range but only in certain areas. For small amplitudes of the input signal, a much higher range of digital values is used than for high-amplitude values.

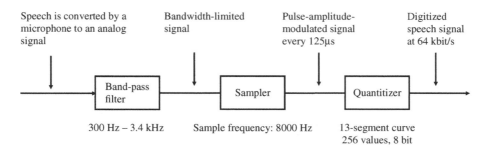

Figure 1.11   Digitization of an analog voice signal

The resulting digital data stream is called a pulse code-modulated (PCM) signal. Which volume is represented by which digital 8-bit value is described in the A-law standard for European networks and in the μ-law standard in North America.

The use of different standards unfortunately complicates voice calls between networks that use different standards. Therefore, it is necessary, for example, to convert a voice signal for a connection between France and the United States.

As the MSC controls all connections, it is also responsible for billing. This is done by creating a billing record for each call, which is later transferred to a billing server. The billing record contains information like the number of the caller and the calling party, cell-ID of the cell from which the call originated, time of call origination, the duration of the call, and so on. Calls for prepaid subscribers are treated differently as the charging is already done while the call is running. The prepaid billing service is usually implemented on an IN system and not on the MSC as is further described in Section 1.11.

MSC Server and Media Gateway

In most of today's mobile voice networks, circuit-switched components have been replaced with IP-based devices. The MSC has been split into an MSC-Server (MSC-S) and an MGW. This is shown in Figure 1.10 and has been specified in 3GPP TS 23.205 [4]. The MSC-Ss are responsible for CC and MM (signaling), and the MGWs handle the transmission of virtual voice circuits (user data).

To establish a voice connection, MSC-Ss and MGWs communicate over the Mc-interface. This interface does not exist in the classical model, as the MSC contained both components. 3GPP TS 29.232 [5] describes this interface on which the H.248/MEGACO (Media Gateway Control) protocol is used [6]. The protocol is used, for example, to establish voice channels to two parties and then to logically connect the two channels in the MGW. The protocol is also used to instruct the MGWs to play announcements to inform users of events such as the called party is currently not available or busy and to establish conference calls between more than two subscribers. To add redundancy and for load balancing reasons, several MSC-Ss and MGWs can be interconnected in a mesh. If an MSC-S fails, an MGW can thus still continue to operate and is then controlled by another server. Thus, a single MSC-S is no longer solely responsible for a single geographical area as was the case in the traditional model.

On the radio network side, the A-interface continues to be used to connect the radio network to the MSC-Ss and MGWs. The connection can be made without any changes in the radio network over the classic E-1-based A-interface or over an IP-based A-interface. In addition, the A-interface has been made more flexible and can now be connected to several media gateways. This adds redundancy toward the radio network as well, as a geographical region can still be served even if a media gateway fails.

The Nc-interface is used to transport voice calls within the core network, for example, to gateways, to other mobiles or fixed networks. The protocol used on this interface is referred to as the BICC protocol and is very similar to the traditional ISUP protocol. This is specified in ITU Q.1901 [7] and 3GPP TS 29.205 [8]. By using an SGW as shown in Figure 1.10, the protocol can be converted into ISUP to be able to forward calls to other core networks that are still based on the classic model. In practice, it can be observed that despite many networks having moved to an IP-based architecture, the gateways between them are still based on the classic architecture.

Virtual speech channels that have been negotiated over the Nc-interface are transmitted between MGWs over the Nb-interface. The combination of the Nb-interface and Nc-interface thus replaces the E-interface of the classic network architecture. A voice channel is transmitted over IP connections either as PCM/G.711, Narrowband-AMR or Wideband-AMR, depending on the type of radio network, configuration of the network and the capabilities of the mobile device. At the borders of the core network, for example, to and from the A-interface to the GSM radio network or to and from a classic fixed-line PSTN network, MGWs can convert media streams, for example, between Narrowband-AMR over IP to G.711/PCM over E-1. This requires, however, that an MGW contains both Ethernet ports and E-1 ports.

Gateways between mobile networks are usually still based on ISUP and circuit-switched links, even though most networks are based on IP technology today. In the future, this is expected to change as advanced speech codecs such as Wideband-AMR can only be used over BICN and IP-based transport links.

Like in classic core networks, the C- and D-interfaces are used in a BICN network to communicate with the HLR. Instead of E-1 links, however, communication is based on IP links today.

1.6.2 The Visitor Location Register (VLR)

Each MSC has an associated Visitor Location Register (VLR), which holds the record of each subscriber that is currently served by the MSC (Figure 1.12). These records are only copies of the original records, which are stored in the HLR (see Section 1.6.3). The VLR is mainly used to reduce the signaling between the MSC and the HLR. If a subscriber roams into the area of an MSC, the data are copied to the VLR of the MSC and are thus locally available for every connection establishment. The verification of the subscriber's record at every connection establishment is necessary, as the record contains information about the services that are active and the services from which the subscriber is barred. Thus, it is possible, for example, to bar outgoing calls while allowing incoming calls to prevent abuse of the system. While the standards allow implementation of the VLR as an independent hardware component, all vendors have implemented the VLR simply as a software component in the MSC. This is possible because MSC and VLR use different SCCP SSNs as shown in Figure 1.12 (see Section 1.4.1) and can thus run on a single physical node.

When a subscriber leaves the coverage area of an MSC, his record is copied from the HLR to the VLR of the new MSC, and is then removed from the VLR of the previous MSC. The

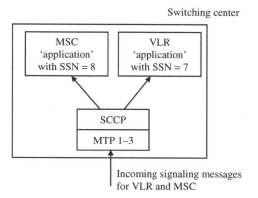

Figure 1.12   Mobile Switching Center (MSC) with integrated Visitor Location Register (VLR)

communication with the HLR is standardized in the D-interface specification, which is shown together with other MSC interfaces in Figure 1.9 and Figure 1.10.

1.6.3   The Home Location Register (HLR)

The HLR is the subscriber database of a GSM network. It contains a record for each subscriber, which contains information about the individually available services.

The International Mobile Subscriber Identity (IMSI) is an internationally unique number that identifies a subscriber and is used for most subscriber-related signaling in the network (Figure 1.13). The IMSI is stored in the subscriber's subscriber identity module (SIM) card and in the HLR and is thus the key to all information about the subscriber. The IMSI consists of the following parts:

- **The Mobile Country Code (MCC).** The MCC identifies the subscriber's home country. Table 1.2 shows a number of MCC examples.
- **The Mobile Network Code (MNC).** This part of the IMSI is the national part of a subscriber's home network identification. A national identification is necessary because there are usually several independent mobile networks in a single country. In the United Kingdom, for example, the following MNCs are used: 10 for O2, 15 for Vodafone, 30 for T-Mobile, 33 for Orange, 20 for Hutchison 3G, etc.

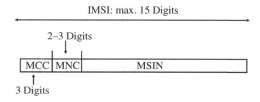

Figure 1.13   The international mobile subscriber identity (IMSI)

Table 1.2   Mobile country codes

MCC	Country
234	United Kingdom
310	United States
228	Switzerland
208	France
262	Germany
604	Morocco
505	Australia

- **The Mobile Subscriber Identification Number (MSIN).** The remaining digits of the IMSI form the MSIN, which uniquely identifies a subscriber within the home network.

As an IMSI is internationally unique, it enables a subscriber to use his phone abroad if a GSM network is available that has a roaming agreement with his home operator. When the mobile device is switched on, the IMSI is retrieved from the SIM card and sent to the MSC. There, the MCC and MNC of the IMSI are analyzed and the MSC is able to request the subscriber's record from the HLR of the subscriber's home network.

For information purposes, the IMSI can also be retrieved from the SIM card with a PC and a serial cable that connects to the mobile device. By using a terminal program such as Hyper-Terminal, the mobile can be instructed to return the IMSI by using the 'at + cimi' command, which is standardized in 3GPP TS 27.007 [9]. Figure 1.14 shows how the IMSI is returned by the mobile device.

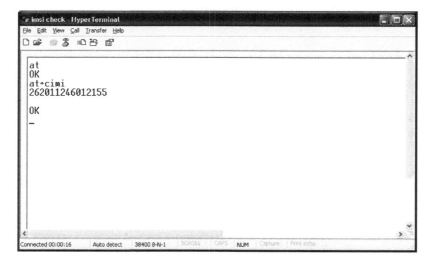

Figure 1.14   A terminal program can be used to retrieve the IMSI from the SIM card

The phone number of the user, which is called the Mobile Subscriber Integrated Services Digital Network Number (MSISDN) in the GSM standards, has a length of up to 15 digits and consists of the following parts:

- The country code is the international code of the subscriber's home country. The country code has one to three digits such as +44 for the United Kingdom, +1 for the United States, and +353 for Ireland.
- The NDC usually represents the code with which the network operator can be reached. It is normally three digits in length. It should be noted that mobile networks in the United States use the same NDCs as fixed-line networks. Thus, it is not possible for a user to distinguish if he is calling a fixed-line or a mobile phone. This impacts both billing and routing, as the originating network cannot deduct which tariff to apply from the NDC.
- The remainder of the MSISDN is the subscriber number, which is unique in the network.

There is usually a 1:1 or 1:N relationship in the HLR between the IMSI and the MSISDN. Furthermore, a mobile subscriber is normally assigned only a single MSISDN. However, as the IMSI is the unique identifier of a subscriber in the mobile network, it is also possible to assign several numbers to a single subscriber.

Another advantage of using the IMSI as the key to all subscriber information instead of the MSISDN is that the phone number of the subscriber can be changed without replacing the user's SIM card or changing any information on it. To change the MSISDN, only the HLR record of the subscriber needs to be changed. In effect, this means that the mobile device is not aware of its own phone number. This is not necessary because the MSC automatically adds the user's MSISDN to the message flow for a mobile-originated call establishment so that it can be presented to the called party.

Many countries have introduced a functionality called mobile number portability (MNP), which allows a subscriber to retain his MSISDN even if he wants to change his mobile network operator. This is a great advantage for subscribers and for the competition between mobile operators, but it also implies that it is no longer possible to discern the mobile network to which the call will be routed from the NDC. Furthermore, the introduction of MNP also increased the complexity of call routing and billing in both fixed-line and mobile networks, because it is no longer possible to use the NDC to decide which tariff to apply to a call. Instead of a simple call-routing scheme based on the NDC, the networks now have to query a MNP database for every call to a mobile subscriber to find out if the call can be routed inside the network or if it has to be forwarded to a different national mobile network.

Apart from the IMSI and MSISDN, the HLR contains a variety of information about each subscriber, such as which services he is allowed to use. Table 1.3 shows a number of 'basic services' that can be activated on a per subscriber basis.

In addition to the basic services described above, the GSM network offers a number of other services that can also be activated on a per subscriber basis. These services are called supplementary services and are shown in Table 1.4.

Most supplementary services can be activated by the network operator on a per subscriber basis and allow the operator to charge an additional monthly fee for some services if desired. Other services, like multiparty, can be charged on a per use basis. Although some network

Table 1.3 Basic services of a GSM network

Basic service	Description
Telephony	If this basic service is activated, a subscriber can use the voice telephony services of the network. This can be partly restricted by other supplementary services that are described below
Short messaging service (SMS)	If activated, a subscriber is allowed to use the SMS
Data service	Different circuit-switched data services can be activated for a subscriber with speeds of 2.4, 4.8, 9.6 and 14.4 kbit/s data calls
FAX	Allows or denies a subscriber the use of the FAX service that can be used to exchange FAX messages with fixed-line or mobile devices

operators made use of this in the early years of GSM, most services are now included as part of the basic monthly fee.

Most services can be configured by the subscriber via a menu on the mobile device. The menu, however, is just a graphical front end for the user and the mobile device translates the user's commands into numerical strings which start with a '*' character. These strings are then sent to the network by using an Unstructured Supplementary Service Data (USSD) message. The codes are standardized in 3GPP TS 22.030 [14] and are thus identical in all networks. As the menu is only a front end for the USSD service, the user can also input the USSD strings himself via the keypad. After pressing the 'send' button, which is usually the button that is also used to start a phone call after typing in a phone number, the mobile device sends the string to the HLR via the MSC, where the string is analyzed and the requested operation is performed. For example, call forwarding to another phone (e.g. 0782 192 8355), while a user is already engaged in another call, call forward busy (CFB) is activated with the following string: **67*07821928355# + call button.

1.6.4 The Authentication Center

Another important part of the HLR is the AuC. The AuC contains an individual key per subscriber (Ki), which is a copy of the Ki on the SIM card of the subscriber. As the Ki is secret, it is stored in the AuC and especially on the SIM card in a way that prevents it from being read directly.

For many operations in the network, for instance, during the establishment of a call, the subscriber is identified by using this key. Thus, it can be ensured that the subscriber's identity is not misused by a third party. Figure 1.16 shows how the authentication process is performed.

The authentication process, as shown in Figure 1.17, is initiated when a subscriber establishes a signaling connection with the network before the actual request (e.g. call establishment request) is sent. In the first step of the process, the MSC requests an authentication triplet from the HLR/AuC. The AuC retrieves the Ki of the subscriber and the authentication algorithm (A3 algorithm) based on the IMSI of the subscriber that is part of the message from the MSC.

Table 1.4 Supplementary services of a GSM network

Supplementary service	Description
Call forward unconditional (CFU)	If this service is activated, a number can be configured to which all incoming calls are forwarded immediately [10]. This means that the mobile device will not even be notified of the incoming call even if it is switched on
Call forward busy (CFB)	This service allows a subscriber to define a number to which calls are forwarded if he is already engaged in a call when a second call comes in
Call forward no reply (CFNRY)	If this service is activated, it is possible to forward the call to a user-defined number if the subscriber does not answer the call within a certain time. The subscriber can change the number to which to forward the call as well as the timeout value (e.g. 25 seconds)
Call forward not reachable (CFNR)	This service forwards the call if the mobile device is attached to the network but is not reachable momentarily (e.g. temporary loss of network coverage)
Barring of all outgoing calls (BAOC)	This functionality can be activated by the network operator if, for example, the subscriber has not paid his monthly invoice in time. It is also possible for the network operator to allow the subscriber to change the state of this feature together with a PIN (personal identification number) so that the subscriber can lend the phone to another person for incoming calls only [11]
Barring of all incoming calls (BAIC)	Same functionality as provided by BAOC for incoming calls [11]
Call waiting (CW)	This feature allows signaling an incoming call to a subscriber while he is already engaged in another call [12]. The first call can then be put on hold to accept the incoming call. The feature can be activated or barred by the operator and switched on or off by the subscriber
Call hold (HOLD)	This functionality is used to accept an incoming call during an already active call or to start a second call [12]
Calling line identification presentation (CLIP)	If activated by the operator for a subscriber, the functionality allows the switching center to forward the number of the caller
Calling line identification restriction (CLIR)	If allowed by the network, the caller can instruct the network not to show his phone number to the called party
Connected line presentation (COLP)	Shows the calling party the MSISDN to which a call is forwarded, if call forwarding is active at the called party side
Connected line presentation restriction (COLR)	If COLR is activated at the called party, the calling party will not be notified of the MSISDN to which the call is forwarded
Multiparty (MPTY)	Allows subscribers to establish conference bridges with up to six subscribers [13]

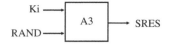

Figure 1.15   Creation of a signed response (SRES)

Extract of a decoded authentication request message
SCCP MSG: Data Form 1
DEST. REF ID: 0B 02 00
DTAP MSG LENGTH: 19
PROTOCOL DISC.: Mobility Management
DTAP MM MSG: Auth. Request
Ciphering Key Seq.: 0
RAND in hex: 12 27 33 49 11 00 98 45 87 49 12 51 22 89 18 81 (16 B = 128 bit)
Extract of a decoded authentication response message
SCCP MSG: Data Form 1
DEST. REF ID: 00 25 FE
DTAP MSG LENGTH: 6
PROTOCOL DISC.: Mobility Management
DTAP MM MSG: Auth. Response
SRES in hex: 37 21 77 61 (4 B = 32 bit)

Figure 1.16   Message flow during the authentication of a subscriber

The Ki is then used together with the A3 algorithm and a random number to generate the authentication triplet, which contains the following values:

- **RAND:** A 128-bit random number.
- **SRES:** The signed response, (SRES), is generated by using Ki, RAND and the authentication A3 algorithm, and has a length of 32 bits (see Figure 1.15).
- **Kc:** The ciphering key, Kc, is also generated by using Ki and RAND. It is used for the ciphering of the connection once the authentication has been performed successfully. Further information on this topic can be found in Section 1.7.7.

RAND, SRES and Kc are then returned to the MSC, which then performs the authentication of the subscriber. It is important to note that the secret Ki key never leaves the AuC.

To speed up subsequent connection establishments, the AuC usually returns several authentication triplets per request. These are buffered by the MSC/VLR and are used during the subsequent connection establishments.

In the next step, the MSC sends the RAND inside an authentication request message to the mobile device. The mobile device forwards the RAND to the SIM card, which then uses the Ki and the authentication A3 algorithm to generate a signed response (SRES*). The SRES* is returned to the mobile device and then sent back to the MSC inside an authentication response

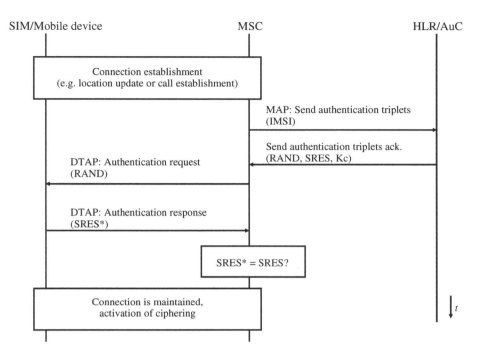

Figure 1.17   Authentication between network and mobile device

message. The MSC then compares SRES and SRES*, and if they are equal, the subscriber is authenticated and allowed to proceed with the communication.

As the secret key, Ki, is not transmitted over any interface that could be eavesdropped on, it is not possible for a third party to correctly calculate a SRES. As a fresh random number is used for the next authentication, it is also pointless to intercept the SRES* and use it for another authentication. A detailed description of the authentication procedure and many other procedures between the mobile device and the core network can be found in 3GPP TS 24.008 [15].

Figure 1.17 shows some parts of an authentication request and an authentication response message. Apart from the format of RAND and SRES, it is also interesting to note the different protocols that are used to encapsulate the message (see Section 1.4.2).

1.6.5   The Short Messaging Service Center (SMSC)

Another important network element is the Short Messaging Service Center (SMSC), which is used to store and forward short messages. The SMS was only introduced about four years after the first GSM networks went into operation as an add-on and has been specified in 3GPP TS 23.040 [16]. Most industry observers were quite skeptical at that time as the general opinion was that if it was needed to convey some information, it was done by calling someone rather than by the more cumbersome typing in a text message on the small keypad. However, they

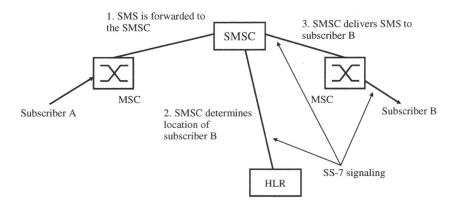

Figure 1.18   SMS delivery principle

were proven wrong and, today, most GSM operators generate a significant amount of their revenue from the short message service, despite a trend to replace SMS messaging with other forms of mobile-Internet-based IM. In Germany, for example, over 59 billion SMS messages were exchanged in 2012 [17].

SMS can be used for person-to-person messaging as well as for providing notification of other events such as a missed call that was forwarded to the voice mail system. The transfer method for both cases is identical.

The sender of an SMS prepares the text for the message and then sends the SMS via a signaling channel to the MSC as shown in Figure 1.18. As a signaling channel is used, an SMS is just an ordinary DTAP SS-7 message and thus, apart from the content, very similar to other DTAP messages, such as a location update message or a setup message to establish a voice call. Apart from the text, the SMS message also contains the MSISDN of the destination party and the address of the SMSC, which the mobile device has retrieved from the SIM card. When the MSC receives an SMS from a subscriber, it transparently forwards the SMS to the SMSC. As the message from the mobile device contains the address of the subscriber's SMSC, international roaming is possible and the foreign MSC can forward the SMS to the home SMSC without the need for an international SMSC database (see Figure 1.18).

To deliver a message, the SMSC analyzes the MSISDN of the recipient and retrieves its current location (the MSC concerned) from the HLR. The SMS is then forwarded to the MSC concerned. If the subscriber is currently attached, the MSC tries to contact the mobile device, and if an answer is received, the SMS is forwarded. Once the mobile device has confirmed the proper reception of the SMS, the MSC notifies the SMSC as well and the SMS is deleted from the SMSC's data storage.

If the subscriber is not reachable because the battery of the mobile device is empty, the network coverage has been lost temporarily, or if the device is simply switched off, it is not possible to deliver the SMS. In this case, the message waiting flag is set in the VLR and the SMS is stored in the SMSC. Once the subscriber communicates with the MSC, the MSC notifies the SMSC to reattempt delivery.

As the message waiting flag is also set in the HLR, the SMS also reaches a subscriber that has switched off the mobile device in London, for example, and switches it on again after a flight to Los Angeles. When the mobile device is switched on in Los Angeles, the visited MSC reports the location to the subscriber's home HLR (location update). The HLR then sends a copy of the user's subscription information to the MSC/VLR in Los Angeles including the message waiting flag and thus the SMSC can also be notified that the user is reachable again.

The SMS delivery mechanism does not include a delivery report for the sender of the SMS by default. The sender is only notified that the SMS has been correctly received by the SMSC. However, if supported by a device, it is also possible to request an end-to-end delivery notification from the SMSC. In practice, there are a number of different ways this is implemented in mobile devices. In some mobile operating systems, delivery reports can be activated in the SMS settings. Confirmations are then shown with a symbol next to the message or are displayed in the status bar. Other operating systems include a separate list of received or pending confirmations.

1.7   The Base Station Subsystem (BSS) and Voice Processing

While most functionality required in the NSS for GSM could be added via additional software, the BSS had to be developed from scratch. This was mainly necessary as earlier generation systems were based on analog transmission over the air interface and thus did not have much in common with the GSM BSS.

1.7.1   Frequency Bands

In Europe, GSM was initially specified only for operation in the 900-MHz band between 890 and 915 MHz in the uplink direction and between 935 and 960 MHz in the downlink direction as shown in Figure 1.19. 'Uplink' refers to the transmission from the mobile device to the network and 'downlink' to the transmission from the network to the mobile device. The bandwidth of 25 MHz is split into 125 channels with a bandwidth of 200 kHz each.

It soon became apparent that the number of available channels was not sufficient to cope with the growing demand in many European countries. Therefore, the regulating bodies assigned an additional frequency range for GSM, which uses the frequency band from 1710 to 1785 MHz for the uplink and from 1805 to 1880 for the downlink. Instead of a total bandwidth of 25 MHz as in the 900-MHz range, the 1800-MHz band offers 75 MHz of bandwidth, which corresponds to 375 additional channels. The functionality of GSM is identical on both frequency bands, with the channel numbers, also referred to as the Absolute Radio Frequency Channel Numbers (ARFCNs), being the only difference (see Table 1.5).

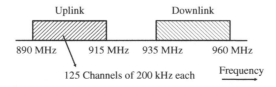

Figure 1.19   GSM uplink and downlink in the 900-MHz frequency band

Table 1.5 GSM frequency bands

Band	ARFCN	Uplink (MHz)	Downlink (MHz)
GSM 900 (primary)	0–124	890–915	935–960
GSM 900 (extended)	975–1023, 0–124	880–915	925–960
GSM 1800	512–885	1710–1785	1805–1880
GSM 1900 (North America)	512–810	1850–1910	1930–1990
GSM 850 (North America)	128–251	824–849	869–894
GSM-R	0–124, 955–1023	876–915	921–960

While GSM was originally intended only as a European standard, the system soon spread to countries in other parts of the globe. In countries outside Europe, GSM sometimes competes with other technologies, such as Code Division Multiple Access CDMA. Today, only a few countries, like Japan and South Korea, are not covered by GSM systems. However, some of the operators in these countries operate Wideband Code Division Multiple Access (WCDMA) UMTS and LTE networks (see Chapter 3 and 4). Therefore, GSM/UMTS subscribers with dual-mode phones can also roam in these countries.

In North America, analog mobile networks continued to be used for some time before second-generation networks, which included the use of the GSM technology, were introduced. Unfortunately, however, the 900-MHz and the 1800-MHz bands were already in use by other systems and thus the North American regulating body chose to open frequency bands for the new systems in the 1900-MHz band and later on in the 850-MHz band. The disadvantage of this approach is that not all North American GSM mobile phones can be used in Europe. Fortunately, the vast majority of current GSM, UMTS and LTE phones sold across the world support the US frequency bands for GSM as well as the European frequency bands, which are also used in most countries in other parts of the world. These quadband devices thus enable a user to truly roam globally.

The GSM standard is also used by railway communication networks in Europe and other parts of the world. For this purpose, GSM was enhanced to support a number of private mobile-radio- and railway-specific functionalities and this version is known as GSM-R. The additional functionalities include the following:

- **The Voice Group Call Service (VGCS).** This service offers a circuit-switched walkie-talkie functionality to allow subscribers that have registered to a VGCS group to communicate with all other subscribers in the area, who have also subscribed to the group. To talk, the user has to press a push to talk button. If no other subscriber holds the uplink, the network grants the request and blocks the uplink for all other subscribers while the push to talk button is pressed. The VGCS service is very efficient, especially if many subscribers participate in a group call, as all mobile devices that participate in the group call listen to the same timeslot in the downlink direction. Further information about this service can be found in 3GPP TS 43.068 [18].
- **The Voice Broadcast Service (VBS).** It is similar to VGCS, with the restriction that only the originator of the call is allowed to speak. Further information about this service can be found in 3GPP TS 43.069 [19].
- **Enhanced Multi-Level Precedence and Preemption (EMLPP).** This functionality, which is specified in 3GPP TS 23.067 [20], is used to attach a priority to a point-to-point, VBS

or VGCS call. This enables the network and the mobile devices to automatically preempt ongoing calls for higher priority calls to ensure that emergency calls (e.g., a person has fallen on the track) are not blocked by lower priority calls and a lack of resources (e.g., because no timeslots are available).

As GSM-R networks are private networks, it has been decided to assign a private frequency band in Europe for this purpose, which is just below the public 900-MHz GSM band. To use GSM-R, mobile phones need to be slightly modified to be able to send and receive in this frequency range. This requires only minor software and hardware modifications. To be also able to use the additional functionalities described above, further extensions of the mobile device software are necessary. More about GSM-R can be found at http://gsm-r.uic.asso.fr [21].

1.7.2   The Base Transceiver Station (BTS)

Base stations, which are also called base transceiver stations (BTSs), are the most visible network elements of a GSM system (Figure 1.20). Compared to fixed-line networks, the base stations replace the wired connection to the subscriber with a wireless connection, which is also referred to as the air interface. The base stations are also the most numerous components of a mobile network. In Germany, for example, Telefonica O2 has over 18,000 GSM base stations and the other three network operators are likely to have deployed similar numbers [22]. Figure 1.20 shows a typical base station antenna.

In theory, a base station can cover an area with a radius of up to 35 km. This area is also called a cell. As a base station can only serve a limited number of simultaneous users, cells are much smaller in practice, especially in dense urban environments. In these environments,

Figure 1.20   A typical antenna of a GSM base station. The optional microwave directional antenna (round antenna at the bottom of the mast) connects the base station with the GSM network. Source: Martin Sauter. Reproduced by permission of Martin Sauter

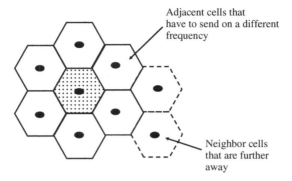

Figure 1.21   Cellular structure of a GSM network

Figure 1.22   Sectorized cell configurations

cells cover areas within a radius between 3 and 4 km in residential and business areas, and down to only several 100 m with minimal transmission power in heavily frequented areas like shopping centers and downtown streets. Even in rural areas, a cell's coverage area is usually less than 15 km as the transmission power of the mobile device of 1 or 2 W is the limiting factor in this case.

As the emissions of different base stations of the network must not interfere with each other, all neighboring cells have to send on different frequencies. As can be seen from Figure 1.21, a single base station usually has quite a number of neighboring sites. Therefore, only a limited number of different frequencies can be used per base station to increase capacity.

To increase the capacity of a base station, the coverage area is usually split into two or three sectors, as shown in Figure 1.22, which are then covered on different frequencies by a dedicated transmitter. This allows a better reuse of frequencies in two-dimensional space than in the case where only a single frequency is used for the whole base station. Each sector of the base station, therefore, forms its own independent cell.

1.7.3   The GSM Air Interface

The transmission path between the BTS and the mobile device is referred to, in the GSM specifications, as the air interface or the Um interface. To allow the base station to communicate with several subscribers simultaneously, two methods are used. The first method is frequency division multiple access (FDMA), which means that users communicate with the base station

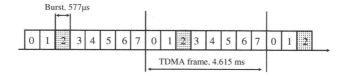

Figure 1.23   A GSM TDMA frame

on different frequencies. The second method used is time division multiple access (TDMA). GSM uses carrier frequencies with a bandwidth of 200 kHz over which up to eight subscribers can communicate with the base station simultaneously as shown in Figure 1.23.

Subscribers are time multiplexed by dividing the carrier into frames with durations of 4.615 milliseconds. Each frame contains eight physically independent timeslots, each for communication with a different subscriber. The time frame of a timeslot is called a burst and the burst duration is 577 microseconds. For example, if a mobile device is allocated timeslot number 2 for a voice call, then the mobile device will send and receive only during this burst. Afterward, it has to wait until the next frame before it is allowed to send again.

By combining the two multiple access schemes it is possible to approximately calculate the total capacity of a base station. For the following example, it is assumed that the base station is split into three sectors and each sector is covered by an independent cell. Each cell is typically equipped with three transmitters and receivers (transceivers). In each sector, $3 \times 8 = 24$ timeslots are thus available. Two timeslots are usually assigned for signaling purposes, which leaves 22 timeslots per sector for user channels. Let us further assume that four or more timeslots are used for the packet-switched GPRS service (see Chapter 2). Therefore, 18 timeslots are left for voice calls per sector, which amounts to 54 channels for all sectors of the base station. In other words, this means that 54 subscribers per base station can communicate simultaneously.

A single BTS, however, provides service to a much higher number of subscribers, as all of them do not communicate at the same time. Mobile operators, therefore, base their network dimensioning on a theoretical call profile model in which the number of minutes per hour that a subscriber statistically uses the system is one of the most important parameters. A commonly used value is 3 for the number of minutes per hour that a subscriber uses the system. This means that a base station is able to provide service to 20 times the number of active subscribers. In this example, a base station with 54 channels is, therefore, able to provide service to about 1080 subscribers.

This number is quite realistic as the following calculation shows: Telefonica O2 Germany had a subscriber base of about 20 million in 2014 [22]. If this value is divided by the number of subscribers per cell, the total number of base stations required to serve such a large subscriber base can be determined. With our estimation above, the number of base stations required for the network would be about 18,500. This value is in line with the numbers published by the operator [22]. In practice, it can be observed today that the voice minutes per subscriber per month are increasing because of falling prices. To compensate, network operators can either increase the number of base stations in the areas of high demand or add additional transceivers to the existing base stations and increase capacity on the backhaul link to the network.

Each burst of a TDMA frame is divided into a number of different sections as shown in Figure 1.24. Each burst ends with a guard time in which no data is sent. This is necessary

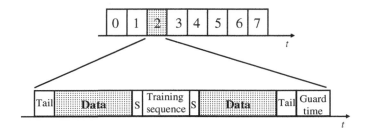

Figure 1.24   A GSM burst

because the distance of the different subscribers relative to the base station can change while they are active. As airwaves propagate 'only' through space at the speed of light, the signal of a faraway subscriber takes a longer time to reach the base station compared to a subscriber that is closer to the base station. To prevent any overlap, guard times were introduced. These parts of the burst are very short, as the network actively controls the timing advance of the mobile device. More about this topic can be found below.

The training sequence in the middle of the burst always contains the same bit pattern. It is used to compensate for interference caused, for example, by reflection, absorption and multi-path propagation. On the receiver side, these effects are countered by comparing the received signal with the training sequence and thus adapting the analog filter parameters for the signal. The filter parameters calculated this way can then be used to modify the rest of the signal and thus to better recreate the original signal.

At the beginning and end of each burst, another well-known bit pattern is sent to enable the receiver to detect the beginning and end of a burst correctly. These fields are called 'tails'. The actual user data of the burst, that is, the digitized voice signal, is sent in the two-user data fields with a length of 57 bits each. This means that a 577-microsecond burst transports 114 bits of user data. Finally, each frame contains 2 bits to the left and right of the training sequence, which are called 'stealing bits'. These bits indicate if the data fields contain user data or are used ('stolen') for urgent signaling information. User data from bursts that carry urgent signaling information are, however, lost. As shown below, the speech decoder is able to cope with short interruptions of the data stream quite well, and thus the interruptions are normally not audible to the user.

For the transmission of user or signaling data, the timeslots are arranged into logical channels. A user data channel for the transmission of digitized voice data, for example, is a logical channel. On the first carrier frequency of a cell, the first two timeslots are usually used for common logical signaling channels while the remaining six independent timeslots are used for user data channels or GPRS. As there are more logical channels than physical channels (timeslots) for signaling, 3GPP TS 45.002 [23] describes how 51 frames are grouped into a multiframe to be able to carry a number of different signaling channels over the same timeslot. In such a multiframe, which is infinitely repeated, it is specified as to which logical channels are transmitted in which bursts on timeslots 0 and 1. For user data timeslots (e.g. voice), the same principle is used. Instead of 51 frames, these timeslots are grouped into a 26-multiframe pattern. For the visualization of this principle, a scheme is shown in Figure 1.25 which depicts

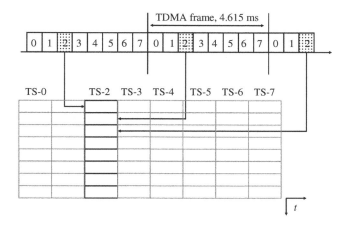

Figure 1.25   Arrangement of bursts of a frame for the visualization of logical channels in Figure 1.26

how the eight timeslots of a frame are grouped into a two-dimensional table. In Figure 1.26, this principle is used to show how the logical channels are assigned to physical timeslots in the multiframe.

Logical channels are arranged into two groups. If data on a logical channel is dedicated to a single user, the channel is called a dedicated channel. If the channel is used for data that needs to be distributed to several users, the channel is called a common channel.

Let us take a look at the dedicated channels first:

- The traffic channel (TCH) is a user data channel. It can be used to transmit a digitized voice signal or circuit-switched data services of up to 14.4 kbit/s.
- The Fast Associated Control Channel (FACCH) is transmitted on the same timeslot as a TCH. It is used to send urgent signaling messages like a handover command. As these messages do not have to be sent very often, no dedicated physical bursts are allocated to the FACCH. Instead, user data are removed from a TCH burst. To inform the mobile device, the stealing bits to the left and right of the training sequence, as shown in Figure 1.24, are used. This is the reason why the FACCH is not shown in Figure 1.26.
- The Slow Associated Control Channel (SACCH) is also assigned to a dedicated connection. It is used in the uplink direction to report signal quality measurements of the serving cell and neighboring cells to the network. The network then uses these values for handover decisions and power control. In the downlink direction, the SACCH is used to send power control commands to the mobile device. Furthermore, the SACCH is used for timing advance control, which is described in Section 1.7.4 and Figure 1.27. As these messages are only of low priority and the necessary bandwidth is very small, only a few bursts are used on a 26-multiframe pattern at fixed intervals.
- The Standalone Dedicated Control Channel (SDCCH) is a pure signaling channel that is used during call establishment when a subscriber has not yet been assigned a TCH. Furthermore, the channel is used for signaling that is not related to call establishment, such as for the location update procedure or for sending or receiving a text message (SMS).

FN	TS-0	TS-1	FN	TS-2	TS-7
0	FCCH	SDCCH/0	0	TCH	TCH
1	SCH	SDCCH/0	1	TCH	TCH
2	BCCH	SDCCH/0	2	TCH	TCH
3	BCCH	SDCCH/0	3	TCH	TCH
4	BCCH	SDCCH/1	4	TCH	TCH
5	BCCH	SDCCH/1	5	TCH	TCH
6	AGCH/PCH	SDCCH/1	6	TCH	TCH
7	AGCH/PCH	SDCCH/1	7	TCH	TCH
8	AGCH/PCH	SDCCH/2	8	TCH	TCH
9	AGCH/PCH	SDCCH/2	9	TCH	TCH
10	FCCH	SDCCH/2	10	TCH	TCH
11	SCH	SDCCH/2	11	TCH	TCH
12	AGCH/PCH	SDCCH/3	12	SACCH	SACCH
13	AGCH/PCH	SDCCH/3	13	TCH	TCH
14	AGCH/PCH	SDCCH/3	14	TCH	TCH
15	AGCH/PCH	SDCCH/3	15	TCH	TCH
16	AGCH/PCH	SDCCH/4	16	TCH	TCH
17	AGCH/PCH	SDCCH/4	17	TCH	TCH
18	AGCH/PCH	SDCCH/4	18	TCH	TCH
19	AGCH/PCH	SDCCH/4	19	TCH	TCH
20	FCCH	SDCCH/5	20	TCH	TCH
21	SCH	SDCCH/5	21	TCH	TCH
22	SDCCH/0	SDCCH/5	22	TCH	TCH
23	SDCCH/0	SDCCH/5	23	TCH	TCH
24	SDCCH/0	SDCCH/6	24	TCH	TCH
25	SDCCH/0	SDCCH/6	25	Free	Free
26	SDCCH/1	SDCCH/6	0	TCH	TCH
27	SDCCH/1	SDCCH/6	1	TCH	TCH
28	SDCCH/1	SDCCH/7	2	TCH	TCH
29	SDCCH/1	SDCCH/7	3	TCH	TCH
30	FCCH	SDCCH/7	4	TCH	TCH
31	SCH	SDCCH/7	5	TCH	TCH
32	SDCCH/2	SACCH/0	6	TCH	TCH
33	SDCCH/2	SACCH/0	7	TCH	TCH
34	SDCCH/2	SACCH/0	8	TCH	TCH
35	SDCCH/2	SACCH/0	9	TCH	TCH
36	SDCCH/3	SACCH/1	10	TCH	TCH
37	SDCCH/3	SACCH/1	11	TCH	TCH
38	SDCCH/3	SACCH/1	12	SACCH	SACCH
39	SDCCH/3	SACCH/1	13	TCH	TCH
40	FCCH	SACCH/2	14	TCH	TCH
41	SCH	SACCH/2	15	TCH	TCH
42	SACCH/0	SACCH/2	16	TCH	TCH
43	SACCH/0	SACCH/2	17	TCH	TCH
44	SACCH/0	SACCH/3	18	TCH	TCH
45	SACCH/0	SACCH/3	19	TCH	TCH
46	SACCH/1	SACCH/3	20	TCH	TCH
47	SACCH/1	SACCH/3	21	TCH	TCH
48	SACCH/1	Free	22	TCH	TCH
49	SACCH/1	Free	23	TCH	TCH
50	Free	Free	24	TCH	TCH
			25	Free	Free

Figure 1.26 Use of timeslots in the downlink direction per 3GPP TS 45.002 [23]

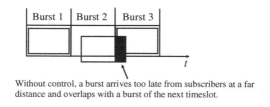

Without control, a burst arrives too late from subscribers at a far
distance and overlaps with a burst of the next timeslot.

Figure 1.27   Time shift of bursts of distant subscribers without timing advance control

Besides the dedicated channels, which are always assigned to a single user, there are a number of common channels that are monitored by all subscribers in a cell:

- The Synchronization Channel (SCH) is used by mobile devices during network and cell searches.
- The Frequency Correction Channel (FCCH) is used by the mobile devices to calibrate their transceiver units, and is also used to detect the beginning of a multiframe.
- The Broadcast Common Control Channel (BCCH) is the main information channel of a cell and broadcasts SYS_INFO messages that contain a variety of information about the network. The channel is monitored by all mobile devices, which are switched on but currently not engaged in a call or signaling connection (idle mode), and broadcasts, among many other things, the following information:
- the MCC and MNC of the cell;
- the identification of the cell which consists of the location area code (LAC) and the cell-ID;
- to simplify the search for neighboring cells for a mobile device, the BCCH also contains information about the frequencies used by neighboring cells. Thus, the mobile device does not have to search the complete frequency band for neighboring cells.
- The Paging Channel (PCH) is used to inform idle subscribers of incoming calls or SMS messages. As the network alone is aware of the location area the subscriber is roaming in, the paging message is broadcast in all cells belonging to the location area. The most important information element of the message is the IMSI of the subscriber or a temporary identification called the Temporary Mobile Subscriber Identity (TMSI). A TMSI is assigned to a mobile device during the network attach procedure and can be changed by the network every time the mobile device contacts the network once encryption has been activated. Thus, the subscriber has to be identified with the IMSI only once and is then addressed with a constantly changing temporary number when encryption is not yet activated for the communication. This increases anonymity in the network and prevents eavesdroppers from creating movement profiles of subscribers.
- The Random Access Channel (RACH) is the only common channel in the uplink direction. If the mobile device receives a message via the PCH that the network is requesting a connection establishment or if the user wants to establish a call or send an SMS, the RACH is used for the initial communication with the network. This is done by sending a channel request message. Requesting a channel has to be done via a 'random' channel because subscribers in a cell are not synchronized with each other. Thus, it cannot be ensured that two devices do not try to establish a connection at the same time. Only when a dedicated channel (SDCCH) has been assigned to the mobile device by the network can there no longer be any collision between different subscribers of a cell. If a collision occurs during the first network access, the colliding messages are lost and the mobile devices do not receive an answer from the

network. Thus, they have to repeat their channel request messages after expiry of a timer that
is set to an initial random value. This way, it is not very likely that the mobile devices will
interfere with each other again during their next connection establishment attempts because
they are performed at different times.
- The Access Grant Channel (AGCH): If a subscriber sends a channel request message on
the RACH, the network allocates an SDCCH or, in exceptional cases, a TCH, and notifies
the subscriber on the AGCH via an immediate assignment message. The message contains
information about which SDCCH or TCH the subscriber is allowed to use.

Figure 1.28 shows how PCH, AGCH and SDCCH are used during the establishment of a
signaling link between the mobile device and the network. The base station controller (BSC),
which is responsible for assigning SDCCH and TCH of a base station, is further described in
Section 1.7.4.
As can also be seen from Figure 1.26, all bursts on timeslots 2−7 are not used for TCHs.
Every 12th burst of a timeslot is used for the SACCH. Furthermore, the 25th burst is also not
used for carrying user data. This gap is used to enable the mobile device to perform signal
strength measurements of neighboring cells on other frequencies. This is necessary so that the
network can redirect the connection to a different cell (handover) to maintain the call while
the user is moving.
The GSM standard offers two possibilities to use the available frequencies. The simplest
case, which has been described already, is the use of a constant carrier frequency (ARFCN) for
each channel. To improve the transmission quality, it is also possible to use alternating frequencies for a single channel of a cell. This concept is known as frequency hopping, and it changes
the carrier frequency for every burst during a transmission. This increases the probability that

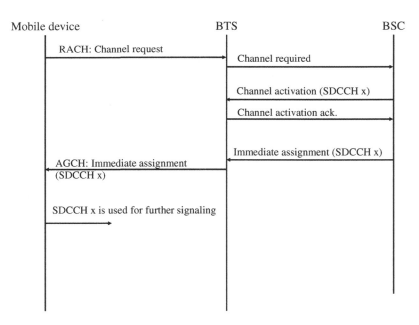

Figure 1.28 Establishment of a signaling connection

only few bits are lost if one carrier frequency experiences a lot of interference from other sources like neighboring cells. In the worst case, only a single burst is affected because the following burst is already sent on a different frequency. Up to 64 different frequencies can be used per base station for frequency hopping. To inform the mobile of the use of frequency hopping, the immediate assignment message used during the establishment of a signaling link contains all the information about the frequencies that are used and the hopping pattern that is applied to the connection.

For carriers that transport the SCH, FCCH and BCCH channels, frequency hopping must not be used. This restriction is necessary because it would be very difficult for mobile devices to find neighboring cells.

In practice, network operators use static frequencies as well as frequency hopping in their networks.

The interface which connects the base station to the network and which is used to carry the information for all logical channels is called the Abis interface. An E-1 connection is usually used for the Abis interface, and owing to its 64 kbit/s timeslot architecture the logical channels are transmitted in a way different from their transmission on the air interface. All common channels as well as the information sent and received on the SDCCH and SACCH channels are sent over one or more common 64 kbit/s E-1 timeslots. This is possible because these channels are only used for signaling data that are not time critical. On the Abis interface, these signaling messages are sent by using the Link Access Protocol (LAPD). This protocol was initially designed for the ISDN D-channel of fixed-line networks and has been reused for GSM with only minor modifications.

For TCHs that use a bandwidth of 13 kbit/s on the Abis interface, only one-quarter of an E-1 timeslot is used. This means that all eight timeslots of an air interface frame can be carried on only two timeslots of the E-1 interface. A base station composed of three sectors, which use two carriers each, thus requires 12 timeslots on the Abis interface plus an additional timeslot for the LAPD signaling. The remaining timeslots of the E-1 connection can be used for the communication between the network and other base stations as shown in Figure 1.29. For this purpose, several cells are usually daisy chained via a single E-1 connection (see Figure 1.29).

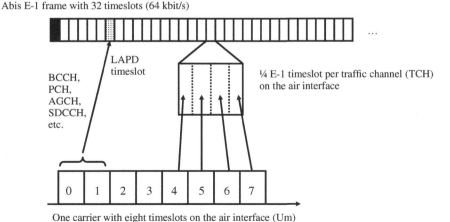

Figure 1.29 Mapping of E-1 timeslots to air interface timeslots

In practice, it can be observed today that physical E-1 links being replaced more and more with virtual connections over IP-based links. This is especially the case if a base station site is used for several radio technologies simultaneously (e.g. GSM, UMTS and LTE).

1.7.4   The Base Station Controller (BSC)

While the base station is the interface element that connects the mobile devices with the network, the BSC is responsible for the establishment, release and maintenance of all connections of cells that are connected to it.

If a subscriber wants to establish a voice call, send an SMS, and so on, the mobile device sends a channel request message to the BSC as shown in Figure 1.28. The BSC then checks if an SDCCH is available and activates the channel in the BTS. Afterward, the BSC sends an immediate assignment message to the mobile device on the AGCH that includes the number of the assigned SDCCH. The mobile device then uses the SDCCH to send DTAP messages that the BSC forwards to the MSC.

The BSC is also responsible for establishing signaling channels for incoming calls or SMS messages. In this case, the BSC receives a paging message from the MSC, which contains the IMSI and TMSI of the subscriber as well as the location area ID in which the subscriber is currently located. The BSC in turn has a location area database that it uses to identify all cells in which the subscriber needs to be paged. When the mobile device receives the paging message, it responds to the network in the same way as in the example above by sending a channel request message.

The establishment of a TCH for voice calls is always requested by the MSC for both mobile-originated and mobile-terminated calls. Once the mobile device and the MSC have exchanged all necessary information for the establishment of a voice call via an SDCCH, the MSC sends an assignment request for a voice channel to the BSC as shown in Figure 1.30.

The BSC then verifies if a TCH is available in the requested cell and, if so, activates the channel in the BTS. Afterward, the mobile device is informed via the SDCCH that a TCH is now available for the call. The mobile device then changes to the TCH and FACCH. To inform the BTS that it has switched to the new channel, the mobile device sends a message to the BTS on the FACCH, which is acknowledged by the BTS. In this way, the mobile also has a confirmation that its signal can be decoded correctly by the BTS. Finally, the mobile device sends an assignment complete message to the BSC which in turn informs the MSC of the successful establishment of the TCH.

Apart from the establishment and release of a connection, another important task of the BSC is the maintenance of the connection. As subscribers can roam freely through the network while a call is ongoing, it can happen that the subscriber roams out of the coverage area of the cell in which the call was initially established. In this case, the BSC has to redirect the call to the appropriate cell. This procedure is called handover. To be able to perform a handover to another cell, the BSC requires signal quality measurements for the air interface. The results of the downlink signal quality measurements are reported to the BSC by the mobile device, which continuously performs signal quality measurements that it reports via the SACCH to the network. The uplink signal quality is constantly measured by the BTS and also reported to the BSC. Apart from the signal quality of the user's current cell, it is also important that the mobile device reports the quality of signals it receives from other cells. To enable the mobile device to perform these measurements, the network sends the frequencies of neighboring cells via the

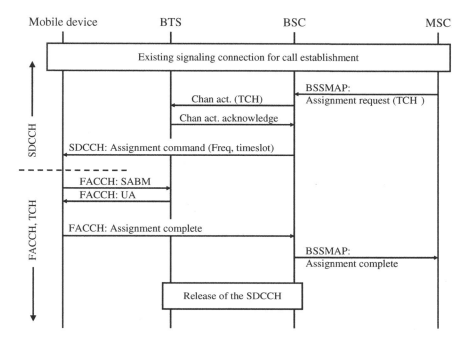

Figure 1.30　Establishment of a traffic channel (TCH)

SACCH during an ongoing call. The mobile device then uses this information to perform the neighboring cell measurements while the network communicates with other subscribers and reports the result via measurement report messages in the uplink SACCH.

The network receives these measurement values and is thus able to periodically evaluate if a handover of an ongoing call to a different cell is necessary. Once the BSC decides to perform a handover, a TCH is activated in the new cell as shown in Figure 1.31. Afterward, the BSC informs the mobile device via the old cell with a handover command message that is sent over the FACCH. Important information elements of the message are the new frequency and timeslot number of the new TCH. The mobile device then changes its transmit and receive frequency, synchronizes to the new cell if necessary and sends a handover access message in four consecutive bursts. In the fifth burst, a Set Asynchronous Balanced Mode (SABM) message is sent, which is acknowledged by the BTS to signal to the mobile device that the signal can be received. At the same time, the BTS informs the BSC of the successful reception of the mobile device's signal with an establish indication message. The BSC then immediately redirects the speech path to the new cell.

From the mobile's point of view the handover is now finished. The BSC, however, has to release the TCH in the old cell and has to inform the MSC of the performed handover before the handover is finished from the network's point of view. The message to the MSC is only informative and has no impact on the continuation of the call.

To reduce interference, the BSC is also in charge of controlling the transmission power for every air interface connection. For the mobile device, an active power control has the advantage that the transmission power can be reduced under favorable reception conditions. Transmission

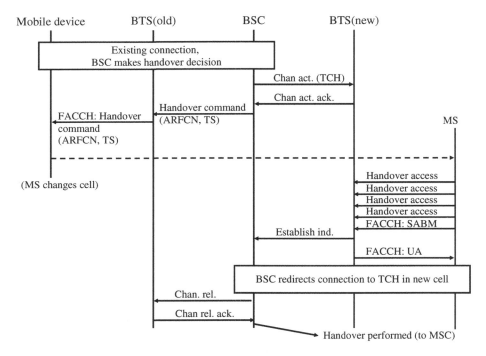

Figure 1.31   Message flow during a handover procedure

power is controlled by using the signal quality measurements of the BTS for the connection. If the mobile device's transmission power has to be increased or decreased, the BSC sends a power control message to the BTS. The BTS in turn forwards the message to the mobile device and repeats the message on the SACCH in every frame. In practice, it can be observed that power control and adaptation is performed every 1–2 seconds. During call establishment, the mobile device always uses the highest allowed power output level, which is then reduced or increased again by the network step by step. Table 1.6 gives an overview of the mobile device power levels. A distinction is made for the 900-MHz versus the 1800-MHz band. While mobile devices operating on the 900-MHz band are allowed to use up to 2 W, connections on the 1800-MHz band are limited to 1 W. For stationary devices or car phones with external antennas, power values of up to 8 W are allowed. The power values in the table represent the power output when the transmitter is active in the assigned timeslot. As the mobile device only sends on one of the eight timeslots of a frame, the average power output of the mobile device is only one-eighth of this value. The average power output of a mobile device that sends with a power output of 2 W is thus only 250 mW.

The BSC is also able to control the power output of the base station. This is done by evaluating the signal measurements of the mobile devices in the current cell. It is important to note that power control can only be performed for downlink carriers that do not broadcast the common channels like frame control header (FCH), SCH and BCCH of a cell. On such carriers, the power output has to be constant to allow mobile devices, which are currently located in other cells of the network, to perform their neighboring cell measurements. This would not be

Table 1.6 GSM power levels and corresponding power output

GSM 900 Power level	GSM 900 Power output	GSM 1800 Power level	GSM 1800 Power output
(0–2)	(8 W)	–	–
5	2 W	0	1 W
6	1.26 W	1	631 mW
7	794 mW	2	398 mW
8	501 mW	3	251 mW
9	316 mW	4	158 mW
10	200 mW	5	100 mW
11	126 mW	6	63 mW
12	79 mW	7	40 mW
13	50 mW	8	25 mW
14	32 mW	9	16 mW
15	20 mW	10	10 mW
16	13 mW	11	6.3 mW
17	8 mW	12	4 mW
18	5 mW	13	2.5 mW
19	3.2 mW	14	1.6 mW
–	–	15	1.0 mW

possible if the signal amplitude varies over time as the mobile devices can only listen to the carrier signal of neighboring cells for a short time.

Owing to the limited speed of radio waves, a time shift of the arrival of the signal can be observed when a subscriber moves away from a base station during an ongoing call. If no countermeasures are taken, this would mean that at some point the signal of a subscriber would overlap with the next timeslot despite the guard time of each burst, which is shown in Figure 1.27. Thus, the signal of each subscriber has to be carefully monitored and the timing of the transmission of the subscriber has to be adapted. This procedure is called timing advance control (Figure 1.30).

The timing advance can be controlled in 64 steps (0–63) of 550 m. The maximum distance between a base station and a mobile subscriber is in theory 64×550 m $= 35.2$ km. In practice, such a distance is not reached very often as base stations usually cover a much smaller area owing to capacity reasons. Furthermore, the transmission power of the mobile device is also not sufficient to bridge such a distance under non-line-of-sight conditions to the base station. Therefore, one of the few scenarios where such a distance has to be overcome is in coastal areas, from ships at sea.

The control of the timing advance already starts with the first network access on the RACH with a channel request message. This message is encoded into a very short burst that can only transport a few bits in exchange for large guard periods at the beginning and end of the burst. This is necessary because the mobile device is unaware of the distance between itself and the base station when it attempts to contact the network. Thus, the mobile device is unable to select an appropriate timing advance value. When the base station receives the burst, it measures the delay and forwards the request, including a timing advance value required for this mobile device, to the BSC. As has been shown in Figure 1.28, the BSC reacts to the connection request by returning an immediate assignment message to the mobile device on the AGCH. Apart from the number of the assigned SDCCH, the message also contains a first timing advance value to be used for the subsequent communication on the SDCCH. Once the connection has been successfully established, the BTS continually monitors the delay experienced for this channel and reports any changes to the BSC. The BSC in turn instructs the mobile device to change its timing advance by sending a message on the SACCH.

For special applications, like coastal communication, the GSM standard offers an additional timeslot configuration to increase the maximum distance to the base station to up to 120 km. This is achieved by only using every second timeslot per carrier, which allows a burst to overlap onto the following (empty) timeslot. While this significantly increases the range of a cell, the number of available communication channels is cut in half. Another issue is that mobile devices that are limited to a transmission power of 1 W (1800-MHz band) or 2 W (900-MHz band) may be able to receive the BCCH of such a cell at a great distance but are unable to communicate with the cell in the uplink direction. Thus, such an extended range configuration mostly makes sense with permanently installed mobile devices with external antennas that can transmit with a power level of up to 8 W.

1.7.5   The TRAU for Voice Encoding

For the transmission of voice data, a TCH is used in GSM as described in Section 1.7.3. A TCH uses all but two bursts of a 26-burst multiframe, with one being reserved for the SACCH, as shown in Figure 1.26, and the other remaining empty to allow the mobile device to perform neighboring cell measurements. As has been shown in the preceding section, a burst that is sent to or from the mobile every 4.615 milliseconds can carry exactly 114 bits of user data. When taking the two bursts, which are not used for user data of a 26-burst multiframe, into account, this results in a raw datarate of 22.8 kbit/s. As is shown in the remainder of this section, a substantial part of the bandwidth of a burst is required for error detection and correction bits. The resulting datarate for the actual user data is thus around 13 kbit/s.

The narrow bandwidth of a TCH stands in contrast to how a voice signal is transported in the core network. Here, the PCM algorithm is used (see Section 1.6.1) to digitize the voice signal, which makes full use of the available 64 kbit/s bandwidth of an E-1 timeslot to encode the voice signal (see Figure 1.32).

A simple solution for the air interface would have been to define air interface channels that can also carry 64 kbit/s PCM-coded voice channels. This has not been done because the scarce resources on the air interface have to be used as efficiently as possible. The decision to compress the speech signal was taken during the first standardization phase in the 1980s because it was foreseeable that advances in hardware and software processing capabilities would allow compression of a voice data stream in real time.

In the mobile network, the compression and decompression of the voice data stream is performed in the Transcoding and Rate Adaptation Unit (TRAU), which is located between the MSC and a BSC and controlled by the BSC (see Figure 1.32). During an ongoing call, the MSC sends the 64 kbit/s PCM-encoded voice signal toward the radio network and the TRAU converts the voice stream in real time into a 13 kbit/s compressed data stream, which is transmitted over the air interface. In the other direction, the BSC sends a continuous stream of compressed voice data toward the core network and the TRAU converts the stream into a 64 kbit/s coded

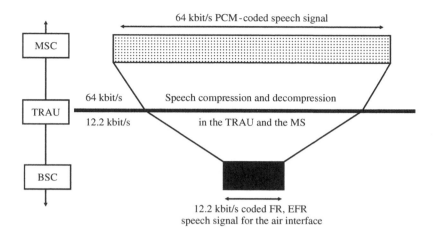

Figure 1.32   GSM speech compression

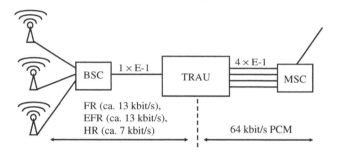

Figure 1.33   Speech compression with a 4:1 compression ratio in the TRAU

PCM signal. In the mobile device, the same algorithms are implemented as in the TRAU to compress and decompress the speech signal (see Figure 1.33).

While the TRAU is a logical component of the BSS, it is most often installed next to an MSC in practice. This has the advantage that four compressed voice channels can be transmitted in a single E-1 timeslot. After compression, each voice channel uses a 16 kbit/s sub-timeslot. Thus, only one-quarter of the transmission capacity between an MSC and BSC is needed in comparison to an uncompressed transmission. As the BSCs of a network are usually located in the field and not close to an MSC, this helps to reduce transmission costs for the network operator substantially as shown in Figure 1.33.

The TRAU offers a number of different algorithms for speech compression. These algorithms are called speech codecs or simply codecs. The first codec that was standardized for GSM is the full-rate (FR) codec that reduces the 64 kbit/s voice stream to about 13 kbit/s.

At the end of the 1990s, the enhanced full-rate (EFR) codec was introduced. The EFR codec not only compresses the speech signal to about 13 kbit/s but also offers a superior voice quality compared to the FR codec. The disadvantage of the EFR codec is the higher complexity of the compression algorithm that requires more processing power. However, the processing power available in mobile devices has increased significantly in the 1990s, and thus modern GSM phones easily cope with the additional complexity.

Besides those two codecs, a half-rate (HR) codec has been defined for GSM that only requires a bandwidth of 7 kbit/s. While there is almost no audible difference between the EFR codec and a PCM-coded speech signal, the voice quality of the HR codec is noticeably inferior. The advantage for the network operator of the HR codec is that the number of simultaneous voice connections per carrier can be doubled. With the HR codec, a single timeslot, which is used for a single EFR voice channel, can carry two TCH (HR)s.

Another speech codec development is the Adaptive Multirate (AMR) algorithm [24] that is used by most devices and networks today. Instead of using a single codec, which is selected at the beginning of the call, AMR allows a change of the codec during a call. The considerable advantage of this approach is the ability to switch to a speech codec with a higher compression rate during bad radio signal conditions to increase the number of error detection and correction bits. If signal conditions permit, a lower rate codec can be used, which only uses every second burst of a frame for the call. This in effect doubles the capacity of the cell as a single timeslot can be shared by two calls in a similar manner to the HR codec. Unlike the HR codec, however, the AMR codecs, which only use every second burst and which are thus called HR AMR codecs, still have a voice quality which is comparable to that of the EFR codec. While AMR is optional for GSM, it has been chosen for the UMTS system as a mandatory feature. In the United States, AMR is used by some network operators to increase the capacity of their network, especially in very dense traffic areas like New York, where it has become very difficult to increase the capacity of the network any further with over half a dozen carrier frequencies per sector already used. Further information about AMR can also be found in Chapter 3.

The latest speech codec development used in practice is AMR-Wideband (AMR-WB) as specified in ITU G.722.2 [25] and 3GPP TS 26.190 [26]. The algorithm allows, as its name implies, to digitize a wider frequency spectrum than is possible with the PCM algorithm that was described earlier. Instead of an upper limit of 3400 Hz, AMR-WB digitizes a voice signal up to a frequency of 7000 Hz. As a consequence, the caller's voice sounds much clearer and more natural on the other end of a connection. A high compression rate is used in practice to reduce the data rate of a voice stream down to 12.65 kbit/s. This way, an AMR-WB data stream can be transmitted in a single GSM timeslot, and also requires no additional capacity in a UMTS network. As AMR-WB is not compatible with the PCM codec used between the BSC and MSC, it is sent transparently between the two nodes. This means that most of the bits in a 64 kbit/s PCM timeslot are unused, as the data rate required by the AMR-WB codec is only 12.65 kbit/s. In practice, AMR-WB is mostly used in UMTS networks today. Therefore, it is described in more detail in Chapter 3.

While the PCM algorithm digitizes analog volume levels by statically mapping them to digital values, the GSM speech digitization is much more complex to reach the desired compression rate. In the case of the FR codec, which is specified in 3GPP TS 46.010 [27], the compression is achieved by emulating the human vocal system. This is done by using a source–filter model (Figure 1.34). In the human vocal system, speech is created in the larynx and by the vocal cords. This is emulated in the mathematical model in the signal creation part, while the filters represent the signal formation that occurs in the human throat and mouth.

On a mathematical level, speech formation is simulated by using two time-invariant filters. The period filter creates the periodic vibrations of the human voice while the vocal tract filter

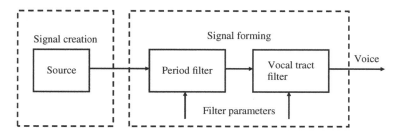

Figure 1.34   Source–filter model of the GSM FR codec

simulates the envelope. The filter parameters are generated from the human voice, which is the input signal into the system. To digitize and compress the human voice, the model is used in the reverse direction as shown in Figure 1.34. As time-variant filters are hard to model, the system is simplified by generating a pair of filter parameters for an interval of 20 milliseconds as shown in Figure 1.35. A speech signal that has previously been converted into an 8- or 13-bit PCM codec is used as an input to the algorithm. As the PCM algorithm delivers 8000 values per second, the FR codec requires 160 values for a 20 milliseconds interval to calculate the filter parameters. As 8 bits are used per value, 8 bits × 160 values = 1280 input bits are used per 20-millisecond interval. For the period filter, the input bits are used to generate a filter parameter with a length of 36 bits. Afterward, the filter is applied to the original input signal. The resulting signal is then used to calculate another filter parameter with a length of 36 bits

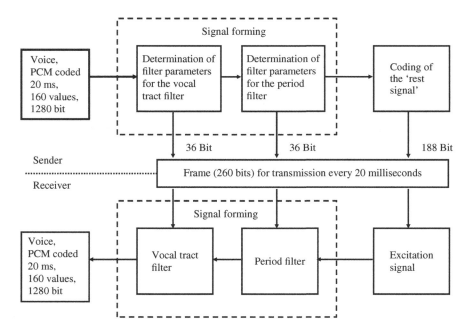

Figure 1.35   Complete transmission chain with the transmitter and receiver of the GSM FR codec

for the vocal tract filter. Afterward, the signal is again sent through the vocal tract filter with the filter parameter applied. The signal, which is thus created, is called the 'rest signal' and coded into 188 bits (see Figure 1.35).

Once all parameters have been calculated, the two 36-bit filter parameters and the rest signal, which is coded into 188 bits, are sent to the receiver. Thus, the original information, which was coded in 1280 bits, has been reduced to 260 bits. In the receiver, the filter procedure is applied in reverse order on the rest signal and thus the original signal is recreated. As the procedure uses a lossy compression algorithm, the original signal and the recreated signal at the other end are no longer exactly identical. For the human ear, however, the differences are almost inaudible.

1.7.6 Channel Coder and Interleaver in the BTS

When a 260-bit data frame from the TRAU arrives at the base station every 20 milliseconds, it is further processed before being sent over the air as shown in Figure 1.36. In the reverse direction, the tasks are performed in the mobile device.

In the first step, the voice frames are processed in the channel coder unit, which adds error detection and correction information to the data stream. This step is very important as the transmission over the air interface is prone to frequent transmission errors due to the constantly changing radio environment. Furthermore, the compressed voice information is very sensitive and even a few bits that might be changed while the frame is transmitted over the air interface create an audible distortion. To prevent this, the channel coder separates the 260 bits of a voice data frame into three different classes as shown in Figure 1.37.

Fifty of the 260 bits of a speech frame are class Ia bits and extremely important for the overall reproduction of the voice signal at the receiver side. Such bits are, for example, the higher order bits of the filter parameters. To enable the receiver to verify the correct transmission of those bits, a three-bit cyclic redundancy check (CRC) checksum is calculated and added to the data stream. If the receiver cannot recreate the checksum with the received bits later on, the frame is discarded.

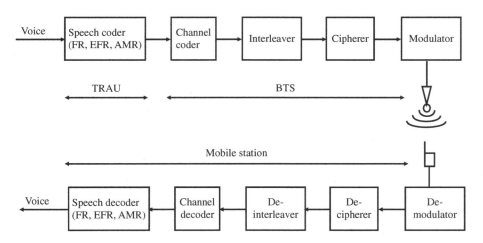

Figure 1.36 Transmission path in the downlink direction between the network and the mobile device

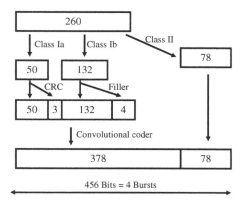

Figure 1.37   GSM channel coder for full-rate speech frames

The other 132 bits of the frame are also quite important and are thus put into class Ib. However, no checksum is calculated for them. To generate the exact amount of bits that are necessary to fill a GSM burst, four filler bits are inserted. Afterward, the class Ia bits, checksum, class Ib bits and the four filler bits are treated by a convolutional coder that adds redundancy to the data stream. For each input bit, the convolutional decoder calculates two output bits. For the computation of the output bits, the coder uses not only the current bit but also the information about the values of the previous bits. For each input bit, two output bits are calculated. This mathematical algorithm is also called a HR convolutional coder.

The remaining 78 bits of the original 260-bit data frame belong to the third class, which is called class II. These are not protected by a checksum and no redundancy is added for them. Errors that occur during the transmission of these bits can neither be detected nor corrected.

As has been shown, the channel coder uses the 260-bit input frame to generate 456 bits on the output side. As a burst on the air interface can carry exactly 114 bits, four bursts are necessary to carry the frame. As the bursts of a TCH are transmitted every 4.6152 milliseconds, the time it takes to transmit the frame over the air interface is about 20 milliseconds. To get to exactly 20 milliseconds, the empty burst and the burst used for the SACCH per 26-burst multiframe has to be included in the calculation.

Owing to the redundancy added by the channel coder, it is possible to correct a high number of faulty bits per frame. The convolutional decoder, however, has one weak point: if several consecutive bits are changed during the transmission over the air interface, the convolutional decoder on the receiver side is not able to correctly reconstruct the original frame. This effect is often observed as air interface disturbances usually affect several bits in a row.

To decrease this effect, the interleaver changes the bit order of a 456-bit data frame in a specified pattern over eight bursts, as shown in Figure 1.38. Consecutive frames are thus interlocked with each other. On the receiver side, the frames are put through the de-interleaver, which puts the bits again into the correct order. If several consecutive bits are changed because of air interface signal distortion, this operation disperses the faulty bits in the frame and the convolutional decoder can thus correctly restore the original bits. A disadvantage of the interleaver, however, is an increased delay in the voice signal. In addition to the delay of 20 milliseconds generated by the FR coder, the interleaver adds another 40 milliseconds as a speech frame is spread over

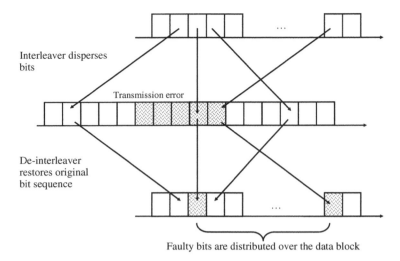

Figure 1.38   Frame interleaving

eight bursts instead of being transmitted consecutively in four bursts. Compared to a voice call in a fixed-line network, a mobile network thus introduces a delay of at least 60 milliseconds. If the call is established between two mobile devices, the delay is at least 120 milliseconds as the transmission chain is traversed twice.

1.7.7   Ciphering in the BTS and Security Aspects

The next module of the transmission chain is the cipherer (Figure 1.39), which encrypts the data frames it receives from the interleaver. GSM uses, like most communication systems, a stream cipher algorithm. To encrypt the data stream, a ciphering key (Kc) is calculated in the AuC and on the SIM card by using a random number (RAND) and the secret key (Ki) as input parameters for the A8 algorithm. Together with the GSM frame number, which is increased for every air interface frame, Kc is then used as input parameter for the A5 ciphering algorithm. The A5 algorithm computes a 114-bit sequence which is XOR combined with the bits of the original data stream. As the frame number is different for every burst, it is ensured that the 114-bit ciphering sequence also changes for every burst, which further enhances security.

To be as flexible as possible, a number of different ciphering algorithms have been specified for GSM. These are called A5/1, A5/2, A5/3 and so on. The intent of allowing several ciphering algorithms was to enable export of GSM network equipment to countries where export restrictions prevent the sale of some ciphering algorithms and technologies. Furthermore, it is possible to introduce new ciphering algorithms into already existing networks to react to security issues if a flaw is detected in one of the currently used algorithms. The selection of the ciphering algorithm also depends on the capabilities of the mobile device. During the establishment of a connection, the mobile device informs the network about the ciphering algorithms that it supports. The network can then choose an algorithm that is supported by the network and the mobile device.

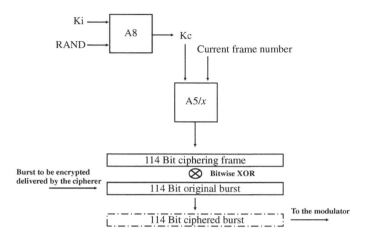

Figure 1.39   Ciphering of an air interface burst

When the mobile device establishes a new connection with the network, its identity is verified before it is allowed to proceed with the call setup. This procedure has already been described in Section 1.6.4. Once the mobile device and subscriber have been authenticated, the MSC usually starts encryption by sending a ciphering command to the mobile device. The ciphering command message contains, among other information elements, the ciphering key, Kc, which is used by the base station for the ciphering of the connection on the air interface. Before the BSC forwards the message to the mobile device, however, the ciphering key is removed from the message because this information must not be sent over the air interface. The mobile device does not need to receive the ciphering key from the network as the SIM card calculates the Kc on its own and forwards the key to the mobile device together with the SRES during the authentication procedure. Figure 1.40 further shows how ciphering is activated during a location update procedure.

With the rising popularity of GSM over the last 20 years, its authentication and encryption procedures have received a lot of scrutiny. From a user point of view, encryption and other security measures must prevent eavesdropping on any kind of communication such as voice conversations, SMS message transfers and signaling in general. Furthermore, it must prevent the theft and misuse of personal authentication data to ensure integrity of the system and to prevent false billing. Also, mobile devices must be protected from third party attacks that attempt to steal or alter personal data from mobile devices that are directly based on the air interface.

At the time of writing, a number of security issues have been found in the GSM security architecture from a user point of view. In this regard, it is important to differentiate between several categories:

1. Theoretical security issues which, at the time of writing, cannot as yet be exploited.
2. Security issues for which practical exploits are likely to have been developed but which require sophisticated and expensive equipment which are not available to the general public.
3. The third group covers security issues which can be exploited with hardware and software available to the general public.

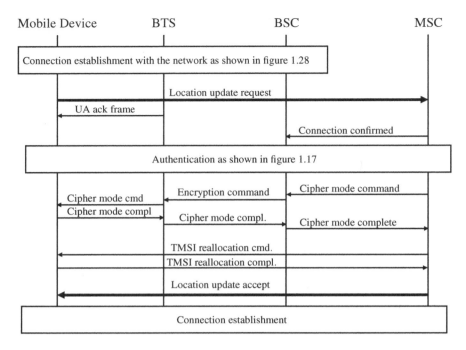

Figure 1.40   Message flow for a location update procedure

The following discussion gives an overview of a number of security issues from the second category, which are described in more detail in Barkan *et al.* [28], the 26C3 [29] and the 28C3 [30]:

- **No encryption on the Abis interface.** The communication link between the base station and the BSC is not ciphered today. Attackers with equipment that is able to intercept E1-based communication over a microwave or cable link can potentially intercept signaling messages and voice calls. In the future, this risk is likely to be reduced with the introduction of encrypted high-speed IP-based communication to multiradio access technology base stations as discussed in Chapters 3 and 4.
- **No mandatory air interface encryption.** Encryption is activated with a ciphering command message by the network. If not activated, all signaling and voice calls are transmitted without protection from eavesdropping. In practice, networks always activate air interface encryption today. Some phones indicate an unencrypted communication link with an open lock symbol.
- **No network authentication.** In practice, this allows attacks that are based on false base stations. By placing such a false base station close to a user and by using transmission power higher than that of any other base station from the network operator, the mobile device will automatically select the false base station and transmit its IMSI in the location update dialog that is further described in Section 1.8.1. The false base station can then use this information to intercept all incoming and outgoing communication by using the user's IMSI itself for communication with the network. By preventing the use of encryption, the

need to get access to the shared secret Ki is dispensed with (cp. Section 1.6.4). Such devices are known as 'IMSI catchers' and further details can be found in [31] and Frick and Bott [32].

Potential protection against such an attack would be to mandate authentication and encryption on the mobile side for every connection establishment. While it would still be possible to collect IMSIs, this would prevent the false base station from eavesdropping on SMS messages and voice calls. At the time of publication, however, such a protection is not implemented in mobile devices.

- **A5/2 Weaknesses.** This encryption algorithm was created to allow the export of GSM systems to countries for which export restrictions concerning security technologies exist. With the processing power of today's computers, it is possible to retrieve the ciphering key Kc within seconds with only little ciphering data collected. As A5/2 is not used in countries where no export restrictions apply, this in itself is not an issue.

- **A5/1 and A5/3 Active attacks.** The weakness of A5/2 can potentially be used for indirect attacks on communication encrypted with more secure A5 algorithms such as A5/1 and A5/3. This requires equipment that can not only intercept a data transfer but that also act as a false base station as described above. In the first step, A5/1 or A5/3 encrypted data are recorded. In the second step, the secret ciphering key, Kc, that was used to encrypt the conversation is recovered by actively contacting the mobile device and instructing it to activate A5/2 ciphering without supplying new keying material. With subsequent frames now being encrypted with A5/2, its weaknesses can be exploited to calculate the secret ciphering key Kc. As no new ciphering material is supplied for this conversation, the recovered Kc is the same as that previously used for the recorded data that were encrypted using A5/1. To counter this attack, the GSM Association recommends that new mobile devices shall no longer support A5/2. This has been implemented in practice and today, only very old mobile devices are still vulnerable.

- **A5/1 Passive attacks.** Researchers have practically demonstrated that passive attacks on A5/1 are possible under the following conditions:
  - A correctly received data stream can be recorded.
  - Empty bits in GSM signaling frames (fillbits) are sent with a repeating bit pattern.
  - A precomputed decryption table with a size of around 4 TB is available.

While computing and storing the decryption table posed an insurmountable challenge even for specialized equipment at the time A5/1 was conceived, it has now become possible to compute the table in a reasonable amount of time and to store the result. The required hardware and open source software are now easily available at low cost, and a practical real-time exploit has been demonstrated during the 28th CCC Congress in December 2011 [30].

This threat can be countered by using the A5/3 encryption algorithm for communication, which at the time of writing is considered to be secure. Today, A5/3 is supported by most new devices appearing in the market but only by a few networks. Further, the mobile device must not support A5/2 to deny an attacker the calculation of the key later on as described above. Another method to protect communication against a passive A5/1 attack is to randomize the fillbits in GSM signaling frames in both the uplink and the downlink directions. This has been standardized a number of years ago in 3GPP TS 44.008, Chapter 5.2. In practice, it can be observed that some devices and networks randomize the fillbits today, but widespread acceptance has still not been reached.

At this point, it is worth noting that the efforts described above were targeted at the ciphering key Kc. No practical methods are known that do not require physical access to the SIM card to break the authentication and key generation algorithms A3 and A8 to get to the shared secret key Ki. This means that should an attacker get the ciphering key Kc of a user, he would still not be able to authenticate during the next network challenge. This means that if the network requires authentication and ciphering for each communication session, it is not possible for an attacker to impersonate another subscriber to receive calls or SMS messages in his place or to make outgoing calls.

1.7.8   Modulation

At the end of the transmission chain, the modulator maps the digital data onto an analog carrier, which uses a bandwidth of 200 kHz. This mapping is done by encoding the bits into changes of the carrier frequency. As the frequency change takes a finite amount of time, a method called Gaussian minimum shift keying (GMSK) is used, which smooths the flanks created by the frequency changes. GMSK has been selected for GSM as its modulation and demodulation properties are easy to handle and implement into hardware and as it interferes only slightly with neighboring channels.

1.7.9   Voice Activity Detection

To reduce the interference on the air interface and to increase the operating time of the mobile device, data bursts are only sent if a speech signal is detected. This method is called discontinuous transmission (DTX) and can be activated independently in the uplink and downlink directions (Figure 1.41). Since only one person speaks at a time during a conversation, one of the two speech channels can usually be deactivated. In the downlink direction, this is managed by the voice activity detection (VAD) algorithm in the TRAU, while in the uplink direction the VAD is implemented in the mobile device.

Simply deactivating a speech channel, however, creates a very undesirable side effect. As no speech signal is transmitted anymore, the receiver no longer hears the background noise on the other side. This can be very irritating, especially for high-volume background noise levels such as when a person is driving a car or sitting in a train. Therefore, it is necessary to generate artificial noise, called comfort noise, which simulates the background noise of the other party for the listener. As the background noise can change over time, the mobile device or the network, respectively, analyzes the background noise of the channel and calculates an approximation for the current situation. This approximation is then exchanged between the

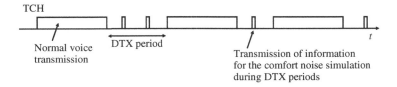

Figure 1.41   Discontinuous transmission (DTX)

mobile device and the TRAU every 480 milliseconds. Additional benefits for the network and mobile device are the ability to perform periodic signal quality measurements of the channel and the ability to use these frames to get an estimation on the current signal timing to adapt the timing advance for the call if necessary. How well this method performs is clear from the audibility as this procedure is used in all mobile device calls today and the simulation of the background noise in most cases cannot be differentiated from the original signal.

Despite using sophisticated methods for error correction, it is still possible that parts of a frame are destroyed beyond repair during the transmission on the air interface. In these cases, the complete 20-millisecond voice frame is discarded by the receiver and the previous data block is used instead to generate an output signal. Most errors that are repaired this way remain undetected by the listener. This trick, however, cannot be used indefinitely. If after 320 milliseconds a valid data block has still not been received, the channel is muted and the decoder keeps trying to decode the subsequent frames. If, during the following few seconds no valid data frame is received, the connection is terminated and the call drops.

Many of the previously mentioned procedures have specifically been developed for the transmission of voice frames. For example, for circuit-switched data connections that are used for fax transmissions or end-to-end encrypted voice calls, a number of modifications are necessary. While it is possible to tolerate a number of faulty bits for voice frames or to discard frames if a CRC error is detected, this is not possible for data calls. If even a single bit is faulty, a retransmission of at least a single frame has to be performed as most applications cannot tolerate a faulty data stream. To increase the likelihood of correctly reconstructing the initial data stream, the interleaver spreads the bits of a frame over a much larger number of bursts than the eight bursts used for voice frames. Furthermore, the channel coder, which separates the bits of a frame into different classes based on their importance, had to be adapted for data calls as well, as all bits are equally important. Thus, the convolutional decoder has to be used for all bits of a frame. Finally, it is also not possible to use a lossy data compression scheme for data calls. Therefore, the TRAU operates in a transparent mode for data calls. If the data stream can be compressed, this has to be performed by higher layers or by the data application itself.

With a radio receiver or an amplifier of a stereo set, the different states of a GSM connection can be made audible. This is possible as the activation and deactivation of the transmitter of the mobile device induce an audible sound in the amplifier part of audio devices. If the GSM mobile device is held close enough to an activated radio or an amplifier during the establishment of a call, the typical noise pattern can be heard, which is generated by the exchange of messages on the signaling channel (SDCCH). At some point during the signaling phase, a TCH is assigned to the mobile device at the point at which the noise pattern changes. As a TCH burst is transmitted every 4.615 milliseconds, the transmitter of the mobile device is switched on and off with a frequency of 217 Hz. If the background noise is low enough or the mute button of the telephone is pressed, the mobile device changes into the DTX mode for the uplink part of the channel. This can be heard as well, as the constant 217- Hz hum is replaced by single short bursts every 0.5 seconds.

For incoming calls, this method can also be used to check that a mobile device has started communication with the network on the SDCCH one to two seconds before ringing. This delay is due to the fact that the mobile device first needs to go through the authentication phase and the activation of the ciphering for the channel. Only afterward can the network forward further information to the mobile device as to why the channel was established. This is also the reason why it takes a much longer time for the alerting tone to be heard when calling a mobile device compared to calling a fixed-line phone.

Some mobile devices possess a number of interesting network monitoring functionalities, which are hidden in the mobile device software and are usually not directly accessible via the phone's menu. These network monitors allow the visualization of many procedures and parameters that have been discussed in this chapter, such as the timing advance, channel allocation, power control, the cell-ID, neighboring cell information, handover and cell reselection. On the Internet, various web pages can be found that explain how these monitors can be activated, depending on the type and model of the phone. As the activation procedures are different for every phone, it is not possible to give a general recommendation. However, by using the manufacturer and model of the phone in combination with terms like 'GSM network monitor', 'GSM netmonitor' or 'GSM monitoring mode', it is relatively easy to discover if and how the monitoring mode can be activated for a specific phone.

1.8   Mobility Management and Call Control

As all components of a GSM mobile network have now been introduced, the following section gives an overview of the three processes that allow a subscriber to roam throughout the network.

1.8.1   Cell Reselection and Location Area Update

As the network needs to be able to forward an incoming call, the subscriber's location must be known. After the mobile device is switched on, its first action is to register with the network. Therefore, the network becomes aware of the current location of the user, which can change at any time because of the mobility of the user. If the user roams into the area of a new cell, it may need to inform the network of this change. To reduce the signaling load in the radio network, several cells are grouped into a location area. The network informs the mobile device via the BCCH of a cell not only of the cell ID but also of the LAC that the new cell belongs to. The mobile device thus has to report only its new location if the new cell belongs to a new location area. Grouping several cells into location areas not only reduces the signaling load in the network but also the power consumption of the mobile. A disadvantage of this method is that the network operator is only aware of the current location area of the subscriber but not of the exact cell. Therefore, the network has to search for the mobile device in all cells of a location area for an incoming call or SMS. This procedure is called paging. The size of a location area can be set by the operator depending on his particular needs. In operational networks, several dozens of cells are usually grouped into a location area (Figure 1.42).

Figure 1.40 shows how a location area update procedure is performed. While idle, the mobile measures the signal strengths of the serving cell and of the neighboring cells. Neighboring cells can be found because their transmission frequency is announced on the broadcast channel (BCCH) of the serving cell. Typical values that a signal is received with are $-100\,$dBm, which indicates that it is very far away from the base station, and $-60\,$dBm, which indicates that it is very close to the base station. This value is also referred to as the received signal strength indication (RSSI). Once the signal of a neighboring cell becomes stronger than the signal of the current cell by a value that can be set by the network operator, the mobile reselects the new cell and reads the BCCH. If the LAC that is broadcast is different from that of the previous cell a location update procedure is started. After a signaling connection has been established,

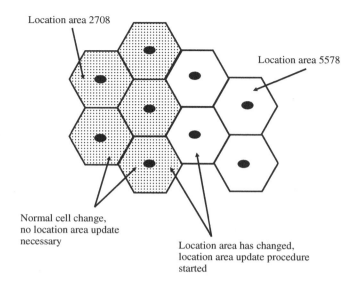

Figure 1.42   Cells in different location areas

the mobile device sends a location update request message to the MSC, which is transparently forwarded by the radio network. Before the message can be sent, however, the mobile device needs to authenticate itself and ciphering is usually activated as well.

Once the connection is secured against eavesdropping, the mobile device is usually assigned a new TMSI by the network, which it uses for the next connection establishment to identify itself instead of the IMSI. By using a constantly changing temporary ID, the identity of a subscriber is not revealed to listeners during the first phase of the call which is not ciphered. Once TMSI reallocation has been performed, the location area update message is sent to the network which acknowledges the correct reception. After receiving the acknowledgment, the connection is terminated and the mobile device returns to idle state.

If the old and new location areas are under the administration of two different MSC/VLRs, a number of additional steps are necessary. In this case, the new MSC/VLR has to inform the HLR that the subscriber has roamed into its area of responsibility. The HLR then deletes the record of the subscriber in the old MSC/VLR. This procedure is called an inter-MSC location update. From the mobile point of view, however, there is no difference compared to a standard location update as the additional messages are only exchanged in the core network.

1.8.2   The Mobile-Terminated Call

An incoming call for a mobile subscriber is called a mobile-terminated call by the GSM standards. The main difference between a mobile network and a fixed-line PSTN network is that the telephone number of the subscriber does not hold any information about where the subscriber is located. In the mobile network, it is thus necessary to query the HLR for the current location of the subscriber before the call can be forwarded to the correct switching center.

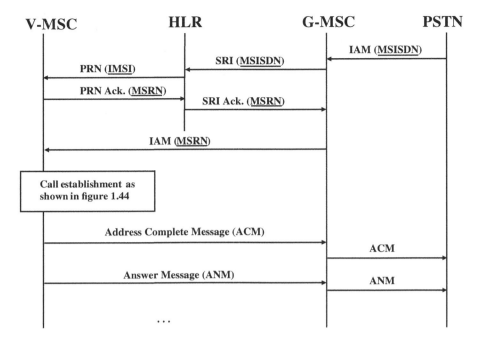

Figure 1.43   Mobile-terminated call establishment, part 1

Figure 1.43 shows the first part of the message flow for a mobile-terminated call initiated from a fixed-line subscriber. From the fixed-line network, the Gateway-Mobile Switching Center (G-MSC) receives the telephone number (MSISDN) of the called party via an ISUP IAM message. The subsequent message flow on this interface is as shown in Figure 1.6 and the fixed-line network does not have to be aware that the called party is a mobile subscriber. The G-MSC in this example is simply a normal MSC with additional connections to other networks. When the G-MSC receives the IAM message, it sends a Send Routing Information (SRI) message to the HLR to locate the subscriber in the network. The MSC currently responsible for the subscriber is also called the subscriber's Visited Mobile Switching Center (V-MSC).

The HLR then determines the subscriber's IMSI by using the MSISDN to search through its database and thus is able to locate the subscriber's current V-MSC. The HLR then sends a Provide Roaming Number (PRN) message to the V-MSC/VLR to inform the switching center of the incoming call. In the V-MSC/VLR, the IMSI of the subscriber, which is part of the PRN message, is associated with a temporary Mobile Station Roaming Number (MSRN) which is returned to the HLR. The HLR then transparently returns the MSRN to the G-MSC.

The G-MSC uses the MSRN to forward the call to the V-MSC. This is possible as the MSRN not only temporarily identifies the subscriber in the V-MSC/VLR but also uniquely identifies the V-MSC to external switches. To forward the call from the G-MSC to the V-MSC, an IAM message is used again, which, instead of the MSISDN, contains the MSRN to identify the subscriber. This has been done as it is possible, and even likely, that there are transit switching centers between the G-MSC and V-MSC, which are thus able to forward the call without querying the HLR themselves.

As the MSRN is internationally unique instead of only in the subscriber's home network, this procedure can still be used if the subscriber is roaming in a foreign network. The presented procedure, therefore, works for both national and international roaming. As the MSRN is saved in the billing record for the connection, it is also possible to invoice the terminating subscriber for forwarding the call to a foreign network and to transfer a certain amount of the revenue to the foreign network operator.

In the V-MSC/VLR, the MSRN is used to find the subscriber's IMSI and thus the complete subscriber record in the VLR. This is possible because the relationship between the IMSI and MSRN was saved when the HLR first requested the MSRN. After the subscriber's record has been found in the VLR database, the V-MSC continues the process and searches for the subscriber in the last reported location area, which was saved in the VLR record of the subscriber. The MSC then sends a paging message to the responsible BSC. The BSC in turn sends a paging message via each cell of the location area on the PCH. If no answer is received, then the message is repeated after a few seconds.

After the mobile device has answered the paging message, an authentication and ciphering procedure has to be executed to secure the connection in a similar way as previously presented for a location update. Only then is the mobile device informed about the details of the incoming call with a setup message. The setup message contains, for example, the telephone number of the caller if the Calling Line Identification Presentation (CLIP) supplementary service is active for this subscriber and not suppressed by the Calling Line Identification Restriction (CLIR) option that can be set by the caller (see Table 1.4).

If the mobile device confirms the incoming call with a call confirmed message, the MSC requests the establishment of a TCH for the voice path from the BSC (see Figure 1.44).

After successful establishment of the speech path, the mobile device returns an alerting message and thus informs the MSC that the subscriber is informed about the incoming call (the phone starts ringing). The V-MSC then forwards this information via the ACM to the G-MSC. The G-MSC then forwards the alerting indication to the fixed-line switch via its own ACM message.

Once the mobile subscriber accepts the call by pressing the answer button, the mobile device returns an Answer Message to the V-MSC. Here, an ISUP answer (ANM) message is generated and returned to the G-MSC. The G-MSC forwards this information again via an ANM message back to the fixed-line switching center.

While the conversation is ongoing, the network continues to exchange messages between different components to ensure that the connection is maintained. Most of the messages are measurement report messages, which are exchanged between the mobile device, the base station and the BSC. If necessary, the BSC can thus trigger a handover to a different cell. More details about the handover process can be found in Section 1.8.3.

If the mobile subscriber wants to end the call, the mobile device sends a disconnect message to the network. After releasing the TCH with the mobile device and after sending an ISUP release (REL) message to the other party, all resources in the network are freed and the call ends.

In this example, it has been assumed that the mobile subscriber is not in the area that is covered by the G-MSC. Such a scenario, however, is quite likely if a call is initiated by a fixed-line subscriber to a mobile subscriber which currently roams in the same region. As the fixed-line network usually forwards the call to the closest MSC to save costs, the G-MSC will, in many cases, also be the V-MSC for the connection. The G-MSC recognizes such a scenario

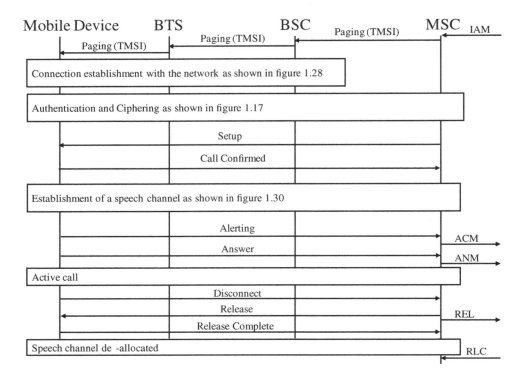

Figure 1.44 Mobile-terminated call establishment, part 2

if the MSRN returned by the HLR in the SRI acknowledge message contains a number, which is from the MSRN pool of the G-MSC. In this case, the call is treated in the G-MSC right away and the ISUP signaling inside the mobile network (IAM, ACM and ANM) is left out. More details about call establishment procedures in GSM networks can be found in 3GPP TS 23.018 [33].

1.8.3   Handover Scenarios

If reception conditions deteriorate during a call because of a change in the location of the subscriber, the BSC has to initiate a handover procedure. The basic procedure and the necessary messages have already been shown in Figure 1.31. Depending on the parts of the network that are involved in the handover, one of the following handover scenarios described in 3GPP TS 23.009 [34] is used to ensure that the connection remains established:

- **Intra-BSC handover.** In this scenario, the current cell and the new cell are connected to the same BSC. This scenario is shown in Figure 1.31.
- **Inter-BSC handover.** If a handover has to be performed to a cell which is connected to a second BSC, the current BSC is not able to control the handover itself as no direct signaling connection exists between the BSCs of a network. Thus, the current BSC requests

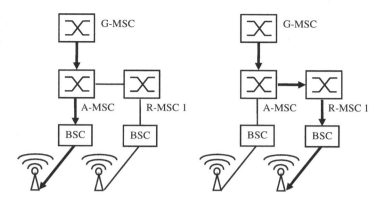

Figure 1.45 Inter-MSC handover

the MSC to initiate a handover to the other cell by sending a handover request message. Important parameters of the message are the cell ID and the LAC of the new cell. As the MSC administers a list of all LACs and cells under its control, it can find the correct BSC and request the establishment of a TCH for the handover in a subsequent step. Once the new BSC has prepared the speech channel (TCH) in the new cell, the MSC returns a handover command to the mobile device via the still existing connection over the current BSC. The mobile device then performs the handover to the new cell. Once the new cell and BSC have detected the successful handover, the MSC can switch over the speech path and inform the old BSC that the TCH for this connection can be released.

- **Inter-MSC handover.** If the current and new cells for a handover procedure are not connected to the same MSC, the handover procedure is even more complicated. As in the previous example, the BSC detects that the new cell is not in its area of responsibility and thus forwards the handover request to the MSC. The MSC also detects that the LAC of the new cell is not part of its coverage area. Therefore, the MSC looks into another table that lists all LACs of the neighboring MSCs. As the MSC in the next step contacts a second MSC, the following terminology is introduced to unambiguously identify the two MSCs: the MSC which has assigned an MSRN at the beginning of the call is called the Anchor-Mobile Switching Center (A-MSC) of the connection. The MSC that receives the call during a handover is called the Relay-Mobile Switching Center (R-MSC) (see Figure 1.45).

To perform the handover, the A-MSC sends an MAP (see Section 1.4.2) handover message to the R-MSC. The R-MSC then asks the responsible BSC to establish a TCH in the requested cell and reports back to the A-MSC. The A-MSC then instructs the mobile device via the still existing connection over the current cell to perform the handover. Once the handover has been performed successfully, the R-MSC reports the successful handover to the A-MSC. The A-MSC can then switch the voice path toward the R-MSC. Afterward, the resources in the old BSC and cell are released.

If the subscriber yet again changes to another cell during the call, which is controlled by yet another MSC, a subsequent inter-MSC handover has to be performed as shown in Figure 1.46.

For this scenario, the current relay-MSC (R-MSC 1) reports to the A-MSC that a subsequent inter-MSC handover to R-MSC 2 is required to maintain the call. The A-MSC then instructs

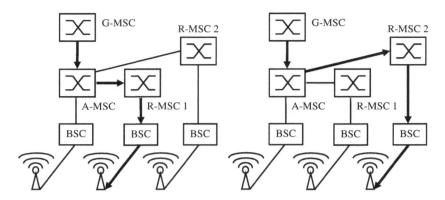

Figure 1.46 Subsequent inter-MSC handover

R-MSC 2 to establish a channel in the requested cell. Once the speech channel is ready in the new cell, the A-MSC sends the handover command message via R-MSC 1. The mobile device then performs the handover to R-MSC 2 and reports the successful execution to the A-MSC. The A-MSC can then redirect the speech path to R-MSC 2 and instruct R-MSC 1 to release the resources. By having the A-MSC in command in all the different scenarios, it is ensured that during the lifetime of a call only the G-MSC, the A-MSC and at most one R-MSC are part of a call. In addition, tandem switches might be necessary to route the call through the network or to a roaming network. However, these switches purely forward the call and are thus transparent in this procedure.

Finally, there is also a handover case in which the subscriber, who is served by an R-MSC, returns to a cell which is connected to the A-MSC. Once this handover is performed, no R-MSC is part of the call. Therefore, this scenario is called a subsequent handback.

From the mobile device point of view, all handover variants are performed in the same way, as the handover messages are identical for all scenarios. To perform a handover as quickly as possible, however, GSM can send synchronization information for the new cell in the handover message. This allows the mobile device to immediately switch to the allocated timeslot instead of having to synchronize first. This can only be done, however, if the current and the new cells are synchronized with each other, which is not possible, for example, if they are controlled by different BSCs. As two cells that are controlled by the same BSC may not necessarily be synchronized, synchronization information is by no means an indication of what kind of handover is being performed in the radio and core network.

1.9   The Mobile Device

Owing to the progress of miniaturization of electronic components during the mid-1980s, it was possible to integrate all components of a mobile device into a single portable device. A few years later, mobile devices had shrunk to such a small size that the limiting factor in future miniaturization was no longer the size of the electronic components. Instead, the space required for user interface components like display and keypad limited a further reduction. Owing to the continuous improvement and miniaturization of electronic components, it is possible to integrate more and more functionalities into a mobile device and to improve the ease of use. While mobile devices were at first used only for voice calls, the trend today is toward feature-rich devices that also include telephony functions. This section, therefore, first describes the architecture of a, from today's perspective, simple voice-centric phone. Afterward, the architecture of a modern smartphone will be discussed.

1.9.1 Architecture of a Voice-Centric Mobile Device

Simple GSM devices for voice and SMS communication can be built with only a few parts today and can therefore be produced very cheaply. The simplest GSM mobile phones can thus be bought today for less than 20 euros.

Figure 1.47 shows the principle architecture of such a device. Its core is based on a baseband processor, which contains a reduced instruction set (RISC), a CPU and a digital signal processor (DSP).

The RISC processor is responsible for the following tasks:

- handling of information that is received via the different signaling channels (BCCH, PCH, AGCH, PCH, and so on);
- call establishment (DTAP);
- GPRS management and GPRS data flow;
- parts of the transmission chain, like channel coder, interleaver and cipherer (dedicated hardware component in some designs);

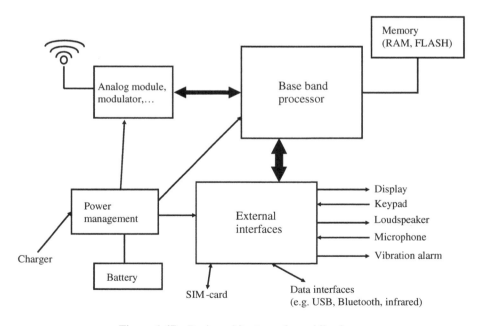

Figure 1.47 Basic architecture of a mobile phone

- mobility management (network search, cell reselection, location update, handover, timing advance, etc.);
- connections via external interfaces like Bluetooth, infrared and Universal Serial Bus (USB);
- user interface (keypad, display, graphical user interface).

As many of these tasks have to be performed in parallel, a multitasking embedded real-time operating system is used on the RISC processor. The real-time component of the operating system is especially important as the processor has to be able to provide data for transmission over the air interface according to the GSM frame structure and timing. All other tasks like keypad handling, display update and the graphical user interface, in general, have a lower priority.

In most devices, the baseband processor is based on the ARM Reduced Instruction Set (RISC) architecture that allows clock speeds of up to 2 GHz today. Such processors are built by several manufacturers that have obtained a license from ARM. For simple devices, ultra low power versions are used with a clock rate of only a few megahertz. Such chips are very power efficient and require only little energy while in sleep mode and while periodically observing the PCH. As a consequence, it is possible to reach standby times of well over a week. Also, the amount of RAM and ROM for such a device is very small by today's standards, usually in the order of only a few hundred kilobytes.

The DSP is another important component of a GSM chipset. Its main task is the decoding of the incoming signal and FR, EFR, HR or AMR speech compression. In GSM, signal decoding starts with the analysis of the training sequence of a burst (see Section 1.7.3). As the DSP is aware of the composition of the training sequence of a frame, the DSP can calculate a filter that is then used to decode the data part of the burst. This increases the probability that the data can be correctly reconstructed. The DSP 56600 architecture with a processor speed of 104 MHz is often used for these tasks.

Figure 1.48 shows the respective tasks that are performed by the RISC processor and the DSP processor. If the transmission chain for a voice signal is compared between the mobile device and the network, it can be seen that the TRAU mostly performs the task of a DSP unit in the mobile device. All other tasks such as channel coding are performed by the BTS, which is thus the counterpart of the RISC CPU of the mobile device.

As hundreds of millions of mobile devices are sold every year, there is a great variety of chipsets available on the market. The chipset is in many cases not designed by the manufacturer of the mobile device. Chipset manufacturers that are not developing their own mobile devices are, for example, Infineon, Broadcom and Mediatek.

1.9.2 Architecture of a Smartphone

Simple GSM voice phones usually have one processor that handles both the modem functionality and the operating system for the user interface. In smartphones, these tasks are performed by independent processors. This has become necessary as each function has become much more complex over time, and the processor architecture required by each function has developed in different directions. In addition, smartphones include many new functionalities that require significant and specialized processing capabilities. Figure 1.49 gives an overview of the typical function blocks of a modern smartphone. Owing to increasing miniaturization, most

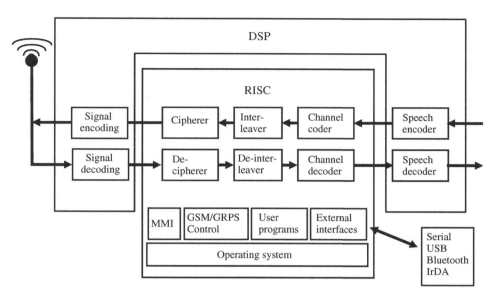

Figure 1.48 Overview of RISC and DSP functionalities

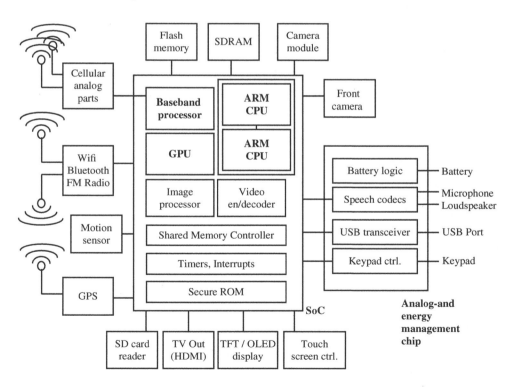

Figure 1.49 Architecture of a modern smartphone

or even all of the functions shown are included in a single chip. Such a combination is often also referred to as a System on a Chip (SoC).

The baseband processor is responsible for communication with a mobile network and does not only support GSM but also UMTS and in many cases, also LTE. A few analog components cannot be included on the SoC because of their size and function and are thus also shown separately in the figure.

The operating system for the user interface, such as Android or iOS, is executed on the application processor that usually consists of one or more ARM processor cores. These ARM CPUs have an instruction set similar to that of the ARM processors of the voice-centric GSM phones discussed earlier. Owing to their higher complexity and higher clock rates, their power consumption is much higher. Baseband processor and application processor operate independently of each other and communicate over a fast serial interface with each other. This is also the case if both units are contained in a single SoC. Other important building blocks contained in a SoC are usually a dedicated Graphics Processing Unit (GPU) and a number of additional supporting functions for memory management, timers, interrupts and dedicated processing units for the external camera module to quickly process images taken by the user for the operating system. Memory chips are usually still physically separate from the SoC. And finally, additional external elements such as a camera module on the back, another one on the front, Wi-Fi, Bluetooth and FM radio chips as well as GPS, SD card readers, TV-Out, the display and the touchscreen are also part of a modern smartphone.

1.10   The SIM Card

Despite its small size, the SIM card, officially referred to as the Universal Integrated Circuit Card (UICC), is one of the most important parts of a GSM network because it contains all the subscription information of a subscriber. Since it is standardized, a subscriber can use any GSM or UMTS phone by simply inserting the SIM card. Exceptions are phones that contain a 'SIM lock' and thus only work with a single SIM card or only with the SIM card of a certain operator. However, this is not a GSM restriction. It was introduced by mobile network operators to ensure that a subsidized phone is used only with SIM cards of their network.

The most important parameters on the SIM card are the IMSI and the secret key (Ki), which is used for authentication and the generation of ciphering keys (Kc). With a number of tools, which are generally available on the Internet free of charge, it is possible to read out most parameters from the SIM card, except for sensitive parameters that are read protected. Figure 1.50 shows such a tool. Protected parameters can only be accessed with a special unlock code that is not available to the end user.

Astonishingly, a SIM card is much more than just a simple memory card as it contains a complete microcontroller system that can be used for a number of additional purposes. The typical properties of a SIM card are shown in Table 1.7.

As shown in Figure 1.51, the mobile device cannot access the information on the Electrically Erasable Programmable Read-Only Memory (EEPROM) directly, but has to request the information from the SIM's CPU. Therefore, direct access to sensitive information is prohibited. The CPU is also used to generate the SRES during the network authentication procedure based on the RAND, which is supplied by the AuC (see Section 1.6.4). It is imperative that the calculation of the SRES is done on the SIM card itself and not in the mobile device to protect the secret Ki key. If the calculation was done in the mobile device itself, this would

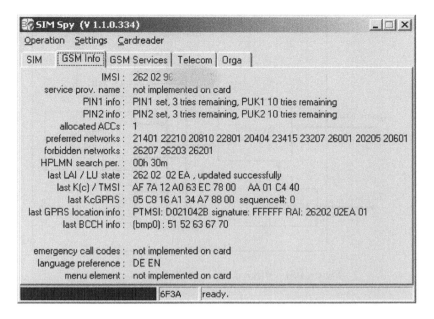

Figure 1.50   Example of a tool to visualize the data contained on a SIM card

Table 1.7   SIM card properties

CPU	8- or 16-bit CPU
ROM	40–100 kB
RAM	1–3 kB
EEPROM	16–64 kB
Clock rate	10 MHz, generated from clock supplied by mobile device
Operating voltage	3 V or 5 V

mean that the SIM card would have to hand over the Ki to the mobile device or any other device upon request. This would seriously undermine security, as tools like the one shown in Figure 1.50 would be able to read the Ki, which then could be used to make a copy of the SIM card.

Furthermore, the microcontroller system on the SIM can also execute programs that the network operator may have installed on the SIM card. This is done via the SIM application toolkit (SAT) interface, which is specified in 3GPP TS 31.111 [36]. With the SAT interface, programs on the SIM card can access functionalities of the mobile device such as waiting for user input or can be used to show text messages and menu entries on the display. Many mobile network operators use this functionality to put an operator-specific menu item into the overall menu structure of the mobile device's graphical user interface. In the menu created by the SIM card program, the subscriber could, for example, request for an overview of the cumulative charges of the current billing cycle. When the subscriber enters the menu, all user

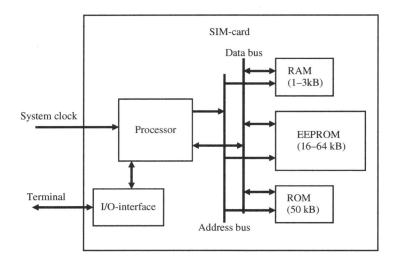

Figure 1.51   Block diagram of SIM card components

inputs via the keypad are forwarded by the mobile device to the SIM card. The program on the SIM card in this example would react to the request by generating an SMS, and then instructing the mobile device to send this SMS to the network. The network replies with one or more SMS messages that contain the balance overview. The SIM card can then extract the information from the SMS messages and present the content to the subscriber.

The SAT can be used for much more complicated programs than the example mentioned earlier. In practice, however, it can be observed that network operators are migrating their services toward smartphone apps.

From a logical point of view, data are stored on a GSM SIM card in directories and files, in a manner similar to the storage on a PC's hard drive. The file and folder structures are specified in 3GPP TS 31.102 [37]. In the specification, the root directory is called the main file (MF), which is somewhat confusing at first. Subsequent directories are called dedicated files (DF), and normal files are called elementary files (EF). As there is only a very limited amount of memory on the SIM card, files are not identified via file and directory names. Instead, hexadecimal numbers with a length of four digits are used, which require only 2 B memory. The standard nevertheless assigns names to these numbers, which are, however, not stored on the SIM card. The root directory, for example, is identified via ID 0x3F00, the GSM directory is identified by ID 0x7F20 and the file containing the IMSI, for example, is identified via ID 0x6F07. To read the IMSI from the SIM card, the mobile device thus has to open the following path and file: 0x3F00 0x7F20 0x6F07.

To simplify access to the data contained on the SIM card for the mobile device, a file can have one of the following three file formats:

- **Transparent.** The file is seen as a sequence of bytes. The file for the IMSI, for example, is of this format. How the mobile device has to interpret the content of the files is again specified in 3GPP TS 31.002 [37].

- **Linear fixed.** This file type contains records of a fixed length and is used, for example, for the file that contains the telephone book records. Each phone record uses one record of the linear fixed file.
- **Cyclic.** This file type is similar to the linear fixed file type but contains an additional pointer that points to the last modified record. Once the pointer reaches the last record of the file, it wraps over again to the first record of the file. This format is used, for example, for the file in which the phone numbers, which have previously been called, are stored.

A number of different access right attributes are used to protect the files on the SIM card. By using these attributes, the card manufacturer can control if a file is read or write only when accessed by the mobile device. A layered security concept also permits network operators to change files which are read only for the mobile device over the air by sending special provisioning SMS messages.

The mobile device can only access the SIM card if the user has typed in the PIN when the phone is started. The mobile device then uses the PIN to unlock the SIM card. SIM cards of some network operators, however, allow deactivating the password protection and thus the user does not have to type in a PIN code when the mobile device is switched on. Despite unlocking the SIM card with the PIN, the mobile device is still restricted to only being able to read or write certain files. Thus, it is not possible, for example, to read or write to the file that contains the secret key Ki even after unlocking the SIM card with the PIN.

Details on how the mobile device and the SIM card communicate with each other have been specified in ETSI TS 102 221 [38]. For this interface, layer 2 command and response messages have been defined which are called Application Protocol Data Units (APDUs). When a mobile device wants to exchange data with the SIM card, a command APDU is sent to the SIM card. The SIM card analyzes the command APDU, performs the requested operation and returns the result in a response APDU. The SIM card only has a passive role in this communication as it can only send response APDUs back to the mobile device.

If a file is to be read from the SIM card, the command APDU contains, among other information, the file ID and the number of bytes to read from the file. If the file is of cyclic or linear fixed type, the command also contains the record number. If access to the file is allowed, the SIM card then returns the requested information in one or more response APDUs.

If the mobile device wants to write some data into a file on the SIM card, the command APDUs contain the file ID and the data to be written into the file. In the response APDU, the SIM card then returns a response as to whether the data were successfully written to the file.

Figure 1.52 shows the format of a command APDU. The first field contains the class of instruction, which is always 0xA0 for GSM. The instruction (INS) field contains the ID of the command that has to be executed by the SIM card.

Table 1.8 shows some commands and their IDs. The fields P1 and P2 are used for additional parameters for the command. P3 contains the length of the following data field which contains the data that the mobile device would like to write on the SIM card.

CLA	INS	P1	P2	P3	Data

Figure 1.52   Structure of a command APDU

Table 1.8 Examples for APDU commands

Command	ID	P1	P2	Length
Select (open file)	A4	00	00	02
Read binary (read file)	B0	Offset high	Offset low	Length
Update binary (write file)	D6	Offset high	Offset low	Length
Verify CHV (check PIN)	20	00	ID	08
Change CHV (change PIN)	24	00	ID	10
Run GSM algorithm (RAND, SRES, Kc, …)	88	00	00	10

Data	SW1	SW2

Figure 1.53 Response APDU

The format of a response APDU is shown in Figure 1.53. Apart from the data field, the response also contains two fields called SW1 and SW2. These are used by the SIM card to inform the mobile device if the command was executed correctly.

For example, to open a file for reading or writing, the mobile device sends a SELECT command to the SIM card. The SELECT APDU is structured as shown in Figure 1.54.

As a response, the SIM card replies with a response APDU that contains a number of fields. Some of them are shown in Table 1.9.

For a complete list of information returned for the example, see [38]. In the next step, the READ BINARY or WRITE BINARY APDU can be used to read or modify the file.

To physically communicate with the SIM card, there are eight contact areas on the top side of the SIM card. Only five of those contacts are required:

- C1: power supply;
- C2: reset;
- C3: clock;
- C5: ground;
- C7: input/output.

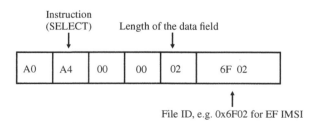

Figure 1.54 Structure of the SELECT command APDU

Table 1.9 Some fields of the response APDU for a SELECT command

Byte	Description	Length
3–4	File size	2
5–6	File ID	2
7	Type of file (transparent, linear fixed, cyclic)	1
9–11	Access rights	3
12	File status	1

As only a single line is used for the input and output of command and status APDUs, the data are transferred in half-duplex mode only. The clock speed for the transmission has been defined as C3/327. At a clock speed of 5 MHz on C3, the transmission speed is thus 13,440 bit/s.

1.11 The Intelligent Network Subsystem and CAMEL

All components that have been described in this chapter are mandatory elements for the operation of a mobile network. Mobile operators, however, usually offer additional services beyond simple postpaid voice services for which additional logic and databases are necessary in the network. Here are a few examples:

- Location-based services (LBS) are offered by most network operators in Germany in different variations. One LBS example is to offer cheaper phone calls to fixed-line phones in the area in which the mobile subscriber is currently located. To be able to apply the correct tariff for the call, the LBS service in the network checks if the current location of the subscriber and the dialed number are in the same geographical area. If so, additional information is attached to the billing record so that the billing system can later calculate the correct price for the call.
- Prepaid services have become very popular in many countries since their introduction in the mid-1990s. Instead of receiving a bill once a month, a prepaid subscriber has an account with the network operator, which is funded in advance with a certain amount of money determined by the subscriber. The amount on the account can then be used for phone calls and other services. During every call, the account is continually charged. If the account runs out of credit, the connection is interrupted. Furthermore, prepaid systems are also connected to the SMSC, the multimedia messaging server (MMS-Server, see Chapter 2) and the GPRS network (see Chapter 2). Therefore, prepaid subscribers can also be charged in real time for the use of these services.

These and many other services can be realized with the help of the IN subsystem. The logic and the necessary databases are located on an SCP, which has already been introduced in Section 1.4.

In the early years of GSM, the development of these services had been highly proprietary because of the lack of a common standard. The big disadvantage of such solutions was that they were customized to work only with very specific components of a single manufacturer. This meant that these services did not work abroad as foreign network operator-used components of other network vendors. This was especially a problem for the prepaid service as prepaid subscribers were excluded from international roaming when the first services were launched.

To ensure the interoperability of intelligent network components between different vendors and in networks of different mobile operators, industry and operators standardized an IN protocol in 3GPP TS 22.078 [39], which is called Customized Applications for Mobile-Enhanced Logic, or CAMEL for short. While CAMEL also offers functionality for SMS and GPRS charging, the following discussion describes only the basic functionality necessary for circuit-switched connections.

CAMEL is not an application or a service, but forms the basis to create services (customized applications) on an SCP, which is compatible with network elements of other vendors and between networks. Thus, CAMEL can be compared with HTTP. HTTP is used for transferring web pages between a web server and a browser. HTTP ensures that any web server can communicate with any browser. Whether the content of the data transfer is a web page or a picture is of no concern to HTTP because this is managed on a higher layer directly by the web server and the web client. Transporting the analogy back to the GSM world, the CAMEL specification defines the protocol for the communication between different network elements such as the MSC and the SCP, as well as a state model for call control.

The state model is called the Basic Call State Model (BCSM) in CAMEL. A circuit-switched call, for example, is divided into a number of different states. For the originator (O-BCSM), the following states, which are also shown in Figure 1.55, have been defined:

- call establishment;
- analysis of the called party's number;
- routing of the connection;
- notification of the called party (alerting);
- ongoing call (active);
- disconnection of the call;
- no answer from the called party;
- called party busy.

For a called subscriber, CAMEL also defines a state model which is called the Terminating Basic Call State Model (T-BCSM). T-BCSM can be used for prepaid subscribers who are currently roaming in a foreign network to control the call to the foreign network and to apply real-time charging.

For every state change in the state model, CAMEL defines a detection point (DP). If a DP is activated for a subscriber, the SCP is informed of the particular state change. Information contained in this message includes the IMSI of the subscriber, the current position (MCC, MNC, LAC and cell ID) and the number that was called. Whether a DP is activated is part of the subscriber's HLR entry. This allows the creation of specific services on a per subscriber basis. When the SCP is notified that the state model has triggered a DP, the SCP is able to influence the way the call should proceed. The SCP can take the call down, change the number that was called or return information to the MSC, which is put into the billing record of the call for later analysis on the billing system.

For the prepaid service, for example, the CAMEL protocol can be used between the MSC and the SCP as follows.

If a subscriber wants to establish a call, the MSC detects during the setup of the call that the 'authorize origination' DP is activated in the subscriber's HLR entry. Therefore, the MSC

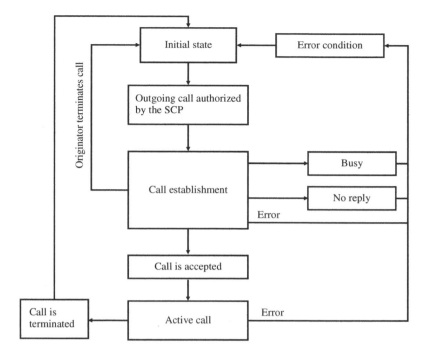

Figure 1.55 Simplified state model for an originator (O-BCSM) according to 3GPP TS 23.078 [40]

sends a message to the SCP and waits for a reply. As the message contains the IMSI of the subscriber as well as the CAMEL service number, the SCP recognizes that the request is for a prepaid subscriber. By using the destination number, the current time and other information, the SCP calculates the price per minute for the connection. If the subscriber's balance is sufficient, then the SCP allows the call to proceed and informs the MSC about the duration for which the authorization is valid. The MSC then continues and connects the call. At the end of the call, the MSC sends another message to the SCP to inform it of the total duration of the call. The SCP then modifies the subscriber's balance. If the time that the SCP initially granted for the call expires, the MSC has to contact the SCP again. The SCP then has the possibility to send an additional authorization to the MSC which is again limited to a determined duration. Other options for the SCP to react are to send a reply in which the MSC is asked to terminate the call or to return a message in which the MSC is asked to play a tone as an indication to the user that the balance on the account is almost depleted.

LBS is another application of CAMEL. Again, the HLR entry of a subscriber contains information about the DPs at which the CAMEL service is to be invoked. For LBS, the 'authorize origination' DP is activated. In this case, the SCP determines, by analyzing the IMSI and the CAMEL service ID, that the call has been initiated by a user that has subscribed to an LBS service. The service on the SCP then deduces from the current location of the subscriber and the NDC of the dialed number which tariff to apply for the connection. The SCP then informs the MSC of the correct tariff by returning a 'Furnish Charging Information' (FCI) message. At the end of the call, the MSC includes the FCI information in the billing record and thus enables the billing system to apply the correct tariff for the call.

Questions

1. Which algorithm is used to digitize a voice signal for transmission in a digital circuit-switched network and at which data rate is the voice signal transmitted?
2. Name the most important components of the GSM NSS and their tasks.
3. Name the most important components of the GSM radio network (BSS) and their tasks.
4. How is a BTS able to communicate with several subscribers at the same time?
5. Which steps are necessary to digitize a speech signal in a mobile device before it can be sent over the GSM air interface?
6. What is a handover and which network components are involved?
7. How is the current location of a subscriber determined for a mobile-terminated call and how is the call forwarded through the network?
8. How is a subscriber authenticated in the GSM network? Why is an authentication necessary?
9. How is an SMS message exchanged between two subscribers?
10. Which tasks are performed by the RISC processor and which tasks are performed by the DSP in a mobile device?
11. How are data stored on the SIM card?
12. What is CAMEL and for which services can it be used?

Answers to these questions can be found on the companion website for this book at http://www.wirelessmoves.com.

References

[1] European Technical Standards Institute (ETSI), http://www.etsi.org.
[2] The 3rd Generation Partnership Project, http://www.3gpp.org.
[3] 3GPP, Mobile Application Part (MAP) Specification, TS 29.002.
[4] 3GPP, Bearer-Independent Circuit-Switched Core Network – Stage 2, TS 23.205.
[5] 3GPP, Media Gateway Controller (MGC) – Media Gateway (MGW) Interface – Stage 3, TS 29.232.
[6] ITU, H.248: Gateway Control Protocol, http://www.itu.int/rec/T-REC-H.248/.
[7] ITU, Q.1901: Bearer Independent Call Control Protocol, http://www.itu.int/rec/T-REC-Q.1901.
[8] 3GPP, Application of Q.1900 Series to Bearer Independent Circuit Switched (CS) Core Network Architecture – Stage 3, TS 29.205.
[9] 3GPP, AT Command Set for 3G User Equipment, TS 27.007.
[10] 3GPP, Call Forwarding (CF) Supplementary Services – Stage 1, TS 22.082.
[11] 3GPP, Call Barring (CB) Supplementary Services – Stage 1, TS 22.088.
[12] 3GPP, Call Waiting (CW) and Call Hold (HOLD) Supplementary Services – Stage 1, TS 22.083.
[13] 3GPP, Multi Party (MPTY) Supplementary Services – Stage 1, TS 22.084.
[14] 3GPP, Man–Machine Interface (MMI) of the User Equipment (UE), TS 22.030.
[15] 3GPP, Mobile Radio Interface Layer 3 Specification; Core Network Protocols – Stage 3, TS 24.008.
[16] 3GPP, Technical Realisation of Short Message Service (SMS), TS 23.040.
[17] Bundesnetzagentur, Annual Report 2012, http://www.bundesnetzagentur.de, 2012.
[18] 3GPP, Voice Group Call Service (VGCS) – Stage 2, TS 43.068.
[19] 3GPP, Voice Broadcast Service (VGS) – Stage 2, TS 43.069.
[20] 3GPP, Enhanced Multi-Level Precedence and Preemption Service (eMLPP) – Stage 2, TS 23.067.
[21] Union Internationale des Chemins de Fer, GSM-R http://gsm-r.uic.asso.fr.
[22] Telefonica O2 Germany, Zahlen und Fakten, January 2014.
[23] 3GPP, Multiplexing and Multiple Access on the Radio Path, TS 45.002.
[24] 3GPP, AMR Speech CODEC: General Description, TS 26.071.

[25] ITU, G.722.2: Wideband Coding of Speech at Around 16 kbit/s using Adaptive Multi-Rate Wideband (AMR-WB), http://www.itu.int/rec/T-REC-G.722.2-200307-I/en.

[26] 3GPP, Speech Codec Speech Processing Functions; Adaptive Multi-Rate-Wideband (AMR-WB) Speech Codec; Transcoding Functions, TS 26.190.

[27] 3GPP, Full Speech Transcoding, TS 46.010.

[28] E. Barkan, E. Biham, and N. Keller, Instant Ciphertext-Only Cryptanalysis of GSM Encrypted Communication, http://cryptome.org/gsm-crack-bbk.pdf, 2003.

[29] 26C3, GSM Related Activities and Presentations, http://events.ccc.de/congress/2009/wiki/GSM, December 2009.

[30] 28C3, GSM Related Activities and Presentations, http://events.ccc.de/category/28c3/, December 2011.

[31] Wikipedia – IMSI-Catcher, http://en.wikipedia.org/wiki/IMSI_catcher, 5 October 2009.

[32] J. Frick and R. Bott, Method for Identifying a Mobile Phone User for Eavesdropping on Outgoing Calls, European Patent Office, EP1051053, http://v3.espacenet.com/publicationDetails/biblio?CC=DE&NR=19920222A1 &KC=A1&FT=D&date=20001109&DB=&locale=, November 2009.

[33] 3GPP, Basic Call Handling: Technical Realization, TS 23.018.

[34] xx

[35] 3GPP, Handover Procedures, TS 23.009.

[36] 3GPP, USIM Application Toolkit, TS 31.111.

[37] 3GPP, Characteristics of the USIM Application, TS 31.102.

[38] ETSI, Smart Cards; UICC-Terminal Interface; Physical and Logical Characteristics, TS 102 221.

[39] 3GPP, Customised Applications for Mobile Network Enhanced Logic (CAMEL): Service Description – Stage 1, TS 22.078.

[40] 3GPP, Customised Applications for Mobile Network Enhanced Logic (CAMEL): Service Description – Stage 2, TS 23.078.

2

General Packet Radio Service (GPRS) and EDGE

In the mid-1980s voice calls were the most important service in fixed and wireless networks. This is the reason why GSM was initially designed and optimized for voice transmission. Since the mid-1990s, however, the importance of the Internet has been constantly increasing. GPRS, the General Packet Radio Service, enhanced the GSM standard to transport data in an efficient manner and enabled wireless devices to access the Internet. With Enhanced Datarates for GSM Evolution (EDGE), further additions were specified to improve speed and latency. The first part of this chapter discusses the advantages and disadvantages of GPRS and EDGE compared to data transmission in classic GSM and fixed-line networks. The second part of the chapter focuses on how GPRS and EDGE have been standardized and implemented. At the end of the chapter, some applications of GPRS and EDGE are discussed and an analysis is presented on how the network behaves during a web-browsing session.

2.1 Circuit-Switched Data Transmission over GSM

As discussed in Chapter 1, the GSM network was initially designed as a circuit-switched network. All resources for a voice or data session are set up at the beginning of the call and are reserved for the user until the end of the call, as shown in Figure 2.1. The dedicated resources assure a constant bandwidth and end-to-end delay time. This has a number of advantages for the subscriber:

- Data that is sent does not need to contain any signaling information such as information about the destination. Every bit simply passes through the established channel to the receiver. Once the connection is established, no overhead, for example, addressing information, is necessary to send and receive the information.
- As the circuit-switched channel has a constant bandwidth, the sender does not have to worry about a permanent or temporary bottleneck in the communication path. This is especially

From GSM to LTE-Advanced: An Introduction to Mobile Networks and Mobile Broadband,
Revised Second Edition. Martin Sauter.
© 2014 John Wiley & Sons, Ltd. Published 2014 by John Wiley & Sons, Ltd.

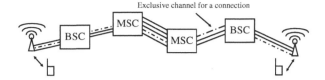

Figure 2.1   Exclusive connections of a circuit-switched system

important for a voice call. As the datarate is constant, any bottleneck in the communication path would lead to a disruption of the voice call.

• Furthermore, circuit-switched connections have a constant delay time. This is the time between sending a bit and receiving it at the other end. The greater the distance between the sender and receiver, the longer the delay time. This makes a circuit-switched connection ideal for voice applications as they are extremely sensitive to a variable delay time. If a constant delay time cannot be guaranteed, a buffer at the receiving end is necessary. This adds additional unwanted delay, especially for applications like voice calls.

While circuit-switched data transmission is ideally suited to voice transmissions, there are a number of significant disadvantages for data transmission with variable bandwidth usage. Web browsing is a typical application with variable or 'bursty' bandwidth usage. For sending a request to a web server and receiving the web page, as much bandwidth as possible is desired to receive the web page as quickly as possible. As the bandwidth of a circuit-switched channel is constant, there is no possibility of increasing the data transmission speed while the page is being downloaded. After the page has been received, no data is exchanged while the subscriber reads the page. The bandwidth requirement during this time is zero and the resources are simply unused and are thus wasted.

2.2   Packet-Switched Data Transmission over GPRS

For bursty data applications, it would be far better to request for resources to send and receive data and release them again after the transmission, as shown in Figure 2.2. This can be done by collecting the data in packets before it is sent over the network. This method of sending data is called 'packet switching'. As there is no longer a logical end-to-end connection, every packet has to contain a header. The header, for example, contains information about the sender (source address) and the receiver (destination address) of the packet. This information is used in the network to route the packets through the different network elements. In the Internet, for example, the source and destination addresses are the Internet Protocol (IP) addresses of the sender and receiver.

To send packet-switched data over existing GSM networks, GPRS was designed as a packet-switched addition to the circuit-switched GSM network. It should be noted that IP packets can be sent over a circuit-switched GSM data connection as well. However, until they reach the Internet service provider they are transmitted in a circuit-switched channel and thus cannot take advantage of the benefits described below. GPRS, on the other hand,

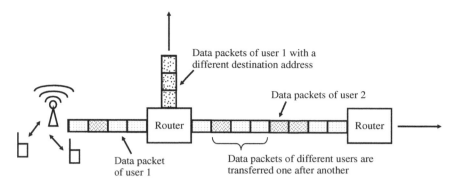

Figure 2.2   Packet-switched data transmission

is an end-to-end packet-switched network and IP packets are sent packet switched from end-to-end.

The packet-switched nature of GPRS also offers a number of other advantages for bursty applications over GSM circuit-switched data transmission:

- By flexibly allocating bandwidth on the air interface, GPRS exceeds the slow datarates of GSM circuit-switched connections of 9.6 or 14.4 kbit/s. Datarates of up to 170 kbit/s are theoretically possible. Multislot class 10 mobile devices (see below) reach speeds of about 85 kbit/s and are thus in the range of fixed-line analog modems that were in widespread use at the time GPRS was introduced.
- With the EDGE update of the GSM system, further speed improvements were made. The enhancements of EDGE for GPRS are called EGPRS in the standards. The term, however, is not widely used in practice and preference has been given to the term EDGE. With an EDGE class 32 mobile device, it is possible to reach transmission speeds of up to 270 kbit/s in today's networks. Despite most network operators having deployed Universal Mobile Telecommunications System (UMTS) and High-Speed Packet Access (HSPA) networks in the meantime in addition to GSM, many network operators have also chosen to upgrade their GPRS networks to EDGE. The extra network capacity and speed benefit 2G customers, offer faster data transmission speeds in buildings with limited 3G and long-term evolution (LTE) coverage and offer higher speeds in rural areas where UMTS or LTE is not available A speed comparison of the different technologies is shown in Figure 2.3.
- GPRS is usually charged by volume and not by time as shown in Figure 2.4. For subscribers this offers the advantage that they pay for downloading a web page but not for the time reading it, as would be the case with a circuit-switched connection. For the operator of a wireless network it offers the advantage that the scarce resources on the air interface are not wasted by 'idle' data calls because they can be used for other subscribers.
- GPRS significantly reduces the call set-up time. Similar to a fixed-line analog modem, a GSM circuit-switched data call took about 20 seconds to establish a connection with the Internet service provider, while GPRS accomplishes the same in less than 5 seconds.

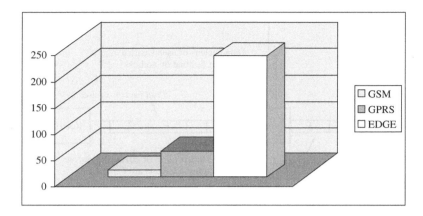

Figure 2.3   GSM, GPRS and EDGE data transmission speed comparison

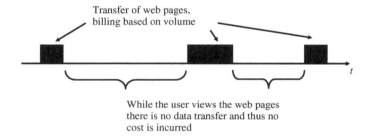

Figure 2.4   Billing based on volume

- Since the subscriber does not pay for the time when no data is transferred, the call does not have to be disconnected to save costs. This is called 'always-on' and enables applications like e-mail programs to poll for incoming e-mails in certain intervals or allows messaging clients to wait for incoming messages.
- When the subscriber is moving, by train for example, the network coverage frequently becomes very bad or is even lost completely for some time. When this happens, circuit-switched connections are disconnected and have to be reestablished manually once network coverage is available again. GPRS connections, on the other hand, are not dropped as the logical GPRS connection is independent of the physical connection to the network. After regaining coverage the interrupted data transfer simply resumes.

GPRS was initially designed to support different types of packet-switching technologies. The great success of the Internet, which uses the IP exclusively for packet switching, has led to IP being the only supported protocol today. Therefore, the terms 'user data transfer', 'user data transmission' or 'packet switching' used in this chapter always refer to 'transferring IP packets'.

2.3 The GPRS Air Interface

2.3.1 GPRS vs. GSM Timeslot Usage on the Air Interface

Circuit-Switched TCH vs. Packet-Switched PDTCH

As discussed in Chapter 1, GSM uses timeslots on the air interface to transfer data between subscribers and the network. During a circuit-switched call, a subscriber is assigned exactly one traffic channel (TCH) that is mapped to a single timeslot. This timeslot remains allocated for the duration of the call and cannot be used for other subscribers even if there is no data transfer for some time.

In GPRS, the smallest unit that can be assigned is a block that consists of four bursts of a packet data traffic channel (PDTCH). A PDTCH is similar to a TCH in that it also uses one physical timeslot. If the subscriber has more data to transfer, the network can assign more blocks on the same PDTCH right away. The network can also assign the following block(s) to other subscribers or for logical GPRS signaling channels. Figure 2.5 shows how the blocks of a PDTCH are assigned to different subscribers.

Instead of using a 26- or 51-multiframe structure as in GSM (see Section 1.7.3), GPRS uses a 52-multiframe structure for its timeslots. Frames 24 and 51 are not used for transferring data. Instead, they are used to allow the mobile device to perform signal strength measurements on neighboring cells. Frames 12 and 38 are used for timing advance calculations as described in more detail later on. All other frames in the 52-multiframe are collected into blocks of four frames (one burst per frame), which is the smallest unit to send or receive data.

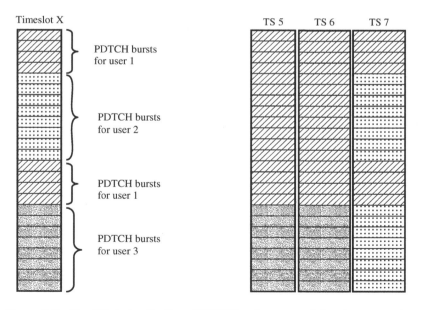

Figure 2.5 Simplified visualization of PDTCH assignment and timeslot aggregation

Timeslot Aggregation

To increase the transmission speed, a subscriber is no longer bound to a single TCH as in circuit-switched GSM. If more than one timeslot is available when a subscriber wants to transmit or receive data, the network can allocate several timeslots (multislot) to a single subscriber.

Multislot Classes

Depending on the multislot class of the mobile device, three, four or even five timeslots can be aggregated for a subscriber at the same time. Thus, the transmission speed for every subscriber is increased, provided that not all of them want to transmit data at the same time. Table 2.1 shows typical multislot classes. Today, most mobile devices on the market support multislot class 10, 12 or 32. As can be seen in the table, multislot class 10 supports four timeslots in the downlink direction and two in the uplink direction. This means that the speed in the uplink direction is significantly less than in the downlink direction. For applications like web browsing, it is not a big disadvantage to have more bandwidth in the downlink than in the uplink direction. The requests for web pages that are sent in the uplink direction are usually quite small, whereas web pages and the embedded pictures require faster speed in the downlink direction. Hence, web browsing benefits from the higher datarates in downlink direction and does not suffer very much from the limited uplink speed. For applications like sending e-mails with file attachments or multimedia messaging server (MMS) messages with large pictures or video content, two timeslots in the uplink direction are a clear limitation and increase the transmission time considerably.

Also important to note in Table 2.1 is that for most classes the maximum number of timeslots used simultaneously is lower than the combined number of uplink and downlink timeslots. For example, for GPRS class 32, which is widely used today, the sum is six timeslots. This means that if five timeslots are allocated by the network in the downlink direction, only one can be allocated in the uplink direction. If the network detects that the mobile devices want to send a larger amount of data to the network, it can reconfigure the connection to use three timeslots in the uplink and three in the downlink, thus again resulting in the use of six simultaneous timeslots. During a web-browsing session, for example, it can be observed that the network assigns two uplink timeslots to the subscriber when the web page request is initially sent. As soon as data to be sent to the subscriber arrives, the network quickly reconfigures the connection to use five timeslots in the downlink direction and only a single timeslot, if required, in the uplink direction.

Table 2.1 Selected GPRS multislot classes from 3GPP (3rd Generation Partnership Project) TS 45.002 Annex B1 [1]

Multislot class	Max. timeslots Downlink	Uplink	Sum
8	4	1	5
10	4	2	5
12	4	4	5
32	5	3	6

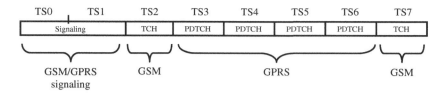

Figure 2.6 Shared use of the timeslots of a cell for GSM and GPRS

In order for the network to know how many timeslots the mobile device supports, it has to inform the network of its capabilities. This so-called mobile station classmark also contains other information such as ciphering capabilities. The classmark information is sent every time the mobile device accesses the network. It is then used by the network together with other information such as available timeslots to decide how many of them can be assigned to the user. The network also stores the classmark sent in the uplink direction and is thus able to assign resources in the downlink direction immediately, without asking the mobile device for its capabilities first.

2.3.2 Mixed GSM/GPRS Timeslot Usage in a Base Station

As GPRS is an addition to the GSM network, the eight timeslots available per carrier frequency on the air interface can be shared between GSM and GPRS. Therefore, the maximum GPRS data rate decreases as more GSM voice/data connections are needed. The network operator can choose how to use the timeslots, as shown in Figure 2.6. Timeslots can be assigned statically, which means that some timeslots are reserved for GSM and some for GPRS. The operator also has the option of dynamically assigning timeslots to GSM or GPRS. If there is a high amount of GSM voice traffic, more timeslots can be used for GSM. If voice traffic decreases, more timeslots can be given to GPRS. It is also possible to assign a minimum number of timeslots for GPRS and dynamically add and remove timeslots depending on voice traffic.

2.3.3 Coding Schemes

Another way to increase the data transfer speed besides timeslot aggregation is to use different coding schemes. If the user is at close range to a base station, the data transmitted over the air is less likely to be corrupted during transmission than if the user is farther away and the reception is weak. As has been shown in Chapter 1, the base station adds error detection and correction to the data before it is sent over the air. This is called coding and the method used to code the user data is called the coding scheme. In GPRS, four different coding schemes (CS-1 to 4) can be used to add redundancy to the user data depending on the quality of the channel [2]. Table 2.2 shows the properties of the different coding schemes.

Although CS-1 and CS-2 are commonly used today, not all GPRS networks use CS-3 and CS-4. This is because data that is carried over one timeslot on the air interface is carried in one-quarter of an E-1 timeslot between base transceiver station (BTS) and base station controller (BSC), which can carry only 16 kbit/s. When the overhead created by the packet header, which is not shown in Table 2.2, is included, CS-3 and CS-4 exceed the amount of

Table 2.2 GPRS coding schemes

Coding scheme	Number of user data bits per block (four bursts with 114 bits each)	Transmission speed per timeslot (kbit/s)
CS-1	160	8
CS-2	240	12
CS-3	288	14.4
CS-4	400	20

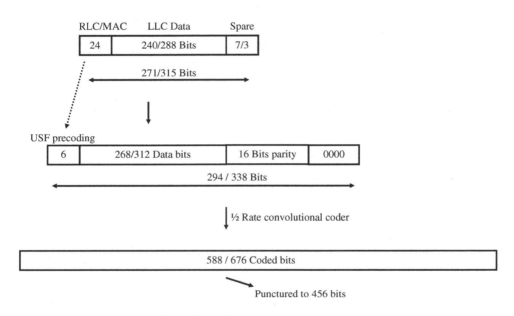

Figure 2.7 CS-2 and CS-3 channel coder

data that can be carried over one-quarter of an E-1 timeslot. To use these coding schemes, it is no longer possible to use fixed mapping, and more capacity is needed on the backhaul link. Thus some network operators have not introduced these coding schemes as it would require a costly change of the backhaul connectivity.

Figure 2.7 shows how CS-2 and CS-3 encode the data before it is transmitted over the air interface. CS-4 does not add any redundancy to the data. Therefore, CS-4 can only be used when the signal quality between the network and the mobile device is very good.

GPRS uses the same 1/2-rate convolutional coder as already used for GSM voice traffic. The use of the convolutional coding in CS-2 and CS-3 results in more coded bits than can be transmitted over a radio block. To compensate for this, some of the bits are simply not transmitted. This is called 'puncturing'. As the receiver knows which bits are punctured, it can insert 0 bits at the correct positions and then use the convolutional decoder to recreate the original data stream. This, of course, reduces the effectiveness of the channel coder as not all the bits that are punctured are 0 bits at the sender side.

2.3.4   Enhanced Datarates for GSM Evolution (EDGE)

To further increase data transmission speeds, an additional modulation and coding scheme, which uses 8 Phase Shift Keying (8 PSK), has been introduced into the standards. The new coding scheme is the basis of the 'enhanced datarates for GSM evolution' package, which is also called EDGE. The packet-switched part of EDGE is also referred to in the standard as enhanced GPRS or EGPRS. In the GPRS context, EGPRS and EDGE are often used interchangeably. By using 8 PSK modulation, EDGE transmits three bits in a single transmission step. This way, data transmission can be up to three times faster compared to GSM and GPRS, which both use Gaussian minimum shift keying (GMSK) modulation, which transmits only a single bit per transmission step. Figure 2.8 shows the differences between GMSK and 8 PSK modulation. While with GMSK the two possibilities 0 and 1 are coded as two positions in the I/Q space, 8 PSK codes the three bits in eight different positions in the I/Q space.

Together with the highest of the nine new coding schemes introduced with EDGE, it is possible to transfer up to 60 kbit/s per timeslot. Similar to CS-3 and CS-4, the use of these coding schemes require an update of the backhaul connection, and in addition, new transceiver elements in the base stations that are 8PSK capable. From the network side, the mobile device is informed of the EDGE capability of a cell by the EDGE capability bit in the GPRS cell options of the system information 13 message, which is broadcast on the Broadcast Common Control Channel (BCCH). From the mobile device side, the network is informed of the mobile device's EDGE capability during the establishment of a new connection. Hence, EDGE is fully backward compatible to GPRS and allows the mixed use of GPRS and EDGE mobile devices in the same cell. EDGE mobile devices are also able to use the standard GMSK modulation for GPRS and can thus be used also in networks that do not offer EDGE functionality.

Another advantage of the new modulation and the nine different coding schemes (MCS) compared to the four different coding schemes of GPRS is the precise use of the best modulation and coding for the current radio conditions. This is done in the mobile device by continuously calculating the current bit error probability (BEP) and reporting the values to the network. The network in turn can then adapt its current downlink modulation and coding to the appropriate value. For the uplink direction, the network can measure the error rate of

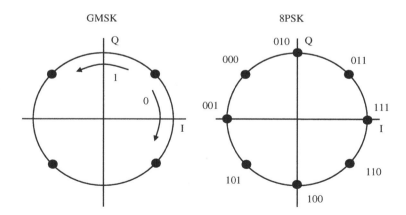

Figure 2.8   GMSK (GPRS) and 8PSK (EDGE) modulation

data that was recently received and instruct the mobile device to change its MCS accordingly. As both network and mobile device can report the BEP very quickly, it is also possible to quickly adapt to changing signal conditions, especially when the mobile device is in a moving car or train. This reduces the error rate and ensures the highest transmission speed in every radio condition. In practice, it can be observed that this control mechanism allows the use of MCS-8 and MCS-9 if reception conditions are good, and a quick fallback to other MCS if radio conditions deteriorate. In practice, transmission speeds of over 270 kbit/s can be reached with a class 32 EDGE mobile device. Table 2.3 gives an overview of the possible modulation and coding schemes and the datarates that can be achieved per timeslot.

Despite the ability to react quickly to changing transmission conditions, it is of course still possible that a block contains too many errors and thus the data cannot be reconstructed correctly. This is even desired to some extent because retransmitting a few faulty blocks is preferred over switching to a slower coding scheme.

To preserve the continuity of the data flow on higher layers, EDGE introduces a number of enhancements in this area as well. To correct transmission errors a method called 'incremental redundancy' has been introduced. As is already the case with the GPRS coding schemes, some error detection and correction bits produced by the convolutional decoder are punctured and therefore not put into the final block that is sent over the air interface. With the incremental redundancy scheme it is possible to send the previously punctured bits in a second or even a third attempt. On the receiver side, the original block is stored and the additional redundancy information received in the first and second retry is added to the information. Usually only a single retry is necessary to be able to reconstruct the original data based on the additional information received. Figure 2.9 shows how MCS-9 uses a 1/3 convolutional decoder to generate three output bits for a single input bit. For the final transmission, however, only one of those three bits is sent. In case the block was not received correctly, the sender will use the second bits that were generated by the convolutional decoder for each input bit to form the retry block. In the unlikely event that it is still not possible for the receiver to correctly decode the data, the sender will send another block containing the third bit. This further increases the probability that the receiver can decode the data correctly by combining the information that is contained in the original block with the redundancy information in the two additional retransmissions.

Table 2.3 EDGE modulation and coding schemes (MCS)

	Modulation	Speed per timeslot (kbit/s)	Coding rate (user bits to error correction bits)	Coding rate with one retransmission
MCS-1	GMSK	8.8	0.53	0.26
MCS-2	GMSK	11.2	0.66	0.33
MCS-3	GMSK	14.8	0.85	0.42
MCS-4	GMSK	17.6	1.00	0.50
MCS-5	8PSK	22.4	0.37	0.19
MCS-6	8PSK	29.6	0.49	0.24
MCS-7	8PSK	44.8	0.76	0.38
MCS-8	8PSK	54.4	0.92	0.46
MCS-9	8PSK	59.2	1.00	0.50

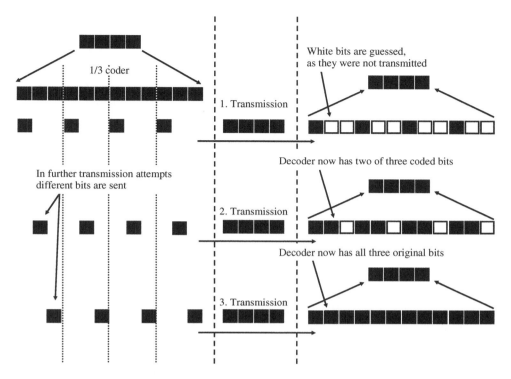

Figure 2.9   MCS-9 convolutional coding and incremental redundancy

Another way of retransmitting faulty blocks is to split them up into two blocks for a retransmission that uses a different MCS. This method is called resegmentation. As can be seen in Table 2.4, the standard defines three code families. If, for example, a block coded with MCS-9 has to be retransmitted, the system can decide to send the content of this block embedded in two blocks, which are then coded by using MCS-6. As MCS-6 is more robust than MCS-9, it is much more likely that the content can be decoded correctly. In practice, it can be observed that the incremental redundancy scheme is preferred over resegmentation.

The interleaving algorithm, which reorders the bits before they are sent over the air interface to disperse consecutive bit errors, has been changed for EDGE as well. GSM voice packets and GPRS data blocks are always interleaved over four bursts as described in Section 1.7.3. As EDGE notably increases the number of bits that can be sent in a burst, it has been decided to decrease the block size for MCS-7, MCS-8 and MCS-9 to fit in two bursts instead of four. This reduces the number of bits that need to be retransmitted after a block error has occurred and thus helps the system to recover more quickly. The block length reduction is especially useful if frequency hopping is used in the system. When frequency hopping is used, every burst is sent on a different frequency to avoid using a constantly jammed channel. Although the approach is good for voice services that can hide badly damaged blocks from the user up to a certain extent, it poses a retransmission risk for packet data if one of the frequencies used in the hopping sequence performs very badly. Thus, limiting the size of MCS-7, MCS-8 and MCS-9 blocks to two bursts helps to cope better with such a situation.

Table 2.4   Resegmentation of EDGE blocks using a different MCS

MCS	Family	Speed (kbit/s)	Resegmentation
MCS-9	A	59.2 (2 × 29.2)	2 × MCS-6
MCS-8	A	54.4 (2 × 29.2 + padding)	2 × MCS-6 (+ padding)
MCS-6	A	29.2 (2 × 14.8)	2 × MCS-3
MCS-3	A	14.8	–
MCS-7	B	44.8 (2 × 22.4)	2 × MCS-5
MCS-5	B	22.4 (2 × 11.2)	2 × MCS-2
MCS-2	B	11.2	–
MCS-4	C	17.6	2 × MCS-1
MCS-1	C	8.8	–

2.3.5   Mobile Device Classes

The GPRS standard defines three different classes of mobile devices, namely class A, class B and class C. Class C mobiles can only be attached to GPRS or GSM at one time. As this is quite inconvenient for a user, class C mobiles are suited only for embedded applications that only need either GSM or GPRS to transfer data.

Today, all mobile devices that are available on the market are class B devices. They can be attached to both GPRS and GSM at the same time, but early GPRS specifications had one important limitation: GSM and GPRS could not be used at the same time. In most networks this meant and still means today that during an ongoing voice call it is not possible to transfer data via GPRS. Similarly, during data transmission no voice call is possible. For outgoing calls this is not a problem. If a GPRS data transmission is ongoing, it will be interrupted when the user starts a telephone call and is automatically resumed once the call is terminated. There is no need to reconnect to GPRS as only the data transfer is interrupted. The logical GPRS connection remains in place during the voice call.

During data transmission, the mobile device is unable to listen to the GSM paging channel. This means that without further mechanisms on the network side, the mobile device is not able to detect incoming voice calls or short messaging service (SMS) messages. When applications generate only bursty data traffic, the probability of missing a paging message is reduced. Once the current data transfer is completed, the PDTCHs are released and the mobile device is again able to listen to the paging channel (PCH). As the paging message is usually repeated after a couple of seconds, the probability of overhearing a paging message and thus missing a call depends on the ratio between active data transmission time and idle time. As this is clearly not ideal, a number of enhancements have been specified to allow the circuit-switched and packet-switched parts of the network to exchange information about incoming calls or SMS messages. This is described in more detail in Section 2.3.6 and ensures that no paging message is lost during an ongoing data transfer. In practice, it can be observed today that most GSM/GPRS networks use one of these mechanisms.

The GPRS standard has also foreseen class A mobile devices that can be active in both GSM and GPRS at the same time. This means that a GPRS data transfer and a GSM voice call can be active at the same time. Today, there are no such devices on the market as the practical implementation would require two sets of independent transceivers in the mobile device. As this has been deemed impractical, a further enhancement was put into the GPRS standard that is referred to as 'dual transfer mode', or DTM for short. DTM synchronizes the circuit- and packet-switched parts of the GSM/GPRS network and thus allows GPRS data transfers during an ongoing GSM voice call with a single transceiver in the mobile device. Even though many mobile devices support DTM today, there is no widespread use of it on the network side. One reason for this may be that network operators are content with their UMTS networks being capable of simultaneous voice and data transmission.

2.3.6 Network Mode of Operation

Similar to GSM, the data transferred over the GPRS network can be both user data and signaling data. Signaling data is exchanged, for example, during the following procedures:

- the network pages the mobile device to inform it of incoming packets;
- the mobile device accesses the network to request for resources (PDTCHs) to send packets;
- modification of resources assigned to a subscriber;
- acknowledgment of correct reception of user data packets.

This can be done in a number of ways.

In GPRS, NOM I signaling for packet- and circuit-switched data is done via the GSM PCH. To make sure that incoming voice calls are not missed by class B mobile devices during an active data transfer, an interface between the circuit-switched part (MSC) and the packet-switched part (Serving GPRS Support Node – SGSN) of the network is used. This interface is called the Gs interface. Paging for incoming circuit-switched calls will be forwarded to the packet-switched part and then sent to the mobile device as shown in Figure 2.10. If a packet data transfer is in progress when paging needs to be sent, the mobile device will be informed via the Packet-Associated Control Channel (PACCH) to which the circuit-switched GSM part of the network does not have access. Alternatively, the paging is done via the PCH. The Gs interface can also be used for combined GSM/GPRS attach procedures and location updates (LU). As it is optional, some, but not all, networks use the functionality today.

GPRS NOM II is simpler than NOM I and is the most commonly used network operation mode today. This is because there is no signaling connection between the circuit-switched and packet-switched parts of the core network, that is, no Gs interface is used.

To overcome the shortcoming of not being able to signal incoming SMS and voice calls during a GPRS data transfer between the circuit-switched and packet-switched core network, a method has been defined for the BSC in the radio network to inform the GPRS Packet Control Unit (PCU), that is described below, of the incoming SMS or call. If an active data transfer is ongoing, the PCU will then send the paging message during the data transfer to the mobile device. The data transfer can then be interrupted, and the mobile device can respond to the paging message to the MSC.

To inform mobile devices which of the two GPRS network modes is used, GPRS uses the GSM BCCH channel and the SysInfo 13 message.

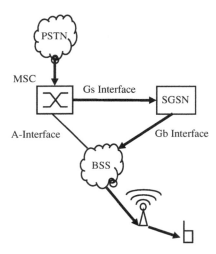

Figure 2.10   Paging for an incoming voice call via the Gs interface

2.3.7   GPRS Logical Channels on the Air Interface

GPRS uses a number of logical channels on the air interface in addition to those which are shared with GSM. They are used for transmitting user data and signaling data in the uplink and downlink direction. The following logical channels, which are shown in Figure 2.11, are mandatory for the operation of GPRS:

- **The PDTCH**: This is a bidirectional channel, which means it exists in the uplink and downlink direction. It is used to send user data across the air interface. The PDTCH is carried over timeslots that are dedicated for GPRS in a 52-multiframe structure that was introduced in Section 2.3.1.
- **The PACCH**: This channel is also bidirectional and is used to send control messages. These are necessary to acknowledge packets that are transported over the PDTCH. When a mobile device receives data packets from the network via a downlink PDTCH, it has to acknowledge them via the uplink PACCH. Similar to the PDTCH, the PACCH is also carried over the GPRS dedicated timeslots in blocks of the 52-multiframe structure that was introduced in Section 2.3.1. In addition, the PACCH is used for signaling messages that assign uplink and downlink resources. For the mobile device and the network to distinguish between PDTCH and PACCH that are carried over the same physical resource, the header of each block contains a logical channel information field as shown in Figure 2.12.
- **The PTCCH**: This channel is used for timing advance estimation and control of active mobile devices. To calculate the timing advance, the network can instruct an active mobile device to send a short burst at regular intervals on the PTCCH. The network then calculates the timing advance and sends the result back in the downlink direction of the PTCCH.

In addition, GPRS shares a number of channels with GSM to initially request for the assignment of resources. Figure 2.13 shows how some of these channels are used and how data are transferred once uplink and downlink resources have been assigned.

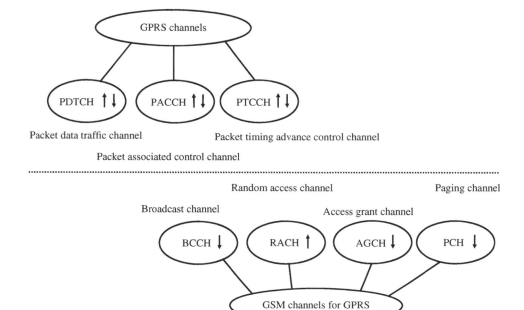

Figure 2.11 PDTCH and PACCH are sent on the same timeslot

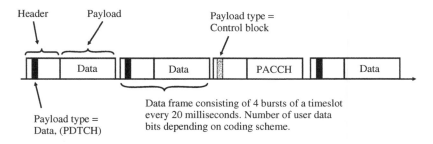

Figure 2.12 GPRS logical channels

- **The Random Access Channel (RACH)**: When the mobile device wants to transmit data blocks to the network, it has to request for uplink resources. This is done the same way as already described in Chapter 1 for voice calls. The only difference is in the content of the channel request message. Instead of asking for a circuit-switched resource, the message asks for packet resources on the air interface.
- **The AGCH**: The network will answer to a channel request on the RACH with an immediate packet assignment message on the AGCH that contains information about the PDTCH timeslot the mobile device is allowed to use in the uplink. As the network is not aware at this stage of the identity of the device, the first uplink transmissions have to contain

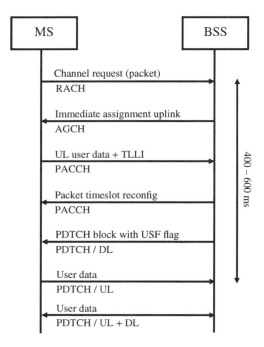

Figure 2.13   Packet resources: requests and assignments

the temporary logical link identifier (TLLI, also known as the packet-temporary mobile subscriber identity, P-TMSI) the mobile device was assigned when it attached to the network. All further GPRS signaling messages are then transmitted over the PACCH, which shares the dedicated GPRS timeslots with the PDTCH. Once data is available for the mobile device in downlink direction, the network needs to assign timeslots in downlink direction. This is done by transmitting a packet timeslot reconfiguration message with information about which timeslots the mobile device can use in uplink and downlink direction.

- **The PCH**: In case the mobile device is in standby state, only the location area of a subscriber is known. As the cell itself is not known, resources cannot be assigned right away and the subscriber has to be paged first. GPRS uses the GSM PCH to do this.
- **The BCCH**: A new system information message (SYS_INFO 13) has been defined on the BCCH to inform mobile devices about GPRS parameters of the cell. This is necessary to let mobile devices know, for example, if GPRS is available in a cell, which NOM is used, if EDGE is available, and so on.

2.4   The GPRS State Model

When the mobile device is attached to the GSM network, it can be either in 'idle' mode as long as there is no connection, or in 'dedicated' mode during a voice call or exchange of signaling information. Figure 2.14 shows the state model introduced to address the needs of a packet-switched connection for GPRS.

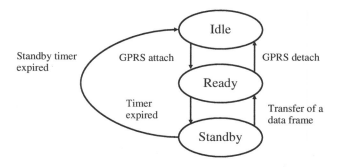

Figure 2.14   The GPRS state model

The Idle State

In this state the mobile device is not attached to the GPRS network at all. This means that the SGSN is not aware of the user's location, no Packet Data Protocol (PDP) context is established and the network cannot forward any packets for the user. It is very unfortunate that the standards body named this state 'idle' because in the GSM circuit-switched 'idle mode' the mobile device is attached to the circuit-switched side of the network and is reachable by the network. Therefore, great care has to be taken not to mix up the packet-switched Idle state with the GSM circuit-switched idle mode.

The Ready State

When the user wants to attach to the GPRS network, the mobile device enters the ready state as soon as the first packet is sent. While in ready state, the mobile device has to report every cell reselection to the network so that the SGSN can update the user's position in its database. This process is called 'cell update'. It enables the network to send any incoming data for a user directly to the mobile device instead of having to page the mobile device first to locate the user's serving cell. The mobile device will remain in the ready state while signaling or user data is transferred and for a certain time afterward. The timer that controls how long the mobile device will remain in this state after the last block of data was transferred is called T3314. The value of this timer is broadcast on the BCCH or PBCCH as part of the GPRS system information. A typical value for this timer that is used in many networks is 44 seconds. The timer is reset to its initial value in both the mobile device and the SGSN whenever data is transmitted. When the timer reaches 0 the logical connection between the mobile device and the network automatically falls back into the standby state, which is further described below.

It is important to note that the ready state of a mobile device is not synonymous with the ability of the mobile device to transfer data to and from the Internet. To transfer user data, a so-called PDP context is necessary, which is further described in Section 2.8.2. Being in ready state simply means that both signaling and possibly user data can be sent to the mobile device without prior paging by the network.

The ready state resembles in some ways the GSM dedicated mode. However, it should be noted that in the GPRS ready state the network is not responsible for the user's mobility as would be the case in the GSM dedicated mode. The decision to select a new cell for an ongoing data transfer is not made by the network (see Section 1.8.3) but by the mobile device. When the signal quality deteriorates during an ongoing data transfer and the mobile device sees a better cell, it will interrupt the ongoing data transfer and change to the new cell. After reading the system information on the BCCH it reestablishes the connection and informs the network of the cell change. The complete procedure takes about two seconds, after which the communication resumes. Data of the aborted connection might have to be resent if it was not acknowledged by the network or the mobile device before the cell change.

To minimize the impact of cell changes, an optional method, requiring the support of both the mobile device and the network has been added to the GPRS standard, which is referred to as network-assisted cell change (NACC). If implemented, it is possible for the mobile device to send a packet cell change notification message to the network when it wants to change into a different cell. The network responds with a packet neighbor cell data message alongside the ongoing user data transfer that contains all necessary parts of the system information of the new cell to perform a quick reselection. Subsequently, the network stops the user data transfer in the downlink direction and instructs the mobile device to switch to the new cell. The mobile device then moves to the new cell and reestablishes the connection to the network without having to read the system information messages from the broadcast channel first. By skipping this step, the data traffic interruption is reduced to a few hundred milliseconds. The network can then resume data transfer in the downlink direction from the point the transmission was interrupted. While there is usually some loss of data during the standard cell change procedure in the downlink, this is not the case with NACC. Thus, this additional benefit also contributes to speeding up the cell change. To complete the procedure, the mobile device asks the network for the remaining system information via a provide system information message while the user data transfer is already ongoing again.

Although the implementation of NACC in the mobile devices is quite simple, there are a number of challenges on the network side. When the old and new cells are in the same location area and controlled from the same radio network node, the procedure is straightforward. If the new and old cells are in different location areas, however, they might be controlled by different network elements. Therefore, an additional synchronization between the elements in the network is necessary to redirect the downlink data flow to the new cell before the mobile device performs the cell reselection. Unfortunately, this synchronization was not included when NACC was first introduced into the GPRS standards in Release 4. As a consequence many of the NACC-capable networks today only support the feature for cells in the same location area.

The Standby State

In case no data is transferred for some time, the ready timer expires and the mobile device changes into the standby state. In this state, the mobile device only informs the network of a cell change if the new cell belongs to a different routing area than the previous one. If data arrives in the network for the mobile device after it has entered the standby state, the data needs to be buffered and the network has to page the subscriber in the complete routing area to get the current location. Only then can the data be forwarded as shown in Figure 2.15. A routing area is a part of a location area and thus also consists of a number of cells. Although it would have been possible to use location areas for GPRS as well, it was decided that splitting location areas into smaller routing areas would enable operators to better fine-tune their networks by being able to control GSM and GPRS signaling messages independently.

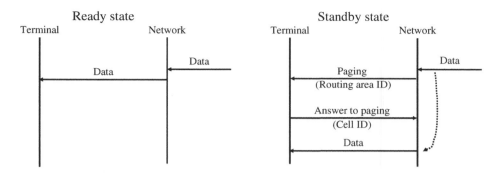

Figure 2.15   Difference between ready and standby states

If the mobile device detects after a cell change that the routing area is different from the previous cell, it starts to perform a routing area update (RAU) which is similar to a GSM location area update. In case the location area has changed as well, the mobile device needs to perform both an LU and an RAU.

The advantage of the standby state for the network is the reduced signaling overhead as not every cell change has to be reported. Thus, scarce resources on the RACH, the AGCH and the PDTCH can be saved. For the mobile device, the advantage of the standby state is that it can stop the continuous monitoring of the AGCH and only infrequently monitor the PCH as described in more detail below. Most operators have set the PCH monitoring interval to around 1.5 seconds (e.g. 6–8 multiframes), which helps to significantly reduce power consumption.

In the uplink direction, there is no difference between ready and standby states. If a mobile device wants to send data while being in standby state, it implicitly switches back to ready state once the first frame is sent to the network.

2.5   GPRS Network Elements

As discussed in the previous paragraphs, GPRS works in a very different way compared to the circuit-switched GSM network. This is why three new network components were introduced into the mobile network and software updates had to be made for some of the existing components. Figure 2.16 gives an overview of the components of a GPRS network, which are described in more detail below.

2.5.1   The Packet Control Unit (PCU)

The BSC has been designed to switch 16 kbit/s circuit-switched channels between the MSC and the subscribers. It is also responsible for the handover decisions for those calls. As GPRS subscribers no longer have a dedicated connection to the network, the BSC and its switching matrix are not suited to handle packet-switched GPRS traffic. Therefore, this task has been assigned to a new network component, the PCU. The PCU is the packet-switched counterpart of the BSC and fulfills the following tasks:

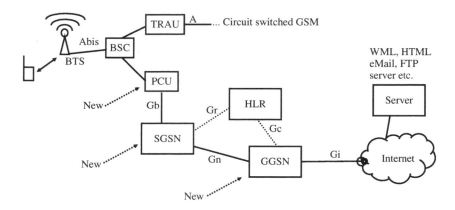

Figure 2.16   GPRS network nodes

- assignment of timeslots to subscribers in the uplink direction when requested for by the mobile device via the RACH or the PRACH;
- assignment of timeslots to subscribers in the downlink direction for data arriving from the core network;
- flow control of data in the uplink and downlink directions and prioritization of traffic;
- error checking and retransmission of lost or faulty frames;
- subscriber paging;
- supervising entity for subscriber timing advance during data transmission.

In order for the PCU to control the GPRS traffic, the BSC turns over control for some of the timeslots to the PCU. This is done by redirecting timeslots in the BSC switching matrix away from the MSC and Transcoding and Rate Adaptation Unit (TRAU) toward the PCU. The BSC then simply forwards all data contained in these timeslots to and from the PCU without any processing.

As GPRS uses GSM signaling channels like the RACH, PCH and AGCH to establish the initial communication, a control connection has to exist between the PCU and the BSC. When the mobile device requests for GPRS resources from the network, the BSC receives a channel request message for packet access. The BSC forwards such packet access request messages straight to the PCU without further processing. It is then the PCU's task to assign uplink blocks on a PDTCH and return an immediate packet assignment command, which contains a packet uplink assignment for the subscriber. The BSC just forwards this return message from the PCU to the BTS without further processing. Once GPRS uplink resources have been assigned to a user by the PCU, further signaling will be handled by the PCU directly over the GPRS timeslots and no longer via the GSM signaling channels.

Monitoring GSM and GPRS Signaling Messages

In GSM, it is quite easy to use a network tracer to monitor all signaling messages being exchanged between the BSC, BTS and the mobile devices communicating over a BTS. All messages use the same logical link access protocol (LAPD) channel that is transmitted over

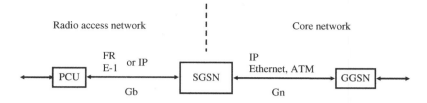

Figure 2.17   Interfaces and protocols of the SGSN on layers 2 and 3

dedicated LAPD timeslots on the Abis interface. The traffic channels used for the voice data are transmitted on different timeslots. To monitor the signaling messages it is only necessary to monitor the LAPD timeslots. Monitoring GPRS signaling messages is a far more complex task, as they can be sent over any GPRS timeslot and in between user data blocks. Therefore, it is also necessary to trace user data and empty packets on the Abis interface, which requires more processing power and memory on the network tracer.

PCU Positioning

The GSM standards allow for a number of different positions of the PCU in the network. The most common implementation is to have the PCU behind the BSC as shown in Figure 2.16. On the basis of the design of the PCU, some network suppliers deliver it as one or more cards that can be plugged into the BSC. Others have chosen to base the PCU on a more powerful computing architecture that is able to process the GPRS traffic of more than just one BSC. In such an architecture, the PCU is implemented in a cabinet physically independent from the BSC. Several BSCs are then connected to a single PCU.

The interface between the PCU and BSC has not been standardized. This means that the PCU and BSC have to be from the same supplier. If a network operator has BSCs from multiple suppliers, he is constrained to also buy the PCUs from the same number of network suppliers.

2.5.2   The Serving GPRS Support Node (SGSN)

The SGSN can be seen as the packet-switched counterpart to the MSC in the circuit-switched core network. As shown in Figure 2.17, it lies between the radio access network and the core network. It is responsible for user plane management and the signaling plane management.

User Plane Management

The user plane combines all protocols and procedures for the transmission of user data frames between the subscriber and external networks like the Internet or a company intranet. All frames that arrive for a subscriber at the SGSN are forwarded to the PCU, which is responsible for the current cell of the subscriber. In the reverse direction the PCU delivers data frames of a subscriber to the SGSN, which in turn will forward them to the next network node, which is called the gateway GPRS support node (GGSN). The GGSN is further described in Section 2.5.3.

IP is used as the transport protocol in the GPRS core network between the SGSN and GGSN. This has the big advantage that on lower layers a great number of different transmission technologies can be used (Figure 2.17). For short distances between the network elements, copper-based twisted pair Gigabit Ethernet links can be used, whereas over longer distances optical links are more likely to be used. By using IP, it is ensured that the capacity of the core network can be flexibly enhanced in the future.

To connect the SGSN with the PCU, the frame relay protocol was selected at first. The decision to not use IP on this interface is somewhat difficult to understand from today's perspective. At that time frame relay was selected because the data frames between SGSN and PCU are usually transported using E-1 links, which are quite common in the GSM BSS. Frame relay, with its similarities to ATM, was well suited for transmitting packet data over 2 Mbit/s E-1 channels and had already been used for many years in wide area network communication systems. The disadvantage of using frame relay, however, is that besides the resulting complicated network architecture, the SGSN has to extract the user data frames from the frame relay protocol and forward them via IP to the GGSN and vice versa.

As ATM and IP have become more common since the introduction of GPRS, the UMTS radio network described in Chapter 2 no longer uses frame relay on this interface but uses ATM and IP instead, which significantly simplifies the network architecture. In the meantime, the 3GPP GPRS standards were also enhanced with an option to use IP and any kind of transmission technology below on the Gb interface instead of frame relay. In practice, it can be observed that many network operators make use of this today.

Although ciphering for circuit-switched traffic is terminated in the BTS, ciphering for packet-switched traffic is terminated in the SGSN as shown in Figure 2.18. This has a number of advantages. In GPRS, the mobile device and not the network has control over cell changes during data transfers. If ciphering were done on the BTS, the network would first have to supply the ciphering information to the new BTS before the data transfer could resume. As this step is not necessary when the ciphering is terminated in the SGSN, the procedure is accelerated. Furthermore, the user data remains encrypted on all radio network links. From a security point of view, this is a great improvement. The link between BTS and BSC is often carried over microwave links, which are not very difficult to spy on. The drawback of this solution is that the processing power necessary for ciphering is not distributed over many BTS but concentrated on the SGSN. Some SGSN suppliers therefore offer not only ciphering in software but also hardware-assisted ciphering. As ciphering is optional, it can be observed that in some networks GPRS ciphering is not enabled as operators can save money by using the processing power they save for additional user data traffic.

Signaling Plane Management

The SGSN is also responsible for the management of all subscribers in its area. All protocols and procedures for user management are handled on the signaling plane.

To be able to exchange data with the Internet, it is necessary to establish a data session with the GPRS network. This procedure is called PDP context activation and is part of the session management (SM) tasks of the SGSN. From the user point of view, this procedure is invoked to get an IP address from the network.

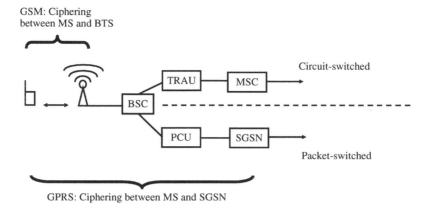

Figure 2.18   Ciphering in GSM and GPRS

Subscribers can change their location in a mobile network frequently. When this happens the SGSN needs to change its routing of packets to the radio network accordingly. This task is done by the GPRS mobility management (GMM) sublayer.

When a subscriber leaves the area of the current SGSN, GMM also contains procedures to change the routing for a subscriber in the core network to the new SGSN. This procedure is called inter-SGSN routing area update (IRAU).

To charge the subscriber for usage of the GPRS network, the SGSN and the GGSN, which is described in more detail in the next paragraph, collect billing information in so-called call detail records (CDRs). These are forwarded to the billing server, which collects all CDRs and generates an invoice for each subscriber once a month. The CDRs of the SGSN are especially important for subscribers that roam in a foreign network. As will be described in Section 2.8.2, the SGSN is the only network node in the foreign network that can generate a CDR for a GPRS session of a roaming subscriber for the foreign operator. For roaming subscribers the CDRs of the SGSN are then used by the foreign operator to charge the home operator for the data traffic the subscriber has generated. For GPRS data traffic generated in the home network, the GGSN usually generates the CDR, and the billing information of the SGSN is not used.

2.5.3   The Gateway GPRS Support Node (GGSN)

Although the SGSN routes user data packets between the radio access network and the core network, the GGSN connects the GPRS network to the external data network. The external data network will in most cases be the Internet. For business applications, the GGSN can also be the gateway to a company intranet [3].

The GGSN is also involved in setting up a PDP context. In fact, it is the GGSN that is responsible for assigning a dynamic or static IP address to the user. The user keeps this IP address while the PDP context is established.

As shown in Figure 2.19, the GGSN is the anchor point for a PDP context and hides the mobility of the user to the Internet. When a subscriber moves to a new location, a new SGSN might become responsible and data packets are sent to the new SGSN (IRAU). In this scenario,

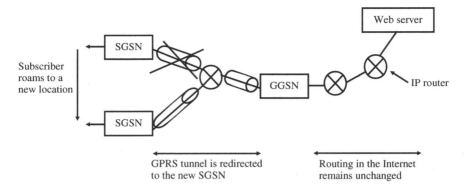

Figure 2.19   Subscriber changes location within the GPRS network

the GGSN has to update its routing table accordingly. This is invisible to the Internet as the GGSN always remains the same. It can thus be seen as the anchor point of the connection, which ensures that despite user mobility, the assigned IP address does not have to be changed.

2.6   GPRS Radio Resource Management

As described earlier, a GPRS timeslot can be assigned to several users at the same time. It is also possible to assign several timeslots to a single subscriber to increase his data transmission speed. In any case, the smallest transmission unit that can be assigned to a user is one block, which consists of four bursts on one timeslot on the air interface for GPRS and two bursts for EDGE MCS 7–9. A block is also called a GPRS radio link control/Medium Access Control (RLC/MAC) frame.

Temporary Block Flows (TBF) in the Uplink Direction

Every RLC/MAC frame on the PDTCH or PACCH consists of an RLC/MAC header and a user data field. When a user wants to send data on the uplink, the mobile device has to request for resources from the network by sending a packet channel request message via the RACH or the PRACH as previously shown in Figure 2.13.

The PCU then answers with an immediate packet assignment message on the AGCH. The message contains information as to the timeslots in which the mobile device is allowed to send data. As a timeslot in GPRS may not only be used exclusively by a single subscriber, a mechanism is necessary to indicate to a mobile device when it is allowed to send on the timeslot. Therefore, the uplink assignment message contains a parameter called the uplink state flag (USF). A different USF value is assigned to every subscriber that is allowed to send on the timeslot. The USF is linked to the so-called temporary flow identity (TFI) of a temporary block flow (TBF). A TBF identifies data to or from a user for the time of the data transfer. Once the data transfer is completed, the TFI is reused for another subscriber. To know when it can use the uplink timeslots, the mobile device has to listen to all the timeslots it has been

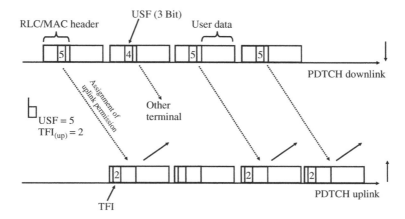

Figure 2.20   Use of the uplink state flag

assigned in the downlink direction. Every block that is sent in the downlink to a subscriber contains a USF in its header as shown in Figure 2.20. It indicates who is allowed to send in the next uplink block. By including the USF in each downlink block, the PCU can dynamically schedule who is allowed to send in the uplink. Therefore, this procedure is also referred to as 'dynamic allocation'.

Mobile devices that support high multislot classes are not able to listen in downlink direction on all the timeslots assigned to it for uplink transmission opportunities. In such cases the 'Extended Dynamic Allocation' scheme is used, which uses a single USF to assign resources on several timeslots for a user for uplink data transfers.

Note that the USF information in the header and data portion of a downlink block is usually not intended for the same user. This is because the assignments of up- and downlink resources are independent. This makes sense when considering web surfing, for example, where it is usually not necessary to already assign downlink resources at the time the universal resource locator (URL) of the web page is sent to the network.

For mobile devices that have an uplink TBF established, the network needs to send control information from time to time. This is necessary to acknowledge the receipt of uplink radio blocks. The logical PACCH that can be sent in a radio block, instead of a PDTCH, is used to send control information. The mobile device recognizes its own downlink PACCH blocks because the header of the block contains its TFI value.

The PCU will continue to assign uplink blocks until the mobile device indicates that it no longer requires blocks in the uplink direction. This is done with the so-called 'countdown procedure'. Every block header in the uplink direction contains a four-bit countdown value. The value is decreased by the mobile device for every block sent at the end of the data transfer. The PCU will no longer assign uplink blocks for the mobile device once this value has reached 0.

While coordinating the use of the uplink in this way is quite efficient, it creates high latency if data is only sent sporadically. This is especially problematic during a web-browsing session for two reasons: As shown at the end of this chapter, high latency has a big impact on the time it takes to establish Transmission Control Protocol (TCP) connections, which are necessary

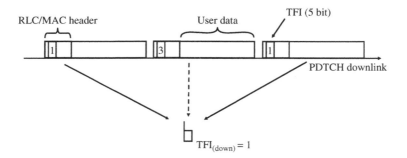

Figure 2.21   Use of the temporary flow identifier (TFI) in the downlink direction

before a web page can be requested. Furthermore, several TCP connections are usually opened to download the different elements like text, pictures and so on, of a web page, so high latency slows down the process in several instances. To reduce this effect, the GPRS standard was enhanced by a method called the 'extended uplink TBF'. In case both network and mobile device support the functionality, the uplink TBF is not automatically closed at the end of the countdown procedure but is kept open by the network until the expiry of an idle timer, which is usually set in the order of several seconds. While the uplink TBF is open, the network continues to assign blocks in the uplink direction to the mobile device. This enables the mobile device to send data in the uplink direction quickly without requesting for a new uplink TBF. The first mobile devices and networks that supported extended uplink TBF appeared on the market in 2005 and a substantial improvement of web page download and delay times can be observed, as discussed at the end of the chapter.

Temporary Block Flows in the Downlink Direction

If the PCU receives data for a subscriber from the SGSN, it will send a packet downlink assignment message to the mobile device similar to the one shown in Figure 2.21 in the AGCH or the PAGCH. The message contains a TFI of a TBF and the timeslots the mobile device has to monitor. The device will then immediately start monitoring the timeslots. In every block it receives, it will check if the TFI included in the header equals the TFI assigned to it in the packet downlink assignment message as shown in Figure 2.22. If they are equal, it will process the data contained in the data portion of the block. If they are not equal, the mobile device discards the received block. Once the PCU has sent all data for the subscriber currently in its queue, it will set the 'final block indicator' bit in the last block it sends to the mobile device. Subsequently, the mobile device stops listening on the assigned timeslots and the TFI can be reused for another subscriber. To improve performance, the network can also choose to keep the downlink TBF established for several seconds so that no TBF establishment is necessary if further data for the user arrives.

To acknowledge blocks received from the network, the mobile device has to send control information via the logical PACCH. For sending control information to the network, it is not necessary to assign an uplink TBF. The network informs the mobile device in the header of downlink blocks as to which uplink blocks it can use to send control information.

[...]	
	RLC/MAC PACKET TIMESLOT RECONFIGURE
000111--	Message Type : 7 = packet timeslot reconfigure
------00	Page Mode : 0 = normal paging
	Global TFI:
--01111-	Uplink Temporary Flow Identifier : 15
00------	Channel Coding Command : Use CS-1 in Uplink
	Global Packet Timing Advance:
----0001	Uplink TA Index : 1
101-----	Uplink TA Timeslot Number : 5
----0001	Downlink TA Index : 1
101-----	Downlink TA Timeslot Number : 5
---0----	Downlink RLC Mode : RLC acknowledged mode
----0---	CTRL ACK : 0 = downlink TBF already established
xxxxxxxx	Downlink Temporary Flow ID: 11
xxxxxxxx	Uplink Temporary Flow ID: 15
	Downlink Timeslot Allocation:
-0------	Timeslot Number 0 : 0
--0-----	Timeslot Number 1 : 0
---0----	Timeslot Number 2 : 0
----0---	Timeslot Number 3 : 0
-----1--	Timeslot Number 4 : 1 = assigned
------1-	Timeslot Number 5 : 1 = assigned
-------1	Timeslot Number 6 : 1 = assigned
0-------	Timeslot Number 7 : 0
	Frequency Parameters:
--000---	Training Sequence Code : 0
xxxxxxxx	ARFCN : 067
[...]	

Figure 2.22 Packet timeslot reconfiguration message according to 3GPP TS 44.060, 11.2.31 [4]

Timing Advance Control

The farther a mobile device is away from a BTS, the sooner it has to start sending its data bursts to the network in order for them to arrive at the BTS at the correct time. As the position of the user can change during the data exchange, it is necessary for the network to constantly monitor how far away the user is from the serving base station. If the user moves closer to the BTS, the network has to inform the mobile device to delay sending its data compared to the current timing. If the user moves farther away, it has to start sending its bursts earlier. This process is called timing advance control.

As we have seen in the previous paragraph, the assignment of uplink and downlink resources is independent of each other. When downloading a large web page, for example, it might happen that a downlink TBF is assigned while no uplink TBF is established because the mobile device has no data to send.

Even though no uplink TBF is established, it is necessary from time to time to send layer 2 acknowledgment messages to the network for the data that has been received in the downlink. To send these messages quickly, no uplink TBF has to be established. In this case, the PCU informs the mobile device in the downlink TBF from time to time as to which block to use to send the acknowledgment. As this only happens infrequently, the network cannot utilize the previous acknowledgment bursts for the timing advance calculation for the following bursts.

Hence, a number of methods have been standardized to measure and update the timing advance value while the mobile device is engaged in exchanging GPRS data.

The Continuous Timing Advance Update Procedure

In a GPRS 52-multiframe, frames 12 and 38 are dedicated to the logical PTCCH uplink and downlink. The PTCCH is further divided into 16 subchannels. When the PCU assigns a TBF to a mobile device, the assignment message also contains an information element that instructs the mobile device to send access bursts on one of the 16 subchannels in the uplink with a timing advance 0. These bursts can be sent without a timing advance because they are much shorter than a normal burst. For more information about the access burst, see Chapter 1. The BTS monitors frames 12 and 38 for access bursts and calculates the timing advance value for every subchannel. The result of the calculation is sent on the PTCCH in the following downlink block. As the PTCCH is divided into 16 subchannels, the mobile device sends an access burst on the PTCCH and receives an updated value every 1.92 seconds.

2.7 GPRS Interfaces

The GPRS standards define a number of interfaces between components. Apart from the PCU, which has to be from the same manufacturer as the BSC, all other components can be selected freely. Thus, it is possible, for example, to connect a Huawei PCU to an Ericsson SGSN, which is in turn connected to a Cisco GGSN.

The Abis Interface

The Abis interface connects the BTS with the BSC. The protocol stack as shown in Figure 2.23 is used on all timeslots of the radio network which are configured as (E)GPRS PDTCHs. Usually, all data on these timeslots is sent transparently over the nonstandardized interface between the BSC and PCU. However, as the link is also used to coordinate the BSC and PCU with each other, it is still not possible to connect the BSC and PCUs of two different vendors. On the lower layers of the protocol stack, the RLC/MAC protocol is used for the radio resource management. On the next protocol layer, the Logical Link Control (LLC) protocol is responsible for the framing of the user data packets and signaling messages of the mobility management and SM subsystems of the SGSN. Optionally, the LLC protocol can also ensure a reliable connection between the mobile device and the SGSN by using an acknowledgment mechanism for the correctly received blocks (acknowledged mode). On the next higher layer, the Subnetwork-Dependent Convergence Protocol (SNDCP) is responsible for framing IP user data to send it over the radio network. Optionally, SNDCP can also compress the user data stream. The LLC layer and all layers above are transparent for the PCU, BSC and BTS as they are terminated in the SGSN and the mobile device, respectively.

The Gb Interface

The Gb interface connects the SGSN with the PCU as shown in Figure 2.23. On layer 1, mostly 2 Mbit/s E-1 connections were used in the past. An SGSN is usually responsible for several

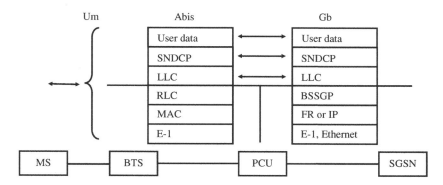

Figure 2.23   GPRS protocol stacks in the radio network

PCUs in an operational network and they were thus usually connected by several 2 Mbit/s connections to the SGSN. On layers 2 and 3 of the protocol stack, the frame relay protocol was used, which was a standard packet-switched protocol used in the telecom world for many years. Frame relay was also a predecessor of the ATM protocol, which has gained a lot of popularity for packet-based long-distance transmission in the telecom world and which was heavily used in early UMTS network as will be described in Chapter 3. Thus, its properties were very well known at the time of standardization, especially for packet-switched data transfer over 2 Mbit/s E-1 connections. The disadvantage was that the user data had to be encapsulated into frame relay packets, which made the overall protocol stack more complex as in the wireless world it is only used on the Gb interface. In newer versions of the 3GPP standards, a feature was specified to also use IP as a lower layer protocol for the Gb interface. Today, the IP protocol has mostly replaced frame relay, and fiber links are used on the physical layer for transporting the IP packets of the Gb interface and data of other interfaces simultaneously over longer distances.

The Gn Interface

This is the interface between the SGSNs and GGSNs of a GPRS core network and is described in detail in 3GPP TS 29.060 [5]. Usually, a GPRS network comprises more than one SGSN because a network usually has more cells and subscribers than can be handled by a single SGSN. Another reason for having several GGSNs in the network is to assign them different tasks. One GGSN, for example, could handle the traffic of postpaid subscribers, while another one could be specialized in handling the traffic of prepaid subscribers. Yet another GGSN could be used to interconnect the GPRS network with companies that want to offer direct intranet access to their employees without sending the data over the Internet. Of course, all of these tasks can also be done by a single GGSN if it has enough processing power to handle the number of subscribers for all these different tasks. In practice, it is also quite common to use several GGSNs that can handle the same kind of tasks for load balancing and redundancy reasons.

On layer 3, the Gn interface uses IP as the routing protocol as shown in Figure 2.24. If the SGSN and GGSN are deployed close to each other, Ethernet over twisted pair or optical cables can be used for the interconnection. If larger distances need to be overcome, alternatives are IP over Carrier Ethernet optical links or IP over ATM, over various transport technologies such as synchronous transfer mode (STM). To increase capacity or for redundancy purposes, several physical links are usually used between two network nodes.

Gn

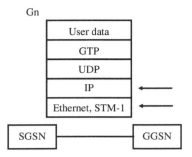

Figure 2.24   The Gn interface protocol stack

User data packets are not sent directly on the IP layer of the Gn interface but are encapsulated into GPRS Tunneling Protocol (GTP) packets. This creates some additional overhead, which is needed for two reasons: Each router in the Internet between the GGSN and the destination makes its routing decision for a packet based on the destination IP address and its routing table. In the fixed-line Internet, this approach is very efficient as the location of the destination address never changes and thus the routing tables can be static. In the GPRS network, however, subscribers can change their location at any time as shown in Figure 2.19 and thus the routing of the packets must be flexible. As there are potentially many IP routers between the GGSN and SGSN, these would have to change their routing tables whenever a subscriber changes its location. To avoid this, the GPRS network does not use the source and destination IP address of the user's IP packet. Instead, the IP addresses of the current SGSN and GGSN are used for the routing process. As a consequence, the user data packets need to be encapsulated into GTP packets to be able to tunnel them transparently through the GPRS network. If the location of a subscriber changes, the only action that needs to be taken in the core network is to inform the GGSN of the IP address of the new SGSN that is responsible for the subscriber. The big advantage of this approach is that only the GGSN has to change its routing entry for the subscriber. All IP routers between the GGSN and SGSN can, therefore, use their static routing tables and no special adaptation of those routers is necessary for GPRS. Figure 2.25 shows the most important parameters on the different protocol layers on the Gn interface. The IP addresses on layer 3 are those of the SGSN and GGSN, while the IP addresses of the user data packet that is encapsulated into a GTP packet belong to the subscriber and the server in the Internet with which the subscriber communicates. This means that such a packet contains two layers on which IP is used.

When the GGSN receives a GTP packet from an SGSN, it removes all headers including the GTP header. Later, the remaining original IP packet is routed via the Gi interface to the Internet.

The Gi Interface

This interface connects the GPRS network to external packet networks, for example, the Internet. From the perspective of the external networks, the GGSN is just an ordinary IP router. As on the Gn interface, a number of different transmission technologies from 'ordinary' twisted pair 100 Mbit/s Ethernet to ATM over STM-1 optical interface can be used. To increase the bandwidth or to add redundancy, several physical interfaces can be used simultaneously.

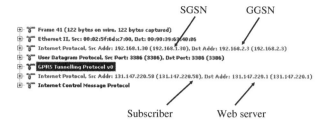

Figure 2.25 GTP packet on the Gn interface

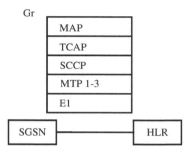

Figure 2.26 The Gr interface

The Gr Interface

This interface connects the SGSN with the Home Location Register (HLR), which contains information about all subscribers on the network (Figure 2.26). It was enhanced with a software upgrade to also act as a central database for GPRS subscriber data. The following list shows some examples:

- GPRS service admission on a per user (International mobile subscriber identity, IMSI) basis;
- GPRS services that a user is allowed to use (Access Point Names, APNs);
- GPRS international roaming permissions and restrictions.

As described in Chapter 1, the HLR is an SS7 service control point (SCP). Therefore, the Gr interface was initially based on E1 trunks, SS7 on layer 3 and Mobile Application Part (MAP) on the application layer, as shown in Figure 2.26. Today, the lower layers have been replaced by IP connectivity. The MAP protocol was also extended to be able to exchange GPRS-specific information. The following list shows some of the messages that are exchanged between SGSN and HLR:

- **Send authentication information:** This message is sent from the SGSN to the HLR when a subscriber attaches to the network for which the SGSN does not yet have authentication information.
- **Update location:** The SGSN informs the HLR that the subscriber has roamed into its area.

- **Cancel location:** When the HLR receives an update location message from an SGSN, it sends this message to the SGSN to which the subscriber was previously attached.
- **Insert subscriber data:** As a result of the update location message sent by the SGSN, the HLR will forward the subscriber data to the SGSN.

The Gc Interface

This interface connects the GGSN with the HLR. It is optional and is not widely used in networks today.

The Gp Interface

This interface is described in 3GPP TS 29.060 [5] and connects GPRS networks of different countries or different operators with each other for GTP traffic as shown in Figure 2.27. It enables a subscriber to roam outside the coverage area of the home operator and still use GPRS to connect to the Internet. The user's data will be tunneled via the Gp interface similarly to the Gn interface from the SGSN in the foreign network to the GGSN in the subscriber's home network and from there to the Internet or a company intranet. From an end-user's perspective, using a GGSN in the home network has the big advantage that no settings in the device need to be changed.

The ability to be able to simply connect to the Internet while roaming abroad without any configuration changes is an invaluable improvement over previous methods. Before GPRS and mobile networks, travelers could use only fixed-line modems. Using fixed-line modems abroad was often difficult or even impossible because of not having a fixed-line connection available where required and because of connectors that differed from country to country. If an adapter was available, the next problem was which dial-in service provider could be called to establish a connection to the Internet. A provider in the visited country was usually not an option. Establishing a modem connection with a provider in the home country was often also difficult because of their use of special numbers that could not be reached from abroad.

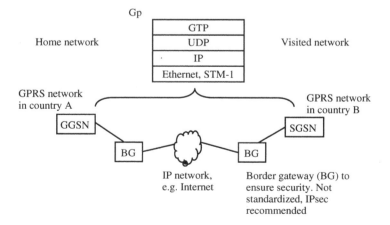

Figure 2.27   The Gp interface

Note that the Gp interface is for GTP traffic only. For signaling with the HLR, the two networks also need an SS7 interconnection so that the visited SGSN can communicate with the HLR in the home network.

It should be noted at this point that it is also possible to use a GGSN in the visited network. This option is referred to as 'local breakout' and can, for example, be used by network operators in Europe to offer cheaper data roaming. The downside of 'local breakout' for data roaming is that the user has to manually change the configuration of his device, that is, he has to manually change the APN. For details see Section 2.8.2.

The Gs Interface

3GPP TS 29.018 [6] describes this interface, which is also optional. It connects the SGSN and the MSC/VLR. The functionality and benefits of this interface in conjunction with GPRS NOM I are discussed in Section 2.3.6.

2.8 GPRS Mobility Management and Session Management (GMM/SM)

Apart from forwarding data packets between GPRS subscribers and the Internet, the GPRS network is also responsible for the mobility management of the subscribers and the SM to control the individual connections between subscribers and the Internet. For this purpose, signaling messages and signaling flows have been defined that are part of the GMM/SM protocol.

2.8.1 Mobility Management Tasks

Before a connection to the Internet can be established, the user has to first connect to the network. This is similar to attaching to the circuit-switched part of the network. When a subscriber wants to attach, the network usually starts an authentication procedure, which is similar to the GSM authentication procedure. If successful, the SGSN sends an LU message to the HLR to update the location information of that subscriber in the network's database. The HLR acknowledges this operation by sending an 'insert subscriber data' message back to the SGSN. As the name of the message suggests, it not only acknowledges the LU but also returns the subscription information of the user to the SGSN so that no further communication with the HLR is necessary as long as the subscriber does not change the location. The SGSN, subsequently, will send an attach accept message to the subscriber. The attach procedure is complete when the subscriber returns an attach complete message to the SGSN. Figure 2.28 shows the message flow for this procedure.

If the subscriber was previously attached to a different SGSN, the procedure is somewhat more complex. In this case, the new SGSN will ask the old SGSN for identification information of the subscriber. Once the subscriber has authenticated successfully, the SGSN will send the LU message as above to the HLR. As the HLR knows that the subscriber was previously attached to a different SGSN, it sends a cancel location message to the old SGSN. It later returns the insert subscriber data message to the new SGSN.

It is also possible to do a combined GSM/GPRS attach procedure in case the Gs interface is available. To inform the mobile device of this possibility, the network broadcasts the GPRS network operation mode on the BCCH. Should the mobile device thus send a request for a

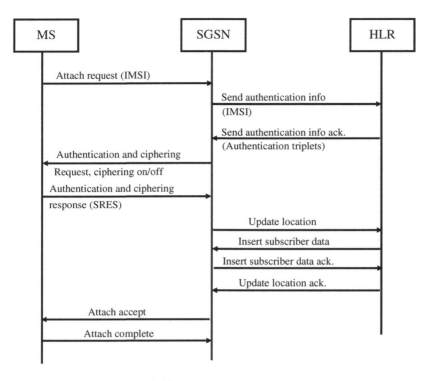

Figure 2.28   GPRS attach message flow

combined attach from the SGSN, it is the task of the new SGSN to inform the new MSC of the location of the subscriber. The new MSC will then send an update location to the HLR for the circuit-switched part of the network. The HLR will then cancel the location in the old MSC and send an insert subscriber data back to the new MSC. Once all operations have been performed, the new MSC sends back an LU accept to the SGSN, which will then finally return the attach accept message to the mobile device. Although this message flow is more complex from the point of view of the core network, it allows the mobile device to attach to both circuit- and packet-switched network parts with only a single procedure. This speeds up the process for the mobile device and reduces the signaling load in the radio network.

Once the attach procedure is complete, the mobile device is authenticated and known to the network. In the circuit-switched part of the network, the user can now go ahead and establish a voice call by dialing a number. In the GPRS packet-switched part of the network, the mobile device can now go ahead and establish a data session. This so-called PDP context activation procedure is described in the next paragraph.

Figure 2.29 shows an example of a GPRS attach message that was traced on the Gb interface. Some interesting parameters are highlighted in bold. As can be seen in the message, the mobile device does not only inform the network about its identity, but it also includes its capabilities such as its multislot capabilities and which frequency bands it supports (850, 900, 1800, 1900 MHz). Although standards evolve quickly, mobile device developers often only implement a subset of functionality at the beginning and add more features over time in

[...]	Mobility Management: ATTACH REQUEST
	MS Network Capability:
1-------	GPRS encryption algorithm GEA/1: 1 = available
[...]	
-----001	**Attach Type : 001bin = GPRS attach**
-100----	GPRS Ciphering Key Sequence Number : 100bin
	DRX Parameter
01000000	Split PG cycle code : 64 = 64
-----011	Non-DRX timer: max. 4 sec non-DRX mode after transfer state
----0---	SPLIT on CCCH: not supported
	Mobile Identity
-----100	Type of identity: TMSI
----0---	Parity: 0 = even
Xxxxxxxx	TMSI: D4CC3EC4h
	Old Routing Area Identification
Xxxxxxxx	**Mobile Country Code: 232**
Xxxxxxxx	**Mobile Network Code: 03**
Xxxxxxxx	**Location area code: 6F32h**
00000001	**Routing area code: 0Fh**
	MS Radio Access Capability
0001----	**Access technology type: 1 = GSM E (900 MHz Band)**
	Access capabilities
---100--	RF power capability: 4h
	A5 bits
-------1	A5/1: 1 = Encryption algorithm available
0-------	A5/2: 0 = Encryption algorithm not available
-1------	A5/3: 1 = Encryption algorithm available
[...]	
------1-	ES IND : 1h = early Classmark Sending is implemented
[...]	
	Multislot capability
Xxxxxxxx	**GPRS multi slot class: 10 (4 downlink + 2 uplink)**
--0-----	GPRS extended dynamic allocation: not implemented
----1101	Switch-measure-switch value: 0
1000----	Switch-measure value: 8
Xxxxxxxx	**Access technology type: 3 = GSM 1800**
Xxxxxxxx	Access capabilities
001-----	RF power capability: 1
----1---	ES IND: 1 = early Classmark Sending is implemented
[...]	

Figure 2.29   GPRS attach message on the Gb interface

new software versions or even only in new models. This flexibility and thus fast time to market are only possible if networks and mobile devices are able to exchange information about their capabilities.

A good example of such an approach is the multislot capability. Early GPRS mobile devices were able to aggregate only two downlink timeslots and use only a single one in the uplink. Current mobile devices support up to five timeslots in the downlink and three in the uplink (multislot class 32).

Once the mobile device is attached, the network has to keep track of the location of the mobile device. As discussed in Chapter 1, this is done by dividing the GSM network into location areas. When a mobile device in idle mode changes to a cell in a different location area, it has to perform a so-called location update (LU). This is necessary so that the network will be able to find the subscriber for incoming calls or SMS messages. In GPRS, the same principle exists. To be more flexible, the location areas are subdivided into GPRS routing areas. If a mobile device in ready or standby state crosses a routing area border, it reports to the SGSN. This procedure is called RAU.

If the new routing area is administered by a new SGSN the process is called IRAU. Although from the mobile device point of view there is no difference between a RAU and IRAU, there is quite a difference from the network point of view. This is because the new SGSN does not yet know the subscriber. Therefore, the first task of the new SGSN is to get the subscriber's authentication and subscription data. As the RAU contains information about the previous routing area, the SGSN can then contact the previous SGSN and ask for this information. At the same time this procedure also prompts the previous SGSN to forward all incoming data packets to the new SGSN in order not to lose any user data while the procedure is ongoing. Subsequently, the GGSN is informed about the new location of the subscriber so that, henceforth, further incoming data is sent directly to the new SGSN. Finally, the HLR is also informed about the new location of the subscriber and this information is deleted in the old SGSN. Further information about this procedure can be found in 3GPP TS 23.060, 6.9.1.2.2 [7].

2.8.2 GPRS Session Management

To communicate with the Internet, a PDP context has to be requested to use after the attach procedure. For the end user, this in effect means getting an IP address from the network. As this procedure is in some ways similar to establishing a voice call, it is sometimes also referred to as 'establishing a packet call'.

Although there are some similarities between a circuit-switched call and a packet-switched call, there is one big difference that is important to remember: For a circuit-switched voice or data call the network reserves resources on all interfaces. A timeslot is reserved for this connection on the air interface, in the radio network and also in the core network. These timeslots cannot be used by anyone else while the call is established even if no data is transferred by the user. When a GPRS packet call is established there are no resources dedicated to the PDP context. Resources on the various interfaces are used only during the time that data is transmitted. Once the transmission is complete (e.g. after the web page has been downloaded), the resources are used for other subscribers. Therefore, the PDP context represents only a logical connection with the Internet. It remains active even if no data is transferred for a prolonged length of time. For this reason a packet call can remain established indefinitely without blocking resources. This is also sometimes referred to as 'always on'.

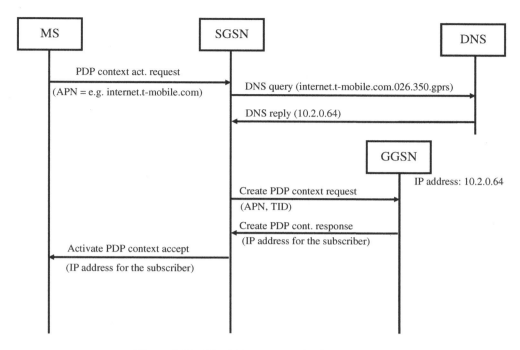

Figure 2.30   The PDP context activation procedure

Figure 2.30 shows the PDP context activation procedure. Initially, the subscriber sends a PDP context activation request message to the SGSN. The most important parameter of the message is the APN. The APN is the reference that GGSN uses as a gateway to an external network. The network operator could have one APN to connect to the Internet transparently, one to offer Wireless Application Protocol (WAP) services, several other APNs to connect to corporate intranets, etc. The SGSN compares the requested APN with the list of allowed APNs for the subscriber that has been received from the HLR during the attach procedure. The APN is a fully qualified domain name like 'internet.t-mobile.com' or simply 'Internet' or 'wap'. The names of the APN can be chosen freely by the GPRS network operator.

In a second step, the SGSN uses the APN to locate the IP address of the GGSN that will be used as a gateway. To do this, the SGSN performs a domain name service (DNS) lookup with the APN as the domain name to be queried. The DNS lookup is identical to a DNS lookup that a web browser has to perform to get the IP address of a web server. Therefore, a standard DNS server can be used for this purpose in the GPRS network. To get an internationally unique qualified domain name, the SGSN adds the mobile country code (MCC) and mobile network code (MNC) to the APN, which is deduced from the subscriber's IMSI. As a top level domain, '.gprs' is added to form the complete domain name. An example of domain name for the DNS query is 'internet.t-mobile.com.026.350.gprs'. Adding the MCC and MNC to the APN by the SGSN enables the subscriber to roam in any country that has a GPRS roaming agreement with the subscriber's home network and use the service without having to modify any parameters. The foreign SGSN will always receive the IP address of the home GGSN from the DNS server, and all packets will be routed to and from the home GGSN and

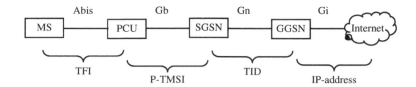

Figure 2.31   Identification of user data packets on different GPRS interfaces

from there to the external network. Of course, it is also possible to use a GGSN in the visited network. To do that, however, the user would have to change the settings in his device, which is very undesirable. Therefore, most operators prefer to always route the traffic back to the home GGSN and thus offer a seamless service to the user.

After the DNS server has returned the GGSN's IP address, the SGSN can then forward the request to the correct GGSN. The APN and the user's IMSI are included in the message as mandatory parameters. To tunnel the user data packets through the GPRS network later on, the SGSN assigns a so-called tunnel identifier (TID) for this virtual connection that is also part of the message. The TID consists of the user's IMSI and a two-digit network service access point identifier (NSAPI). This allows a mobile device to have more than one active PDP context at a time. This is quite useful, for example, to separate Internet access from network operator internal services such as MMS.

If the GGSN grants access to the external network (e.g. the Internet) it will assign an IP address out of an address pool for the subscriber. For special purposes it is also possible to assign a fixed IP address for a subscriber. Subsequently, the GGSN responds to the SGSN with a PDP context activation response message that contains the IP address of the subscriber. Furthermore, the GGSN will store the TID and the subscriber's IP address in its PDP context database. This information is needed later on to forward packets between the subscriber and the Internet and, of course, for billing purposes.

Once the SGSN receives the PDP context activation response message from the GGSN, it also stores the context information in its database and forwards the result to the subscriber. The subscriber then uses the IP address to communicate with the external network.

Different IDs are used for packets of a certain user on each network interface due to the different nature of the protocols and due to the different packet sizes. On the GPRS air interface, with its small data frames of only 456 bits or 57 bytes, which even includes the overhead for error detection and correction, the three-bit TFI is used to route the frame to the correct mobile device. In the radio network the P-TMSI/TLLI is used to identify the packets of a user. Finally in the core network, the GPRS TID is used as identification. Figure 2.31 shows the different interfaces and IDs used on them at a glance.

2.9   Session Management from a User's Point of View

From a user's point of view a PDP context is established in two cases:

The first scenario that requires a PDP context is when applications that run on mobile devices, such as mobile web browsers, instant messaging clients, weather apps, etc., want to access the Internet. Activation of a PDP context usually happens after the mobile devices are switched on and the required configuration data is stored in the device. As the average smartphone user today uses mobile apps that occasionally send and receive data in the background, the PDP context usually remains active until the device is switched off again. Many mobile devices today can also offer external devices, such as notebooks, access to the Internet. This is usually done over Wi-Fi or a Universal Serial Bus (USB) cable. Depending on the network operator, the same or a different PDP context is used for this purpose and the configuration is also stored in the mobile device. This is necessary as the external device just uses the offered Wi-Fi or USB connection without the knowledge of how the Internet connection is established by the mobile device.

Another possibility to connect a notebook to the Internet over GPRS, UMTS and LTE is to use USB data sticks. For this kind of connectivity, the configuration information for the GPRS/UMTS/LTE connection establishment is not stored in the USB data stick but in the notebook. Usually, network operators or the device manufacturers include a connection manager software that is automatically installed when the data stick is used for the first time. It is also possible, however, to use the legacy modem protocol stack of the notebook's operating system for establishing a connection. To better understand this concept, the following section now describes how the modem protocol stack was initially used to connect to a fixed-line Internet service provider over a circuit-switched modem connection. Afterward, the difference between a circuit-switched modem connection and a GPRS/UMTS/LTE connection is described and which additional settings are required in the notebook.

Figure 2.32 shows the equipment that was required to establish a circuit-switched modem connection between a notebook and an Internet service provider in the fixed-line network. On the notebook side, a separate modem was required which later became an integrated device in many notebooks. Modems communicated with external devices with the standardized AT command set, which is an ASCII command and response language. For example, to dial a telephone number, the notebook sent a dial command, which included the telephone number (e.g. 'ATD 0399011782') to the modem. The modem then tried to establish a circuit-switched data connection with the other end. If successful, the modem returned a connect message (e.g. 'CONNECT'). It then entered a transparent mode and forwarded all data that was sent by the notebook to the other end of the connection, instead of interpreting it as a command.

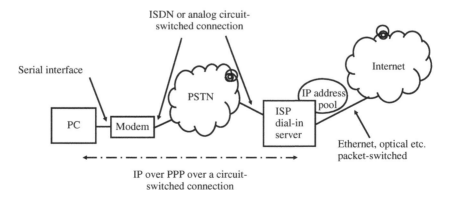

Figure 2.32   IP over PPP for Internet connections

Once the connection was established, it was up to the connected device to interpret the data that was sent and received. In the case of an Internet connection, the Point-to-Point Protocol (PPP) was widely used to send and receive IP packets over modem connections. An example of a service that uses PPP is the 'dial-up network' of the Microsoft Windows operating system. After the call is established, the PPP is responsible for establishing an IP connection with the PPP server on the other side. Usually, a username and password were exchanged before the server at the other side accepted the PPP connection and returned an IP address. The PPP stack on the notebook then connected to the IP stack and encapsulated all IP packets in PPP frames for transmission over the serial interface via the modem to the other side (Figure 2.32).

For GPRS, using this approach in the same way was not possible as there are a number of differences to a circuit-switched dial-up connection:

- GPRS is a packet-switched connection to the Internet.
- There is no telephone number to dial.
- There is no PPP server at the other side of the connection.

To add GPRS functionality to notebooks and other devices, their operating systems could have been extended to support GPRS as part of the network stack. This would have meant a significant amount of software development on many different devices and operating systems. As this seemed impractical, the following approach has been chosen: The already existing PPP stack of the dial-up network on the user's device, as described before, is used for GPRS with slight modifications in the configuration. This means that no software changes had to be made to existing devices. As the PPP stack of the user's device requires a PPP server on the other side, it was decided to implement the PPP server inside the USB stick to terminate the PPP connection. The PPP server in the USB stick then converts the PPP commands and PPP data frames into GPRS commands and GPRS data frames. This approach is shown in Figure 2.33.

To establish a packet-switched GPRS connection from a user's device, the following parameters have to be configured in the dial-up network settings:

The APN is a GPRS-specific parameter and needs to be sent from the user's device to the mobile device during the connection establishment. This is done via a new AT command that has been standardized so that all mobile device vendors that offer GPRS via a serial connection can implement the command in the same way. The AT command,

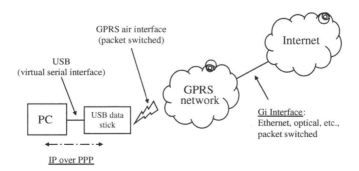

Figure 2.33   PPP termination in the USB data stick for GPRS

Figure 2.34   The advanced settings dialog box for entering the APN

for example, to use `internet.t-mobile.com` as the APN for the connection is, `AT+CGDCONT=1,"IP","internet.t-mobile.com"`. To send the command during the connection establishment, it has to be entered in the dial-up networking configuration in the advanced modem settings dialog box as shown in Figure 2.34.

The next step in the process is to instruct the USB data stick to connect to the Internet via GPRS and not via a circuit-switched connection. This is done by using `*99***1#` as the telephone number in the dial-up networking instead of an ordinary telephone number. When the USB data stick receives this unusual telephone number with the ATD command, it establishes a GPRS connection instead of a circuit-switched data connection. It does this by starting the internal PPP server and sending a PDP context activation request message to the SGSN with the APN that was previously given to it via the `AT+CGDCONT` command. Once it receives a PDP context activation acknowledge message, it will forward the IP address to the external device via the PPP connection. Also, the IP address of the DHCP server that was sent in the activation message is forwarded to the external device so that it can translate domain names to IP addresses. Subsequently, the USB data stick can send and receive IP packets over the connection.

Although these steps might be straightforward to perform for network experts, a simpler solution is necessary for the average user. Therefore, many mobile device suppliers or network operators offer connection manager applications to simplify the process. These applications hide the complexity of creating a new modem and dial-up connection entry and the use of the AT+CGDCONT and ATD *99***1# commands as described above. In addition, the connection manager software usually shows the amount of data transmitted and the current radio signal strength, and, depending on the device manufacturer, offers configuration options such as network and access technology selection.

Another approach rather than to use a USB data stick for Internet connectivity with a notebook is by using a virtual network driver. To the notebook's operating system, the mobile device looks like an Ethernet network adapter. A proprietary connection manager software is then necessary to establish the GPRS/UMTS/LTE session and to configure the received IP address, the DNS server's IP address and the default gateway settings.

2.10   Small Screen Web Browsing over GPRS and EDGE

Although GPRS and EDGE are bearers for IP packets, they have some properties that distinguish them from fixed-line connections. These include longer latency, varying latency if used in moving environments and even loss of service for some time if the user moves outside the coverage area of the network. Furthermore, many devices, especially in emerging markets, that use GPRS and EDGE for communication have limited abilities such as small screens and relatively low processing power when compared to contemporary smartphones, notebooks or desktop PCs. Therefore, a number of applications for which the fixed-line Internet is widely used have been adapted for mobile environments. Web browsing is certainly the most popular Internet application. It benefits from a fast connection and depends on the reliability of the bearer, especially if web pages are big and thus take some time to be transferred, during which no transfer interruptions should occur. Modern web pages are also designed to make use of the big displays and processing power of notebooks or desktop workstations.

From the beginning, GPRS and later also EDGE offered data rates that were usually lower than those of mainstream fixed-line connections for which most web pages were optimized. Speed increased over time but so did those of fixed-line connections. As a result, content of web pages kept increasing at the same rate, thereby always countering the advances made with GPRS and EDGE. As a consequence, methods had to be found to cope with these advancements. In the following two sections, WAP 1.1 and 2.0 are described which were used in the past for this purpose. Afterward, contemporary methods to use GPRS and EDGE for web browsing when faster networks are not available are described.

2.10.1   WAP 1.1 Used in Early GPRS Devices

In the very early days of GPRS, the WAP standard was created by the Wap Forum, which was later consolidated into the Open Mobile Alliance (OMA) forum. Basically, the standard adopted the concepts of HyperText Transfer Protocol (HTTP) and hypertext markup language (HTML) and adapted them for use in a mobile environment. iMode was a rivaling standard to WAP, initially designed by NTT DoCoMo in Japan.

The initial WAP 1.1 standard was designed for web browsing in, from today's point of view, very constrained devices, which are not used anymore today. The discussion in this section is therefore mostly of historical nature.

Special attention was given to the following limitations:

- Very limited bandwidth of the connection, which has an impact on the speed a page can be downloaded.
- Very limited processing power of the mobile device, which has an impact on how quickly pages can be rendered on the screen.
- Reliability of the connection. Pages should be loaded as quickly as possible to reduce the effects of transmission interruptions and lost network coverage on the user experience.

HTML and its successor XHTML are used today to describe how web pages are to be rendered in a personal computer (PC)-based web browser. Although the text and layout of a page are directly embedded in the document, pictures and other elements are usually referenced and have to be requested separately. As these languages are quite complex and offer many possibilities that cannot be used in mobile devices owing to the small displays and limited processing capabilities, WAP 1.1 defined its own page description language, which is called the wireless markup language (WML). Using a WAP browser on a mobile device was sometimes also called WAP browsing. Figure 2.35 shows a simple WML description to show a text on the display.

Although at first the WML source looks quite similar to HTML, there are some differences apart from the limited functionality. The main difference was the use of so-called 'cards' inside a single page. Inside the text of each card a link to other cards can be included so that a user can navigate between the cards. The advantage of this approach was to download several cards that are related to each other in a single transaction rather than having to access the network every time the user clicked on a link. This was helpful to break down long texts into several cards and bind them together with a referring link at the bottom of each card. Devices with small displays benefited from this approach as the user didn't have to scroll down a lot of text but could click on a link. Separating a long text into several cards also sped up the embedded WAP browser as only a single card of the downloaded document needed to be rendered when downloading a page. Apart from formatted text and hyperlinks, WML also supported references to images.

```
<?xml version= "1.0"?>
<!DOCTYPE wml PUBLIC "-//WAPFORUM//DTD WML 1.1//EN"
"http://www.wapforum.org/DTD/wml.dtd">
<wml>
 <card id="main" title="First Card">
  <p mode="wrap"> This is a WML page including only this
  sentence.</p>
 </card>
</wml>
```

Figure 2.35   Simple WML page

WAP 1.1 only supported black and white images in the Wireless Application Bitmap Protocol (WBMP) format as mobile devices at the time were limited to black and white screens. Like HTML, WML also supported 'forms' elements to enable users to type in text that could then be used on a server for processing.

To transfer WML pages, the Wireless Session Protocol (WSP) was created instead of using the well-known HTTP. To be able to request for WML pages from ordinary web servers, a gateway was necessary which acted as a translator between the WSP and the HTTP world. The concept of the WAP gateway, which was sometimes also referred to as WAP proxy, is shown in Figure 2.36. Between the mobile device and the WAP gateway, WSP was used for requesting the web page, while HTTP was used to request the WML page from a web server on the Internet.

The differences between WSP and HTTP lie in the speed with which a page is requested and the subsequent data transfer happens:

- While HTTP uses the session-oriented TCP on layer 4 of the protocol stack, WSP used User Datagram Protocol (UDP) and the Wireless Transaction Protocol (WTP) to compensate for some of the functionality of TCP that cannot be found in UDP. To simplify the mobile device configuration, UDP port 9201 was standardized for the use of UDP/WTP. This way, WAP 1.1 avoided the three-way TCP handshake for the session establishment, which saved a lot of time in wireless environments with long round-trip delay (RTD) times. To ensure that no data is lost during the download of a WAP page, WTP included a packet number in each packet. Thus, the receiver could reassemble the arriving UDP/WTP packets in the correct order and could ask for a retransmission of missing packets.
- Every time an ordinary web browser requests a web page, it includes a lot of information in the request such as a list of supported file formats. This takes a lot of space and increases the overall transaction time for every request. WSP used a different approach. At the beginning of a WAP session, the WAP client registered with the WAP gateway and informed the gateway once about its capabilities. When requesting for WAP pages afterward, only the URL needed to be included in the request. This reduced the size of the page request from over 1000 bytes, which are typically needed for a HTTP request, to about 100 bytes for a WSP request.

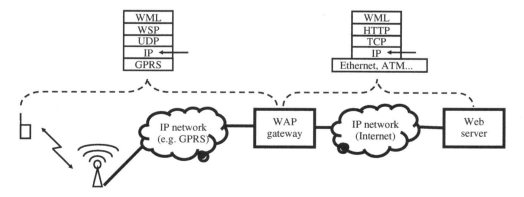

Figure 2.36   Different protocol stacks on the two sides of the WAP gateway

2.10.2   WAP 2.0

After the adoption of the WAP 1.1 standard, capabilities of both the networks and the mobile devices have improved significantly. The processing power of the mobile device had increased, which allowed, among other things, fast data downloads by aggregating several GPRS time-slots. Mobile device manufacturers had also moved to color displays, and their size and resolution had significantly increased compared with early WAP phones as well. As a consequence, the WAP standard was adopted and WAP 2.0, which was approved in 2002, took advantage of the enhanced capabilities of mobile devices and networks. Again, technology has moved forward and today, mobile devices usually include web browsers with full HTML capabilities. The following description of WAP 2.0 is therefore, again, interesting mainly from a historical perspective.

Instead of using a proprietary language for describing WAP pages, WAP 2.0 made use of a subset of XHTML called XHTML Mobile Profile. As XHTML is backward compatible with HTML, ordinary web pages could be viewed as well. It still made sense to adapt standard web pages for the smaller screens of mobile devices, but development of these pages had been greatly simplified by this change as standard development tools could be used.

WAP 2.0 browsers also supported additional graphics formats, such as the widely used graphics interchange format (GIF). Again, standard graphic formats were used compared to the previous proprietary approach. This reduced the overhead of the page creation as pictures no longer had to be converted into a special format, and color images were now used instead of black and white images.

The protocol stack between the mobile device and the WAP gateway had also changed with WAP 2.0. As networks offered higher bandwidths, it was decided to use TCP and HTTP instead of the proprietary WSP and WTP. The concept of the WAP gateway was preserved so that operators could continue to charge for WAP usage as before, such as rate per page or by applying different tariffs depending on whether the user accessed content provided by the network operator or from external servers on the Internet. Apart from the billing and control functionality, which is transparent for the connection, the WAP 2.0 gateway acted as a simple HTTP proxy. TCP port 8080 had been standardized for this use. From the point of view of a mobile device, the WAP gateway was no longer strictly necessary, and most mobile devices allowed skipping of the gateway settings in the WAP set up completely. If no gateway was configured, the WAP browser directly communicated with the server in the Internet. Since some operators charged more for WAP traffic via the gateway as compared to the direct Internet access via a different APN, it made sense in these cases to remove the gateway settings in the mobile device and to change the APN. If the gateway entry could not be removed, another option to save money was to set the IP address of the gateway to that of a public HTTP proxy that supported TCP port 8080. Also, some phones allowed setting up the gateway port so that any free HTTP proxy could be used.

Changing the proprietary gateway approach in WAP 1.1 to standard HTTP proxy functionality had one disadvantage: instead of registering once with the gateway and including all capability information only at the beginning of a session, the HTTP proxy concept required sending the capability information in every request. As HTTP requests with capability information easily used more than 1000 bytes compared with the small WAP 1.1 requests, which used around 100 bytes, the data volume consumed during a WAP session increased significantly, especially when viewing only small pages. From a technical point of view, there was only a small impact to the page load time due to the enhanced transmission speeds of GPRS and EDGE. Depending on the amount, an operator charged per kilobyte transferred via GPRS; however, the increased request size could have resulted in a more notable impact.

2.10.3 Small Screen Web Browsing with Network Side Compression

Today, high-end mobile devices have built-in web browsers that can download and display standard web pages. The bigger the screen, the better a web page can be displayed. The main downside of this approach is the time it takes to download a standard web page over GPRS and EDGE. As a consequence, this approach results in a much degraded user experience for many web pages today.

An alternative to this approach, and also to light-weight WAP 2.0 browsers discussed in the previous section, is web browsers that use a network side compression server and only download compressed content to the mobile device. Opera Mini is such a web browser and Figure 2.37 shows what web browsing looks like in practice with this approach. By compressing standard web pages on a network-based server before they are downloaded to the mobile device, it is not necessary to rely on web pages specially adapted for small devices. With intelligent zooming mechanisms a standard web page can be shown in overview mode and the user can then zoom into any part of the page. Text and pictures are automatically adapted to the screen size with a mechanism referred to as 'reflow'. This way, no horizontal scrolling is necessary.

In practice, this approach offers an excellent web-browsing experience, especially on devices with a smaller screen size. As only a small amount of data is transmitted, web pages are

Figure 2.37 Small screen web browsing with Opera Mini. Source: Reproduced by permission of Opera Software © 2010

displayed in a matter of only a few seconds. Even if used only over GPRS or EDGE, the experience is often similar to that of using a standard web browser on a small device over a fast UMTS connection. In addition, the quality of experience while moving, for example in trains or cars, is excellent as standard web pages are downloaded very quickly and bad coverage or even loss of coverage has much less impact on the overall experience compared to downloading an uncompressed web page with a standard browser that takes longer and is frequently interrupted by changing coverage conditions.

2.10.4 Small Screen Web Browsing – Quality of Experience

The quality of experience of a small screen web-browsing session from the user's point of view mainly consists of short page load times and a high click success rate. The click success rate is defined in this context as the percentage of pages that start getting displayed within a certain amount of time after the user has selected a link to another page. A typical maximum value for this reaction time is seven seconds. For an excellent experience, a reaction time of not more than three to four seconds is essential and can be reached with the network side compression discussed in the previous section. Beyond that time, the user will get the impression that there is a problem and either abort or repeat the request.

The transmission speed on the air interface is one factor that influences the user's quality of experience. In addition, there are a number of other factors that have to be optimized in the network and handset to increase the user's quality of experience:

- Screen size: As web browsers are included even in very small phones today, user experience is limited by small displays that require the user to scroll through pages a lot more often than on bigger displays. A good approach for a page design optimized for mobile devices is to find a compromise that suits both big and small mobile device displays.
- A fast processor can render pages a lot faster than a slower one. A processor architecture that offers high processing power and a good power-saving mode once the page has been rendered greatly increases the user experience while preserving battery power. Processor speed also has an influence on how fast pages can be scrolled up and down.
- Good integration of the web browser into the mobile device operating system and wireless stack. As discussed in Chapter 1, different companies produce different parts of the overall software of a mobile device. The web browser, for example, is one of the most visible parts of the mobile device software that is often outsourced by phone manufacturers to third-party companies. To quickly react to user input, the browser should be integrated very tightly with the phone's operating system and especially the GPRS stack.
- Sufficient capacity in the radio network. Especially during busy hours, resources are scarce as voice calls usually take precedence over GPRS traffic if the radio network is not dimensioned correctly. This reduces the available bandwidth that has to be shared by all GPRS users of a cell and thus increases the page load times.
- Good network coverage is essential for a good web-browsing experience, especially if the mobile device is used in cars, trains, buses or the subway. While roads are usually covered quite well, the same cannot be often said of train tracks, especially inside tunnels. Also, subways are either only partly covered like the stations themselves or not at all.
- To minimize the impact of a cell change, the network and mobile device should implement the NACC functionality to decrease the time the data transfer is interrupted. As cell changes

happen quite often in moving environments, this optimization has a big impact on the click success rate.

- Sufficient capacity on the Gi and Gp interfaces of the GPRS network as these points of interconnection with other networks can quickly become a bottleneck if not dimensioned properly.
- The size of the web page also has an influence on the quality of experience, especially in moving environments with bad network coverage. The smaller the page to be downloaded, the higher the chance that the complete page is downloaded before a coverage degradation interrupts the ongoing data transfer.

2.11   The Multimedia Messaging Service (MMS) over GPRS

Apart from Internet connectivity, mobile network operators also offer services on their own. With the launch of GPRS in the early 2000s, the MMS was introduced. MMS is advertised by mobile network operators as the multimedia successor of the text-based SMS that can transport not only text but also pictures, music and videos. The architectures of SMS and MMS, however, are fundamentally different. The SMS service is based on the SS-7 signaling channels of the network and is thus fully integrated into the GSM system and the GSM standards. MMS, on the other hand, is based on IP and has many similarities with the Internet's e-mail system. As can be seen in Figure 2.38 the MMS system uses the GPRS network only as a transparent IP network. It, therefore, does not rely on GPRS and can be used with any other wireless IP network technology such as UMTS or Code Division Multiple Access (CDMA).

When a mobile device wants to send an MMS message, it establishes an IP connection to the MMS server via the GPRS network. The PDP context activation procedure that is required to get an IP address has already been described before. Although many network operators use the same APN for MMS as used for a transparent connection to the Internet, others use a dedicated APN. This helps operators to charge separately for the MMS traffic and to separate their own services from the Internet.

Although in the early years of MMS the system used the UDP protocol, the WSP and a WAP gateway to compress the data stream, this has changed in the meantime. Today, MMS clients

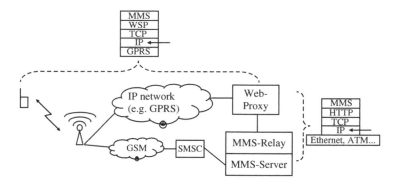

Figure 2.38   MMS architecture overview as defined in 3GPP TS 23.140 [8]

no longer use compression and instead directly communicate with the MMS system in the network over the TCP protocol and a web proxy.

If an MMS is exchanged between two subscribers of the same network, the mobile phone number (MSISDN) is used to identify the recipient. Once the MMS has been sent to the MMS server it needs to notify the recipient that a new MMS message is waiting. This is done by sending an SMS message to the recipient's mobile device. If the recipient has a non-MMS-capable mobile device the SMS contains a text message that informs the user of the MMS, which he can then access via a web page. If the recipient has registered his handset with the MMS server to be MMS capable by having previously sent an MMS message, the SMS is formatted differently. Instead of a text message intended for the subscriber, the SMS is formatted in a standardized way so that the mobile device recognizes on its own that there is a new MMS message waiting on the server. Depending on the mobile device settings, the MMS message is then either retrieved automatically or the mobile device will inform the user that a new MMS message is waiting in the network. The user notification prior to the MMS download can be useful in cases where receiving an MMS is not free, for example, when roaming abroad. If the mobile device is allowed to download the MMS automatically, there is no user interaction and the user is only notified once the MMS message has been fully retrieved.

If the recipient is a subscriber of a different network, the MMS server cannot deliver the MMS directly but has to forward the message to the MMS server of the home network of the recipient. This is only possible if the MMS servers are interconnected.

The MMS standard also allows direct MMS delivery to e-mail addresses. As further described below, this is easily possible as the MMS format is quite similar to an e-mail. In this case, no SMS notification to the receiver is necessary as the MMS is sent in a slightly modified way directly to the e-mail account of the recipient.

To be able to send and receive MMS messages a range of settings are necessary in the mobile device. As MMS is an IP-based service, the first step in sending an MMS is activating a PDP context. Hence, the MMS configuration on the mobile device has to contain an APN and optionally a login name and password. As a proxy server is used between the mobile device and the MMS server, it is also necessary to specify the IP address of the gateway. Finally, the mobile device also has to be configured with the MMS server address, which is specified as a URL. Table 2.5 shows the MMS settings for one of the mobile operators in the United Kingdom.

Similar to an e-mail, MMS messages contain not only text but also attachments like pictures, sound files and video sequences. Unlike an e-mail, however, the user can decide during the creation of an MMS in which order, at which position and for how long pictures and texts are shown on the display and when and for how long sounds and videos are played. On the network side, the MMS server has the possibility of adapting the multimedia elements to the

Table 2.5 Sample MMS settings

APN	orangemms
Username	Orange
Password	Multimedia
IP address of the web proxy	192.168.224.10
MMS server URL	http://mms.orange.co.uk

```
<smil>
  <head>
    <layout>
      <root-layout height="80" width="101"/>        ⎫
        <region id="Image" fit="meet" height="40"    ⎪
                    left="0" top="0" width="101"/>    ⎬  Page
        <region id="Text" fit="meet" height="40"      ⎪  layout
                    left="0" top="40" width="101"/>   ⎭
    </layout>
  </head>

  <body>
    <par dur="10000ms">                              ⎫  First page with a picture
      <img region="Image" src="cid:AA"/>             ⎬  and some text
      <text region="Text" src="cid:AC"/>             ⎪
    </par>                                            ⎭

    <par dur="10000ms">                              ⎫
      <text region="Text" src="cid:AD"/>             ⎬  Second page, text only
    </par>                                            ⎭

  </body>
</smil>
```

Figure 2.39   SMIL description of the layout of an MMS message

capabilities of the recipient's mobile device [9]. This might be necessary, for example, if the receiving mobile device is unable to process an MMS message beyond a certain size.

While creating an MMS, the MMS software in the mobile device converts the user's design of the message into a text-based layout and event description language, which is called synchronized multimedia integration language (SMIL, pronounced 'Smile'). SMIL was standardized by the World Wide Web Consortium (http://www.w3c.org) and has many similarities with HTML, which is used to describe web pages. Figure 2.39 shows a sample SMIL description of an MMS. The general framework first of all describes the layout, number of pages and the presentation flow. The content of each page such as texts, pictures, sounds and video is not part of the SMIL description. These are referenced via 'src = ' tags and are included in the message after the SMIL description.

Similar to an e-mail, the different parts of the MMS including the SMIL description, texts, and pictures are not sent independent of each other but together in a single transaction. To be able to differentiate the different parts of the message, the MMS standard makes use of the Multipurpose Internet Mail Extension (MIME) protocol that is also used in the e-mail standard. As can be seen in Figure 2.40 the MIME header contains a general description of the information that is to be transferred. Subsequently, the SMIL description of the MMS and the content of the message such as texts and pictures are attached to form the complete body of the message. To separate the different parts of the message, a boundary marker is inserted between the different elements.

The boundary markers also contain the reference tags, which have been set in the SMIL description, as well as the description of the content type of the next part of the message. The following formats have been specified for MMS messages in 3GPP TS 26.140 [10]:

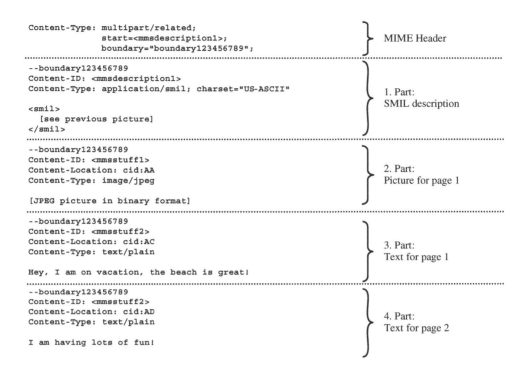

Figure 2.40 MIME boundaries of the different parts of an MMS message

- **Pictures:** JPEG, GIF, WBMP. The maximum guaranteed image size is 160×120 pixels. This corresponds to the display resolution of small mobile devices. If pictures are sent with a higher resolution they might have to be downsized in the receiving mobile device or by the MMS server [9]. Nevertheless, it makes sense to include pictures with a higher resolution in the MMS as the user might transfer the picture to a PC later on via Bluetooth, for example, to benefit from the higher resolution of the PC's display.
- **Text:** ASCII 8-bit, UTF-8 or UTF-16.
- **Audio:** Adaptive Multirate (AMR).
- **Video:** 3GPP MPEG-4 format. Many PC audio/video players offer plugins for this format so that received videos can be transferred to PCs for playback as well.

To be able to send an MMS, an overall header that includes general information like the address of the receiver and the subject line, is necessary. Again, MMS leverages already existing standards and uses a standard e-mail header. The only adaptation that was made to shrink the header size was to not use plain text parameter names. Instead, each parameter name (e.g. To: or Subject:) was given a binary representation. The parameter value, however, is not compressed. To be able to include some additional information that is not necessary for sending e-mails, a number of MMS-specific extension tags were defined. They are all

```
From: <insert address>
Date: Thu, 10 Juni 2004 10:49:55 +0100

To: +4916014867651/TYPE=PLMN
CC: <John Doe> jdoe@cm-networks.de        [optional]

Subject: Still kicking!                   [optional]      MMS header
MIME-Version: 1.0                         [optional]

X-MMS-Version: 1.0
X-MMS-Message-Type: m-send-req
X-MMS-Transaction-ID: 867634563
X-MMS-Read-Reply: Yes                     [optional]

Content-Type: multipart/related;
              start=<mmsdescription1>;
              boundary="boundary123456789";

--boundary123456789
Content-ID: <mmsdescription1>
Content-Type: application/smil; charset="US-ASCII"      See previous picture

<smil>
  [see previous picture]
</smil>

[…]
```

Figure 2.41   Uncompressed view of an MMS header

named X-MMS ... and some of them are shown in Figure 2.41. As MMS messages are usually exchanged between mobile subscribers, an MSISDN is used in the To: field of the header. For clear identification, the /Type=PLMN suffix is added to the MSISDN. PLMN stands for public land mobile network, the technical term for ground-based mobile telecommunication network. As can also be seen in the figure, the From: field of the header does not contain the sender's identification. This was done on purpose to prevent the sender from using a random originator address and thus conceal his true identity. The MMS server therefore has to query the GPRS network for the identification of the user.

While there are many similarities between MMS and e-mail, they use different protocols for the actual transfer of the message. While e-mail uses the Simple Mail Transfer Protocol (SMTP) it was decided to use the HTTP POST method for MMS. This protocol was initially designed to send the user-supplied content of input boxes on web pages to the web server for processing. Using HTTP POST for the MMS transfer has the advantage that no handshake procedure is necessary on the application layer when the MMS server is contacted, which speeds up the MMS transaction.

As discussed above, the MMS sever uses an SMS message to notify the receiver of a waiting message. The message contains a URL that identifies the stored MMS. To save time, it was decided not to use the POP3 or IMAP protocols which are used for retrieving e-mail but to retrieve the MMS via the HTTP GET protocol. HTTP GET was initially designed to allow web browsers to request for web pages from a server.

Owing to the many similarities between MMS and e-mail messages, it is quite simple to forward an MMS to an e-mail address as well. As SMIL is not understood by e-mail programs, the

MMS server removes the SMIL description of the message and sends the user's multimedia content (text, pictures, etc.) as file attachments. More sophisticated MMS server implementations even use HTML formatting for the presentation of the converted MMS and allow the recipient to return an e-mail to the sender, which is then converted back into an MMS message and delivered to the mobile device.

As MMS is a pure IP application and only uses open standards, the bearer over which it is transferred is no longer relevant. This means that theoretically MMS messages could also be used for messaging purposes between PCs in the Internet. Practically this does not make a lot of sense as the enhancements over the e-mail standard that were put into the MMS specification do not entail many advantages for fixed-line use. In Europe, MMS is used in GSM, UMTS and LTE networks, which are based on the same core architecture. The use of IP and open standards also allows using the MMS system with mobile networks that use a different network architecture, like CDMA. Unlike SMS messages, which require a special gateway to allow message exchange between networks that use different standards, MMS messages can be exchanged between different network standards like UMTS and CDMA with only minor adaptations for some multimedia elements.

2.12   Web Browsing via GPRS

2.12.1   Impact of Delay on the Web-Browsing Experience

While a high bandwidth connection is certainly one of the most important factors for a good web-browsing experience, the RTD time of the connection must also not be underestimated [11, 12]. The RTD time is defined in this context as the time it takes to receive a response to a transmitted frame. This RTD can be measured, for example, with the `ping` command. The following example shows how delay impacts the web-browsing experience. When requesting a new page, the following delays are experienced before the page can be downloaded and displayed:

- The URL has to be converted into the IP address of the web server that hosts the requested page. This is done via a DNS query that causes a delay of one RTD of the connection.
- Once the IP address of the server has been determined, the web browser needs to establish a TCP connection. This is done via a three-way handshake. During this handshake, the client sends a synchronization packet to the server, which is answered by a synchronization-ack packet. This in turn is acknowledged by the client by sending an acknowledgment packet. As three packets are sent before the connection is established, the whole operation causes a delay of 1.5 times the RTD of the connection. As the first packet containing user data is sent right after the acknowledgment packet, the time is, however, reduced to approximately a single RTD time.
- Only after the TCP connection has been established can the first packet be sent to the web server, which usually contains the actual request of the web browser. The server then analyzes the request and sends back a packet that contains the beginning of the requested web page. As the request (e.g. 300–500 bytes) and the first response packet (1200–1400 bytes) are quite large, the network requires more time to transfer those packets than the simple RTD time.

There are three different RTD times experienced in GPRS and EDGE networks. If there is no TBF established at the time the web browser starts the request, then the mobile device first needs to set up a physical connection in the uplink and downlink directions. This typically results in an RTD time of the DNS query of 650–750 milliseconds in EDGE networks.

As the network expects further data to be transferred, the downlink TBF is usually kept open by the network, which reduces the RTD time. Early EDGE capable networks had a round-trip time of about 550 milliseconds for any subsequent communication. In current networks the round-trip time has been further reduced to 180–250 milliseconds (Section 2.6), the exact value depending on the radio network equipment vendor. This has been achieved with the extended uplink TBF functionality discussed in Section 2.6 and by enhancing internal procedures for TBF assignments.

The third delay time can be experienced when large frames are sent and received. As transmitting large frames takes some time by itself, the time before the first packet that contains a part of the web page arrives is somewhat longer than the RTD time. Figure 2.42 shows an IP packet trace without the initial DNS query.

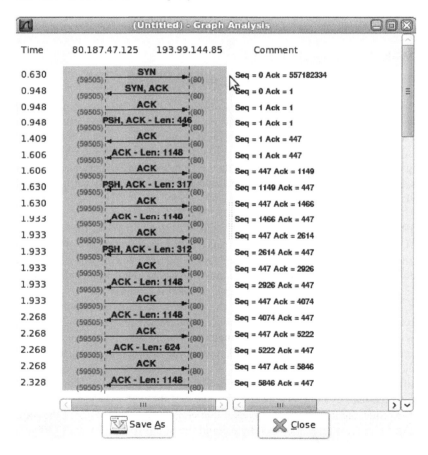

Figure 2.42 IP packet flow and delay times during the download of a web page. Source: Reproduced by permission of WireShark © 2010.

As per the processes presented above, the time between entering a URL and being shown the first part of the web page can be estimated as follows:

$$\text{Total Delay (EDGE)} = \text{Delay DNS query} + \text{Delay TCP Establish}$$
$$+ \text{Delay Request/Response}$$
$$= 750\,\text{ms} + 320\,\text{ms} + 660\,\text{ms}$$
$$= 1730\,\text{ms}$$

After this time, the page is not fully loaded but the browser usually starts to present the text of the main page. If the user then clicks on a link to a different page of the same website, no further DNS query is necessary as the IP address of the server is still stored in the DNS cache. Hence, the delay time for loading subsequent pages is reduced by about 750 milliseconds.

Compared to the page download delay of less than 100 milliseconds of a high-speed fixed-line DSL connection, the EDGE delay appears to be rather long. This difference is caused by the fact that no uplink and downlink connections have to be established prior to the transfer of data when using a DSL connection. The higher bandwidth of the DSL connection also allows downloading the remainder of the page faster than via EDGE. Nevertheless, a delay of less than two seconds before the first parts of a page are displayed is still acceptable. Together with the higher bandwidth offered by EDGE as compared to the standard GPRS network, it was possible for some time to use an EDGE network for full-screen web browsing. However, the size of web pages for PC-based web browsers has continued to increase, thereby limiting the use of EDGE for full-screen web browsing today.

An excellent source for further information about EDGE and performance aspects can be found in Benko *et al.* [11].

2.12.2 Web Browser Optimization for Mobile Web Browsing

As discussed earlier, GPRS and EDGE connections behave differently when used for web surfing compared to fixed-line connections. As most users access the Internet via fixed-line connections, web browsers are usually optimized for short latency and very high bandwidth connections. However, some browsers allow changing their network access settings, which can help improve the download times. The Firefox browser, for example, allows changing the values for pipelining of web requests. Pipelining is a method that was introduced with HTTP 1.1 to request for several elements of a page at once in a single TCP connection instead of only sending the next request once the previous element was downloaded. In this way, it is possible to reduce the effect of the higher latency of the wireless connection.

Pipelining is activated in the Firefox browser as follows:

- In the URL field, `about:config` has to be entered to get the list of all parameters that can be changed.
- Pipelining is activated by setting the `network.http.pipelining` parameter to TRUE.
- A good value for the number of accumulated requests that should be set in `network.http.pipelining.maxrequests` is eight.

When comparing the page download time of what can be considered a medium-sized web page today with a limited number of pictures and active content, a big difference can be observed with and without pipelining in an EDGE network. Although the delay to display the first parts of such a page is almost unchanged, the time to download the complete page including all images can typically be reduced from 60 to 40 seconds. Despite this reduction, however, page load time is still significantly longer than what can be achieved over UMTS and LTE.

Questions

1. What are the differences between circuit-switched and packet-switched data transmission?
2. What are the advantages of the data transmission over GPRS compared to GSM?
3. Why are different modulation and coding schemes used?
4. What is the difference between the GPRS ready state and the GPRS standby state?
5. Does the GPRS network perform a handover if a cell change is required while data is transferred?
6. Which are the new network elements that have been introduced with GPRS and what are their responsibilities?
7. What is a temporary block flow?
8. What actions are performed during an IRAU?
9. Why is IP used twice in the protocol stack of the Gn interface?
10. Why is it not necessary to change any settings on the mobile device for GPRS when roaming abroad?
11. What is the difference between a GPRS attach and a PDP context activation?
12. Why is an APN necessary for the PDP context activation procedure?
13. How are MMS messages sent and received via GPRS?
14. Name the different parts of an MMS message.

Answers to these questions can be found on the companion website for this book at http://www.wirelessmoves.com.

References

[1] 3GPP, Multiplexing and Multiple Access on the Radio Path, TS 45.002, Annex B1.
[2] 3GPP, Radio Access Network: Channel Coding, TS 45.003.
[3] Y.-K. Chen and Y.-B. Lin, IP Connectivity for Gateway GPRS Support Node, *IEEE Wireless Communications Magazine*, **12**, 37–46, February 2005.
[4] 3GPP, General Packet Radio Service (GPRS); Mobile Station (MS) – Base Station System (BSS) Interface; Radio Link Control/Medium Access Control (RLC/MAC) protocol, TS 44.060.
[5] 3GPP, General Packet Radio Service (GPRS); GPRS Tunneling Protocol (GTP) across the Gn and Gp Interface, TS 29.060.
[6] 3GPP, General Packet Radio Service (GPRS); Serving GPRS Support Node (SGSN) – Visitors Location Register (VLR); Gs Interface Layer 3 Specification, TS 29.018.
[7] 3GPP, General Packet Radio Service (GPRS); Service Description; Stage 2, TS 23.060.
[8] 3GPP, Multimedia Messaging Service (MMS); Functional Description; Stage 2, TS 23.140.
[9] S. Coulombe and G. Grassel, Multimedia Adaptation for the Multimedia Messaging Service, *IEEE Communications Magazine*, 120–126, July 2004.
[10] 3GPP, Multimedia Messaging Service (MMS); Media Formats and Codecs, TS 26.140.
[11] P. Benko, G. Malicsko, and A. Veres, A Large-scale, Passive Analysis of End-to-End TCP Performance over GPRS, *IEEE Infocom Conference*, Hong Kong, 2004.
[12] R. Chakravorty, J. Cartwright, and I. Pratt, Practical Experience with TCP over GPRS, *IEEE Globecom*, Taipei, Taiwan, 2002.

3

Universal Mobile Telecommunications System (UMTS) and High-Speed Packet Access (HSPA)

The Universal Mobile Telecommunications System (UMTS) is a third generation wireless telecommunication system and followed in the footsteps of Global System for Mobile Communications (GSM) and General Packet Radio Service (GPRS). Since GSM was standardized in the 1980s, huge progress had been made in many areas of telecommunication. This allowed system designers at the end of the 1990s to design a new system that went far beyond the capabilities of GSM and GPRS. UMTS combines the properties of the circuit-switched voice network with the properties of the packet-switched data network and offers a multitude of new possibilities compared to the earlier systems. UMTS was not defined from scratch and reuses a lot of GSM and GPRS. Therefore, this chapter first gives an overview of the advantages and enhancements of UMTS compared to its predecessors, which have been described in the previous chapters. After an end-to-end system overview, the focus of the chapter is then on the functionality of the UMTS radio access network. New concepts like the Radio Resource Control (RRC) mechanisms as well as changes in mobility, call control and session management are also described in detail.

Over the years, the UMTS radio network system has been significantly enhanced and now offers broadband speeds far beyond the original design. These high-speed enhancements are referred to as High-Speed Packet Access (HSPA). An overview of the various enhancements over time is given in Section 3.1 and a more detailed look at the key features for broadband Internet access follows in the second part of this chapter. At the end of the chapter, an analysis is made of the performance of HSPA in deployed networks today and how the functionalities described in this chapter are used in practice.

From GSM to LTE-Advanced: An Introduction to Mobile Networks and Mobile Broadband,
Revised Second Edition. Martin Sauter.
© 2014 John Wiley & Sons, Ltd. Published 2014 by John Wiley & Sons, Ltd.

3.1 Overview, History and Future

The trends and developments seen in fixed-line networks are also appearing in mobile networks albeit with a delay of about 5 years. In fixed networks, the number of people using the network not only for voice telephony but also to connect to the Internet is increasing as steadily as the transmission speeds. When the Internet first became popular, circuit-switched modems were used to establish a dial-up connection to the network. While the first modems that were used in the middle of the 1990s featured speeds of around 14.4 kbit/s, later models achieved around 50 kbit/s in the downlink (network to user) direction. Another incredible step forward was made around the year 2002 when technologies like cable and Asymmetric Digital Subscriber Line (ADSL) modems reached the mass market and dramatically increased transmission speeds for end users. With these technologies, transmission speeds of several megabits per second are easily achieved. In many countries, fixed-line network operators are now rolling out high-speed connections based on optical fibers that are either directly deployed into the basements of buildings (fiber to the building) or to street-side equipment cabinets (fiber to the curb). The last few meters into buildings or apartments are then bridged with copper technologies such as Ethernet or Very-high-bitrate Digital Subscriber Line (VDSL). This way a shorter length of copper cable is used for the last leg of the connection, which helps to enable speeds of 25–100 Mbit/s per line and beyond. TV cable operators are also upgrading their networks to achieve similar speeds.

In the mobile world, GPRS (see Chapter 2) with its packet-oriented transmission scheme was the first step toward the mobile Internet. With datarates of about 50 kbit/s in the downlink direction for operational networks, similar speeds to those of contemporary fixed-line modems were achieved. New air interface modulation schemes like EDGE have since increased the speed to about 150–250 kbit/s per user in operational networks. However, even with EDGE, some limitations of the radio network such as the timeslot nature of a 200-kHz narrowband transmission channel, GSM medium access schemes and longer transmission delays compared to fixed-line data transmission could not be overcome. Therefore, further increase in transmission speed was difficult to achieve with the GSM air interface.

Since the first GSM networks went into service at the beginning of the 1990s, the increase in computing power and memory capacity has not stopped. According to Moore's law, the number of transistors in integrated circuits grows exponentially. Therefore, the processing power of today's processors used in mobile networks is in orders of magnitude more than that of the processors of the early GSM days. This in turn enables the use of much more computing-intensive air interface transmission methods that utilize the scarce bandwidth on the air interface more effectively than the comparatively simple GSM air interface.

For UMTS, these advances were consistently used. Although voice communication was the most important application for a wireless communication system when GSM was designed, it was evident at the end of the 1990s that data services would play an increasingly important role in wireless networks. Therefore, the convergence of voice and high-speed data services into a single system has been a driving force in the UMTS standardization from the beginning.

As will be shown in this chapter, UMTS is as much an evolution as it is a revolution. While the UMTS radio access network (UTRAN) was a completely new development, many components of the GSM core network were reused, with only a few changes, for the first step of UMTS. New core and radio network enhancements were then specified in subsequent steps. Today, this process still continues, not only with UMTS but also with the Long-Term Evolution (LTE) and LTE-Advanced radio access networks that followed in the footsteps of UMTS. These developments, together with the evolved packet core (EPC) are discussed in Chapter 4.

The Third Generation Partnership Project (3GPP) is responsible for evolving the GSM, UMTS and LTE standards and refers to the different versions as 'Releases'. Within a certain timeframe all enhancements and changes to the standards documents are collected and once frozen, a new version is officially released. Each 3GPP release of the specifications includes many new features for each of the three radio access technologies, some large and some small. As it is impossible to discuss all of them in this book the following sections give a quick introduction to the most important features of each release. These are then discussed in more detail in the remainder of this chapter. For a complete overview of all features of each release refer to [1].

3.1.1   3GPP Release 99: The First UMTS Access Network Implementation

Initially, 3GPP specification releases were named after the year of ratification, while later on a version number was used. This is why the first combined 3GPP GSM/UMTS release was called Release 99, while subsequent versions were called Release 4, Release 5, Release 6 and so on. At the time of publication, 3GPP is in the process of working on Release 12, which combines GSM, UMTS, LTE and LTE-Advanced.

Release 99 contains all the specifications for the first release of UMTS. The main improvement of UMTS compared to GSM in this first step was the completely redesigned radio access network, which the UMTS standards refer to as the UMTS Terrestrial Radio Access Network (UTRAN). Instead of using the time- and frequency-multiplexing method of the GSM air interface, a new method called Wideband Code Division Multiple Access (WCDMA) was introduced. In WCDMA, users are no longer separated from each other by timeslots and frequencies but are assigned a unique code. Furthermore, the bandwidth of a single carrier was significantly increased compared to GSM, enabling a much faster data transfer than previously possible. This allowed a Release 99 UTRAN to send data with a speed of up to 384 kbit/s per user in the downlink direction and also up to 384 kbit/s in the uplink direction. In the first few years, however, uplink speeds were limited to 64–128 kbit/s.

For the overall design of the UTRAN, the concept of base stations and controllers was adopted from GSM. While these network elements are called base transceiver station (BTS) and base station controller (BSC) in the GSM network, the corresponding UTRAN network elements are called Node-B and Radio Network Controller (RNC). Furthermore, the mobile station (MS) also received a new name and is now referred to as the User Equipment (UE). In this chapter, the UE is commonly referred to as the mobile device.

In Europe and Asia, 12 blocks of 5 MHz each have been assigned to UMTS for the uplink direction in the frequency range between 1920 and 1980 MHz. This frequency range is just above the range used by DECT cordless phones in Europe. For the downlink direction, that is, from the network to the user, another 12 blocks of 5 MHz each have been assigned in the frequency range between 2110 and 2170 MHz.

In North America, no dedicated frequency blocks were initially assigned for third generation networks. Instead, UMTS networks shared the frequency band between 1850 and 1910 MHz in the uplink direction and between 1930 and 1990 MHz in the downlink direction with 2G networks such as GSM and CDMA networks. Later, additional spectrum has been assigned for

3G in the range of 1710–1755 MHz in the uplink direction and 2110–2155 MHz in the downlink direction. While the downlink of this frequency allocation overlaps with the downlink in Europe and Asia, a completely different frequency range is used for the uplink.

Over time, many more bands have been opened to wireless networks around the world. An up-to-date list of these can be found in the latest version of 3GPP TS 25.101 [2]. Except for the bands in the 700–900 MHz range, which are discussed in more detail in the following text, most other bands have not received widespread support. At first, this was mostly due to each band requiring additional circuitry in the mobile device. This makes the design and production of devices more costly and the receiver less sensitive. Today, another reason for limited UMTS band support is the uptake of LTE, which is deployed in new frequency bands instead.

Despite being in use for many years the technology for the GSM circuit-switched core network was chosen as the basis for voice and video calls in UMTS. It was decided not to specify major changes in this area but to rather concentrate on the access network. The changes in the circuit core network to support UMTS Release 99 were therefore mainly software enhancements to support the new Iu(cs) interface between the Mobile Switching Center (MSC) and the UTRAN. While the Iu(cs) interface is quite similar to the GSM A-Interface on the upper layers, the lower layers were completely redesigned and were based on ATM. Furthermore, the Home Location Register (HLR) and Authentication Center (AuC) software were enhanced to support the new UMTS features.

The GPRS packet core network, which connects users to the Internet or a company intranet, was adopted from GSM with only minor changes. No major changes were necessary for the packet core because GPRS was a relatively new technology at the time of the Release 99 specification, and was already ideally suited to a high-speed packet-oriented access network. Changes mostly impact the interface between the SGSN and the radio access network, which is called the Iu(ps) interface. The biggest difference to its GSM/GPRS counterpart, the Gb interface, is the use of ATM instead of Frame Relay on lower layers of the protocol stack. In addition, the SGSN software was modified to tunnel GTP user data packets transparently to and from the RNC instead of analyzing the contents of the packets and reorganizing them onto a new protocol stack as was previously done in GSM/GPRS.

As no major changes were necessary in the core network it was possible to connect the UMTS radio access network (UTRAN) to a GSM and GPRS core network. The MSCs and SGSNs only required a software update and new interface cards to support the Iu(cs) and Iu(ps) interfaces. Figure 3.1 shows the network elements of a combined GSM and UMTS network.

Such combined networks simplify the seamless roaming of users between GSM and UMTS. This is still important today as UMTS networks in most countries are nowhere near as ubiquitous as GSM.

Seamless roaming from UMTS to GSM and vice versa requires dual mode mobile devices that can seamlessly handover ongoing voice calls from UMTS to GSM if a user leaves the UMTS coverage area. Similar mechanisms were implemented for data sessions. However, owing to the lower speed of the GSM/GPRS network, the process for data sessions is not seamless.

While UMTS networks can be used for voice telephony, the main goal of the new radio access technology was the introduction of fast packet data services. When the first networks started to operate in 2002, mobile network operators were finally able to offer high-speed Internet access for business and private customers. Release 99 networks could achieve maximum downlink speeds of 384 kbit/s and 128 kbit/s in the uplink direction. While this might seem to be slow

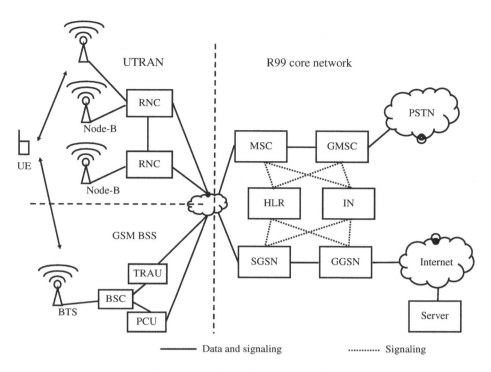

Figure 3.1   Common GSM/UMTS network: Release 99

from today's perspective, it was an order of magnitude faster than could be achieved with GPRS networks at the time. Accessing the Internet was almost as fast as over a 1 Mbit/s ADSL line – a standard speed at the time. Since then, speeds in fixed and wireless networks have increased significantly. Mobile network operators have upgraded their hardware and software with features that are described in the following sections. The general network design as shown in Figure 3.1, however, has remained the same.

3.1.2   3GPP Release 4: Enhancements for the Circuit-Switched Core Network

A major enhancement for circuit-switched voice and data services has been specified with 3GPP Release 4. Up to and including Release 99, all circuit-switched voice calls were routed through the GSM and UMTS core network via E1 connections inside 64 kbit/s timeslots. The most important enhancement of Release 4 was a new concept called the *Bearer-Independent Core Network*, or Bearer-Independent Core Network (BICN) for short. Instead of using circuit-switched 64 kbit/s timeslots, traffic is now carried inside Internet Protocol (IP) packets. For this purpose, the MSC has been split into an MSC-Server (MSC-S), which is responsible for Call Control (CC) and Mobility Management (MM), and a Media Gateway (MG), which is responsible for handling the actual bearer (user traffic). The MG is also responsible for the transcoding of the user data for different transmission methods. This way it is possible, for example, to receive voice calls via the GSM A-Interface via E-1 64 kbit/s timeslots at the MSC MG, which will then convert the digital voice data stream onto a packet-switched IP connection toward another MG in the network. The remote MG will then again convert the incoming user data packets to send it, for example, to a remote party via the UMTS radio access network (Iu(cs) interface) or back to a circuit-switched E-1 timeslot in case a connection is established to the fixed-line telephone network. Further details on the classic and IP-based circuit-switching of voice calls can be found in Chapter 1.

The introduction of this new architecture was driven by the desire to combine the circuit- and packet-switched core networks into a single converged network for all traffic. As the amount of packet-switched data continues to increase so does the need for investment into the packet-switched core network. By using the packet-switched core network for voice traffic as well, network operators can reduce their costs. At the time of publication, many network operators have finished the transition of their circuit-switched core networks to Release 4 MSC-Ss and MGs. Figure 3.2 shows what this setup looks like in practice.

3.1.3   3GPP Release 5: IMS and High-Speed Downlink Packet Access

A further step toward an all IP wireless network is the IP Multimedia Subsystem (IMS). The groundwork for IMS was laid in 3GPP Release 5. Subsequent versions of the standard have extended it with new functionalities. Instead of using the circuit-switched part of the radio

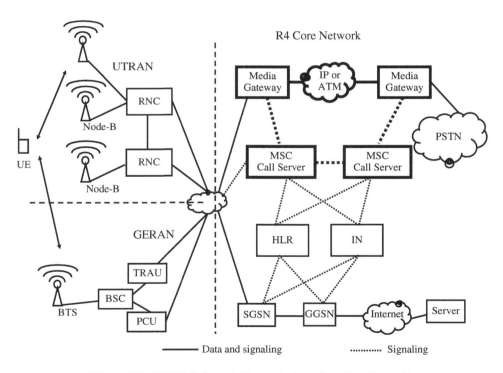

Figure 3.2   UMTS Release 4 (Bearer-Independent Core Network)

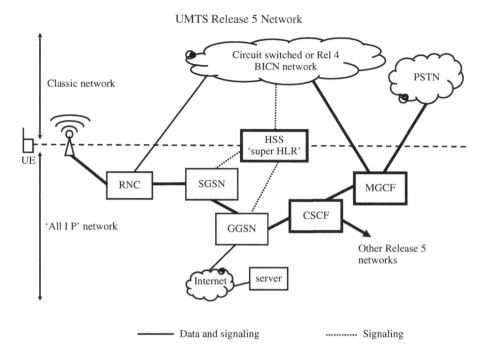

UMTS Release 5 Network

Figure 3.3   UMTS Release 5 IMS architecture

network, the IMS handles voice calls and other services via the packet-switched part of the network as shown in Figure 3.3.

The core of the IMS consists of a number of nodes that form the Call Session Control Function (CSCF). The CSCF is basically a Session Initiation Protocol (SIP) architecture that was initially developed for the fixed-line world and is one of the core protocols for most Voice over Internet Protocol (VoIP) telephony services available in the market today. The CSCF takes the concept one step further and enhances the SIP standard with a number of functionalities necessary for mobile networks. This way, the IMS can transport voice calls over IP not only in the core network but from end to end, that is, from mobile device to mobile device. While the CSCF is responsible for the call setup and call control, the user data packets which, for example, include voice or video conversations, are directly exchanged between the end-user devices.

With the UMTS radio access network, it became possible to implement an IP-based mobile voice and video transmission architecture. This was not only because UMTS offered sufficient bandwidth on the air interface for such applications but also because of the way cell changes are handled for active packet connections. With GPRS, roaming from one cell to another (mobility management) for packet-switched connections is controlled by the mobile device. This results in an interruption of packet traffic of 1–3 seconds at every cell change. For voice or video calls this is not acceptable. With UMTS, the mobility management for active packet-switched connections is now controlled by the network. This ensures uninterrupted packet traffic even while the user is roaming from one cell to another.

As the IMS is an IP-based system it cannot directly communicate with circuit-switched telephony systems that are still dominant in wireless networks. Nevertheless, it has to be ensured that every user can talk to every other user regardless of the kind of telephony architecture that is in use at each end. As can be seen in Figures 3.2 and 3.3 this is achieved by using MGs, which convert between IMS VoIP, BICN, and the classic circuit-switched timeslot transmission.

Unfortunately, the IMS suffers from a number of difficulties that have so far prevented it from becoming an alternative to the Release 99 or Release 4 MSC architecture. Some of these are as follows:

- The IMS is not only designed for voice calls but is intended as a general service platform for all kinds of new multimedia services such as video calling, picture exchange, instant messaging and presence. As a consequence, the system is very complex.
- To be accepted by users, an IMS system must be capable of handing over an ongoing session to a circuit-switched bearer when the user leaves the UMTS coverage area. This requires a complicated interworking with the circuit-switched MSCs in the network and a client-based software on a mobile device that can seamlessly hand over an ongoing connection.
- On mobile devices the IMS client must be seamlessly integrated into the overall device. For the user it should be transparent if the IMS is used for a call or the circuit-switched network if no fast wireless network is available.
- The IMS was designed to be a platform that third party service developers could use to integrate their IP-based services into the mobile network environment. Since its inception, however, services are offered by Internet companies directly to a global rather than a national subscriber base via the Internet. One of the reasons for this is that Internet-based companies mainly see services on mobile devices connected to the Internet as an extension to services they initially designed for the fixed-line Internet and are thus not interested in using mobile specific platforms, which are tied to many different network operators.

Currently, the IMS system is not used for speech calls but mainly for the Rich Communication Service (RCS) that has been introduced by a number of network operators. In addition, the IMS is the basis for the Voice over LTE (VoLTE) system. As LTE supports only IP-based data traffic in the core and access network, a significant effort is underway to make VoLTE a mass market system to remove the need for an LTE-capable device to fall back to UMTS or GSM for voice calls. Chapter 4, on LTE, therefore contains more details on IMS and VoLTE.

The most important new functionality introduced with 3GPP Release 5 was a new data transmission scheme called High-Speed Downlink Packet Access (HSDPA) to increase data transmission speeds from the network to the user. While 384 kbit/s was the maximum speed in Release 99, HSDPA increased speeds per user, under ordinary conditions, to several megabits per second. Top speeds are highly dependent on a number of conditions:

- The maximum throughput capability of the mobile device.
- The sophistication of the receiver and antenna of the mobile device.
- The capability of the network.
- The radio conditions at the place of use. This includes the signal level received from a base station and the interference of neighbor cell transmissions on the same frequency.
- The bandwidth of the backhaul link between the base station and the rest of the network.
- The number of other users in the cell that are actively exchanging data at the same time.

In an ideal radio environment speeds up to 14.4 Mbit/s can be reached if supported by the network and the mobile device. As described later, mobile devices have to be very close to the base station to have signal conditions that allow receiving data at such high speeds. However, even under less ideal radio conditions and with a cell being used by several subscribers simultaneously, a throughput of several megabits per second can still be reached.

Many in the telecommunication industry see HSDPA in combination with smartphones and 3G data dongles for notebooks as 'the' combination that helped UMTS to gain mass market adoption and widespread use.

3.1.4   3GPP Release 6: High-Speed Uplink Packet Access (HSUPA)

The HSPA functionality continued to evolve in 3GPP Release 6. This revision of the specification brought the introduction of methods to increase uplink speeds, which have remained the same since Release 99. This feature set, referred to as High-Speed Uplink Packet Access (HSUPA) in public, enables uplink datarates of 2–3 Mbit/s for a single user under ideal conditions today. Taking realistic signal conditions into account, the number of users per cell and mobile device capabilities, HSUPA enabled devices can still achieve significantly higher uplink speeds than was possible with Release 99. Furthermore, HSUPA also increases the maximum number of users that can simultaneously send data via the same cell and thus further reduces the overall cost of the network. The combination of HSDPA and HSUPA is sometimes also referred to as HSPA. The details on achievable speeds and behavior in practice can be found later in this chapter.

3.1.5   3GPP Release 7: Even Faster HSPA and Continued
Packet Connectivity

One of the shortcomings of UMTS and HSPA compared to GSM is the high power consumption during transmission gaps, for example, between the downloads of two web pages. Even though no user data is transmitted or received during this time, a significant amount of energy is required to send control information to keep the link established and to scan for new incoming data. Only after some time, usually in the order of 5–15 seconds, does the system put the connection into a more power-efficient state. But even this state still requires a significant amount of power and the battery continues to be drained until the point where the network finally puts the air interface connection into a sleep state. In a typical setup, this happens after an additional 10–60 seconds. It then takes around 1–3 seconds to wake up from this state, which the user notices, for example, when he clicks on a link on a web page after the air interface connection gets into sleep mode. Reducing power consumption and a fast return to full active state have been the goals of Release 7 feature package referred to as Continuous Packet Connectivity (CPC). A detailed analysis of the issues that appear in practice and how these are addressed with CPC are described later in this chapter.

In addition, 3GPP Release 7 once again increased the maximum possible data transfer speeds in the downlink direction with the introduction of:

- the use of several antennas and Multiple Input Multiple Output (MIMO) transmission schemes;
- 64 Quadrature Amplitude Modulation (64-QAM).

The maximum speeds reached with these enhancements under ideal signal conditions are 21 Mbit/s with 64 QAM modulation and 28 Mbit/s with MIMO.

In the uplink direction, the HSUPA functionality was also extended in this release. In addition to the Quadrature Phase Shift Keying (QPSK) modulation scheme, 16-QAM is now also specified for uplink operation, which further increases peak datarates to 11.5 Mbit/s under very good signal conditions.

The rising transmission speeds over the past few years has significantly increased bandwidth and processing requirements in the network. To ease this burden, it was decided to simplify the transmission path of user data through the UMTS network as much as possible. As a result, the SGSN (see Chapter 2) can be removed from the user data transmission path with the direct-tunnel functionality.

In cities, UMTS cells usually cover an area with a diameter of less than 1 km. This is due to the higher frequencies used compared to GSM and because of the high population density in such areas. In some countries, for example Australia, where the population density is very low in rural areas, UMTS is used by some operators in the 850 MHz band, which allows for very large coverage areas. However, where the landscape is flat, the maximum cell range of 60 km is no longer sufficient, so that in Release 7 the 3GPP standards were extended to allow for cell ranges of up to 180 km.

3.1.6   3GPP Release 8: LTE, Further HSPA Enhancements and Femtocells

In 3GPP Release 8, a number of features were introduced that had a significant impact on wireless networks. First, Release 8 introduced the successor of the UMTS radio network, the EUTRAN, and the successor architecture of the core network, the EPC. Together they are commonly known as LTE, although technically this is not quite correct. As LTE is a revolution in many ways it is discussed separately in Chapter 4.

In the UMTS domain, this release of the standard also contains some notable enhancements to keep pace with the rising data traffic. To reach even higher data speeds, it is now possible to aggregate two adjacent UMTS carriers to get a total bandwidth of 10 MHz. This is referred to as Dual-Cell or Dual-Carrier operation. Also, the simultaneous use of 64-QAM and MIMO has entered the standards for single carrier operation. Under ideal radio conditions, a peak throughput of 42 Mbit/s in the downlink direction can be reached.

For VoIP applications, an enhancement has been specified that allows the network to prepare a circuit-switched channel in GSM in order to make the handover of a packet-switched VoIP call when the user roams out of the UMTS or LTE coverage area. In previous releases this required the mobile device to communicate simultaneously with the UMTS or LTE network and the GSM network. As such devices do not exist in practice the Single Radio Voice Call Continuity (SR-VCC) feature was introduced. This is discussed in more detail in Chapter 4 in the section on Voice over LTE (VoLTE).

One small, but important functionality specified in 3GPP Release 8 is 'In Case of Emergency' (ICE). Devices that implement this functionality allow the user to store information on the subscriber identity module (SIM) card that can be accessed in a standardized way in emergency situations where the user of the phone is unable to identify himself or to contact his relatives. Unfortunately, the feature has not found widespread adoption so far.

And finally, 3GPP Release 8 laid the groundwork for Femtocells management, referred to as Home Node-Bs in the standard and Self-Organizing Network (SON) functionality, to ease deployment and maintenance of base stations.

3.1.7   3GPP Release 9: Digital Dividend and Dual Cell Improvements

From an LTE point of view, 3GPP Release 9 is mostly a maintenance Release, which includes features and corrections that were not considered essential for the first launch of LTE. The SON and Femtocell architecture (Home Node-B and Home eNode-B) specifications started in Release 8 have been continued in this release.

For UMTS, Release 9 brought a number of further speed enhancements in both uplink and downlink directions. In the uplink direction, aggregation of two adjacent 5 MHz carriers has been specified in a similar way as in the downlink direction in the previous release. This in effect again doubles the theoretical peak uplink datarate to over 20 Mbit/s. In the downlink direction, dual-carrier operation can now be combined with MIMO operation, increasing the peak throughput to 84 Mbit/s. Furthermore, a new work item removed the dual-carrier limitation that the two carriers had to be adjacent to each other. With this addition, carriers can now be in different frequency bands. In Europe, the bands selected for this operation are 900 and 2100 MHz. Mobility is allowed only in one of the two bands, that is, neighbor cell measurements, and the decision to change cells is made with respect to one of the two bands.

In many parts of the world frequencies previously used for analog television have been freed up by the introduction of digital standards. These frequency bands in the 700 MHz region in the United States and the 800 MHz region in Europe, respectively, which are sometimes also referred to as the digital dividend bands, have been reassigned for wireless broadband Internet use. 3GPP Release 9 contains enhancements necessary for the use of LTE in the European digital dividend band. In practice, it can be observed that this spectrum has become very popular as network operators in many countries have deployed LTE networks using this spectrum.

As has been shown in Chapter 1, pre-Release 9 GSM and GPRS security mechanisms have not been changed in quite some time and vulnerabilities have been discovered. With this release, 3GPP has added an additional ciphering algorithm, A5/4. Together with doubling the ciphering key (CK) length to 128 bits, it is considered to be a major security upgrade.

3.1.8   3GPP Releases 10 and 11: LTE-Advanced

After the 'maintenance' performed in 3GPP Release 9, Releases 10 and 11 once again contain a significant new number of features of which all but one have not yet been deployed in practice today. Which of the features described below will be used in practice is not yet clear.

To further increase the peak data rate of a device, 3GPP Release 10 introduced the aggregation of several carriers (Carrier Aggregation, CA) in the same or in different frequency bands. From a European point of view, this feature would allow several 20-MHz carriers in different bands to be bundled together. This could, for example, be a 20-MHz carrier in the 1800-MHz band and another 20-MHz carrier in the 2600-MHz band, respectively. Also, a combination of a carrier in the 800-MHz digital dividend band, in which most carriers are restricted to a 10-MHz channel, and a 20-MHz carrier in the 1800-MHz or 2600-MHz carrier is thus feasible.

Although this feature is of little interest in Europe so far due to the availability of continuous spectrum of 20 MHz in many bands, this feature is of particular interest in the North American market and also in a number of Asian countries. Here, many network operators are limited to 10-MHz carriers because of the limited amount of spectrum in new bands that were assigned to network operators for LTE a few years ago. Thus, higher data rates, as in Europe, are only achievable by combining carriers in several bands. Network operators in North America such as Verizon and AT&T are thus keen to bundle their 10-MHz LTE channels in the 700-MHz band with another channel in the 1700 (uplink) / 2100 (downlink) band.

While LTE devices are usually equipped with two antennas for MIMO reception at the sender and receiver side (2×2 MIMO), the use of four and eight antennas on each side (4×4 and 8×8 MIMO) has now also been specified. In practice, however, this requires new and significantly more complex and bulky antenna systems in base stations. From the mobile device side, it is also challenging to physically include more than the already existing two antenna chains because of size limitations.

As the combination of bandwidth, modulation and the number of simultaneous MIMO streams that have been defined so far exceeds the capabilities of current mobile devices and also the physical limits of average and good transmission conditions, Releases 10 and 11 have thus specified methods to add further capacity to networks by complementing LTE macro cells that cover areas in a radius of several hundred meters by small and low-power cells that only cover a fraction of this area. This new network structure is referred to as a heterogeneous network structure (HetNet) as the coverage area of the macro cells and local cells overlap. To avoid or minimize the resulting interference, a number of methods have been specified for these cells to coordinate their transmissions with each other or to be controlled by a central base station.

Releases 10 and 11 also specify a number of further carrier aggregation steps for HSPA to bundle four or eight 5-MHz carriers. The uplink specification was extended to include MIMO and 64-QAM, as already specified for the downlink direction. It is questionable, however, if these enhancements will be implemented in practice as only few network operators have more than 10–15 MHz of bandwidth in the bands used for UMTS (e.g. 2100 MHz in Europe). Furthermore, due to the quick adoption of LTE across the globe that can be observed today, it is unlikely that network operators are keen to further develop their HSPA networks. Compared to the amount of bandwidth already used by LTE networks today, HSPA enhancements in the amount of spectrum that is currently used with this technology offer only negligible capacity gains.

3.2 Important New Concepts of UMTS

As described in the previous paragraphs, UMTS on the one hand introduces a number of new functionalities compared to GSM and GPRS. On the other hand, many properties, procedures and methods of GSM and GPRS, which are described in Chapter 1 and 2, have been kept. Therefore, this chapter first focuses on the new functionalities and changes that the Release 99 version of UMTS has introduced compared to its predecessors. In order not to lose the end-to-end nature of the overview, references are made to Chapters 1 and 2 for methods and procedures that UMTS continues to use. In the second part of this chapter, advancements are discussed that were introduced with later releases of the standard such as HSPA. These enhancements complement the Release 99 functionality but do not replace it.

3.2.1   The Radio Access Bearer (RAB)

An important new concept that is introduced with UMTS is the Radio Access Bearer (RAB), which is a description of the transmission channel between the network and a user. The RAB is divided into the radio bearer on the air interface and the Iu bearer in the radio network (UTRAN). Before data can be exchanged between a user and the network it is necessary to establish an RAB between them. This channel is then used for both user and signaling data. An RAB is always established by request of the MSC or SGSN. In contrast to the establishment of a channel in GSM, the MSC and SGSN do not specify the exact properties of the channel. Instead, the RAB establishment requests contain only a description of the required channel properties. How these properties are then mapped to a physical connection is up to the UTRAN. The following properties are specified for an RAB:

- service class (conversational, streaming, interactive or background);
- maximum speed;
- guaranteed speed;
- delay;
- error probability.

The UTRAN is then responsible for establishing an RAB that fits the description. The properties not only have an impact on the bandwidth of the established RAB but also on parameters like coding scheme, selection of a logical and physical transmission channel as well as on the behavior of the network in the event of erroneous or missing frames on different layers of the protocol stack. The UTRAN is free to set these parameters as it sees fit; the standards merely contain examples. As an example, for a voice call (service class conversational) it does not make much sense to repeat lost frames. For other services, like web browsing, such behavior is beneficial as delay times are shorter if lost packets are only retransmitted in the radio network instead of end to end.

3.2.2   The Access Stratum and Nonaccess Stratum

UMTS aims to separate functionalities of the core network from the access network as much as possible, in order to be able to independently evolve the two parts of the network in the future. Therefore, UMTS strictly differentiates between functionalities of the Access Stratum (AS) and the Nonaccess Stratum (NAS) as shown in Figure 3.4.

The AS contains all functionalities that are associated with the radio network ('the access') and the control of active connections between a user and the radio network. The handover control, for example, for which the RNC is responsible in the UTRAN, is part of the AS.

The NAS contains all functionalities and protocols that are used directly between the mobile device (UE) and the core network. These have no direct influence on the properties of the established RAB and its maintenance. Furthermore, NAS protocols are transparent to the access network. Functionalities like call control, mobility, and session management as well

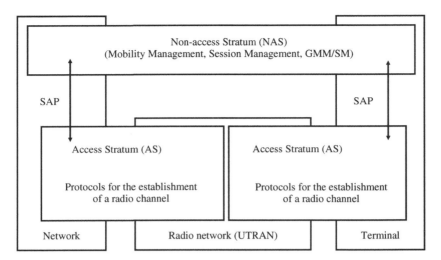

Figure 3.4 Separation of protocols between the core and radio network into Access Stratum (AS) and Nonaccess Stratum (NAS)

as supplementary services (e.g. SMS), which are controlled via the MSC and SGSN, are considered NAS functionalities.

While the NAS protocols have no direct influence on an existing RAB, it is nevertheless necessary for NAS protocols like call control or session management to request the establishment, modification or termination of a bearer. To enable this, three different service access points (SAPs) have been defined between the NAS and the AS:

- notification SAP (Nt, e.g., for paging);
- dedicated control SAP (DC, e.g., for RAB setup);
- general control SAP (GC, e.g., for modification of broadcast messages, optional).

3.2.3 Common Transport Protocols for CS and PS

In GSM networks, data is transferred between the different nodes of the radio network with three different protocols. The most important task of these protocols is to split incoming data into smaller frames, which can be transferred over the air interface. While these protocols are described in more detail in Chapter 1 (GSM) and 2 (GPRS), a short overview is given below:

- **Circuit-switched data (e.g., voice calls).** The Transcoding and Rate Adaptation Unit (TRAU) converts the pulse code-modulated (PCM)-coded voice data, which it receives from the MSC, via optimized codecs like enhanced full-rate (EFR), half-rate (HR) or Adaptive Multirate (AMR). These codecs are much more suitable for data transmission over the air interface as they compress voice data much better than PCM. This data is then sent transparently through the radio network to the BTS. Before the data is sent over the air interface, the BTS only has to perform some additional channel coding (e.g., increase of redundancy by adding error detection and correction bits).

- **Signaling data (circuit-switched signaling as well as some GPRS channel request messaging and paging)**. This data is transferred via the Link Access Protocol (LAPD) protocol, which is already known from the Integrated Services Digital Network (ISDN) world and which has been extended for GSM.
- **Packet-switched user and signaling data for GPRS**. While user and signaling data are separated in GSM, GPRS combines the two data streams into a single lower layer protocol called Release Complete RLC/MAC.

In UMTS, these different kinds of data streams are combined into a single lower layer protocol called the RLC/MAC protocol. Giving this protocol the same name as a protocol in the GPRS network was intentional. Both protocols work quite similarly in areas like breaking up large data packets from higher layers into smaller chunks for transmission over the air interface. However, due to the completely different transmission methods of the UMTS air interface compared to GSM/GPRS, there are also big differences, which are discussed in the next Section.

3.3   Code Division Multiple Access (CDMA)

To show the differences between the UMTS radio access network and its predecessors, the next paragraph gives another short overview of the basic principles of the GSM/GPRS network and its limitations at the time Release 99 UMTS networks were rolled out. As discussed in Chapter 2, some of those limitations have been reduced or overcome in the meantime and are now not as severe anymore as in the description below.

In GSM, data for different users is simultaneously transferred by multiplexing them on different frequencies and timeslots (Frequency and Time Division Multiple Access, FTDMA). A user is assigned one of eight timeslots on a specific frequency. To increase the number of users that can simultaneously communicate with a base station the number of simultaneously used frequencies can be increased. However, it must be ensured that two neighboring base stations do not use the same frequencies as they would otherwise interfere with each other. As the achievable speed with only a single timeslot is limited, GPRS introduced the concept of timeslot bundling on the same carrier frequency. While this concept enabled the network to transfer data to a user much faster than before, there were still a number of shortcomings, which were resolved by UMTS.

With GPRS, it was only possible to bundle timeslots on a single carrier frequency. Therefore, it was theoretically possible to bundle up to eight timeslots. In an operational network, however, it was rare that a mobile device was assigned more than four to five timeslots, as some of the timeslots of a carrier were used for the voice calls of other users. Furthermore, on the mobile device side, most phones could only handle four or five timeslots at a time in the downlink direction.

A GSM base station was initially designed for voice traffic, which only required a modest amount of transmission capacity. This is why GSM base stations were usually connected to the BSC via a single 2 Mbit/s E-1 connection. Depending on the number of carrier frequencies and sectors of the base station, only a fraction of the capacity of the E-1 connection was used. The remaining 64 kbit/s timeslots were used for other base stations. Furthermore, the processing capacity of GSM base stations was only designed to support the modest requirements for voice processing rather than the computing-intensive high-speed data transmission capabilities required today.

At the time UMTS was first rolled out, the existing GPRS implementations assigned resources (i.e., timeslots) in the uplink and downlink directions to the user only for exactly the time they were required. In order for uplink resources to be assigned, the mobile device had to send a request to the network. A consequence of this was unwanted delays ranging from 500 to 700 milliseconds when data needed to be sent.

Likewise, resources were only assigned in the downlink direction if data had to be sent from the core network to a user. Therefore, it was necessary to assign resources before they could be used by a specific user, which took another 200 milliseconds.

These delays, which are compared in Figure 3.5 to the delays experienced with ADSL and UMTS, were tolerable if a large chunk of data had to be transferred. For short and bursty data transmissions as in a web-browsing session, however, the delay was negatively noticeable.

UMTS Release 99 solved these shortcomings as follows:

To increase the data transmission speed per user, UMTS increased the bandwidth per carrier frequency from 200 to 5 MHz. This approach had advantages over just adding more carriers (dispersed over the frequency band) to a data transmission, as mobile devices can be manufactured much more cheaply when only a single frequency is used for the data transfer.

The most important improvement of UMTS was the use of a new medium access scheme on the air interface. Instead of using a FTDMA scheme as per GSM, UMTS introduced code multiplexing to allow a single base station to communicate with many users at the same time. This method is called Code Division Multiple Access (CDMA).

Contrary to the frequency and time multiplexing of GSM, all users communicate on the same carrier frequency and at the same time. Before transmission, a user's data is multiplied by a code that can be distinguished from codes used by other users. As the data of all users is sent at the same time, the signals add up on the transmission path to the base station. The base station

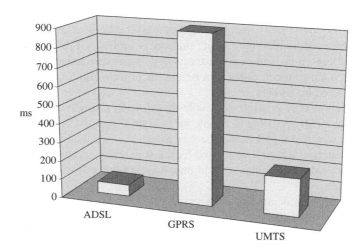

Figure 3.5 Round-trip delay (RTD) time of UMTS (Release 99) compared to ADSL and early GPRS implementations

uses the inverse of the mathematical approach that was used by the mobile device, as the base station knows the code of each user.

This principle can also be described within certain boundaries with the following analogy:

- **Communication during a lecture**. Usually there is only one person speaking at a time while many people in the room are just listening. The bandwidth of the 'transmission channel' is high as it is only used by a single person. At the same time, however, the whispering of the students creates a slight background noise that has no impact on the transmission (of the speaker) due to its low volume.
- **Communication during a party**. Again, there are many people in a room but this time they are all talking to each other. Although the conversations add up in the air the human ear is still able to distinguish the different conversations from each other. Most conversations are filtered out by the ear as unwanted background noise. The more people that speak at the same time, the higher the perceived background noise for the listeners. To be understood, the speakers have to reduce their talking speed. Alternatively, speakers could increase their volume to be able to be heard over the background noise. This, however, means that the background noise for others would increase substantially.
- **Communication in a disco**. In this scenario, the background noise, that is, the music is very loud and no other communication is possible.

These scenarios are analogous to a UMTS system as follows: If only a few users communicate with a base station at the same time, each user will experience only low interference on the transmission channel. Therefore, the transmission power can be quite low and the base station will still be able to distinguish the signal from other sources. This also means that the available bandwidth per user is high and can be used if necessary to increase the transmission speed. If data is sent faster, the signal power needs to be increased to get a more favorable signal to noise ratio. As only a few users are using the transmission channel in this scenario, increasing the transmission speed is no problem as all others are able to compensate.

If many users communicate with a base station at the same time, all users will experience a high background noise. This means that all users have to send at a higher power to overcome the background noise. As each user in this scenario can still increase the power level, the system remains stable. This means that the transmission speed is not only limited by the 5-MHz bandwidth of the transmission channel but also by the noise generated by other users of the cell. Even though the system is still stable, it might not be possible to increase the data transmission speed for some users who are farther away from the base station as they cannot increase their transmission power anymore and thus cannot reach the signal-to-noise ratio required for a higher transmission speed (Figure 3.6).

Transmission power cannot be increased indefinitely as UMTS mobile devices are limited to a maximum transmission power of 0.25 W. Unless the access network continuously controls and is aware of the power output of the mobile devices, a point would be reached at which too many users communicate with the system. As the signals of other users are perceived as noise from a single user's point of view, a situation could occur when a mobile device cannot increase its power level anymore to get an acceptable signal-to-noise ratio. If, on the other hand, a user is close to a base station and increases its power above the level commanded by the network, it could interfere with the signals of mobile devices that are further away and thus weaker.

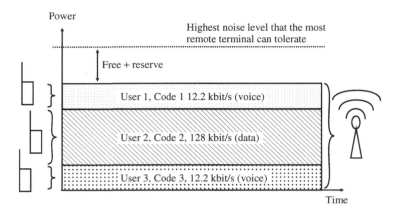

Figure 3.6 Simultaneous communication of several users with a base station in uplink direction (axis not to scale and number of users per base station are higher in a real system)

From a mathematical point of view, CDMA works as follows:

The user data bits of the individual users are not transferred directly over the air interface but are first multiplied with a vector, which, for example, has a length of 128. The elements of the resulting vector are called chips. A vector with a length of 128 has the same number of chips. Instead of transmitting a single bit over the air interface, 128 chips are transmitted. This is called 'spreading' as more information, in this example, 128 times more, is sent over the air interface compared to the transmission of the single bit. On the receiver side the multiplication can be reversed and the 128 chips are used to deduce if the sent bit represents a 0 or 1. Figure 3.7 shows the mathematical operations for two mobile devices that transmit data to a single receiver (base station).

The disadvantage of sending 128 chips instead of a single bit might seem quite serious but there are also two important advantages: Transmission errors that change the values of some of the 128 chips while being sent over the air interface can easily be detected and corrected. Even if several chips are changed because of interference, the probability of correctly identifying the original bit is still very high. As there are many 128-chip vectors, each user can be assigned a unique vector that allows calculation of the original bit out of the chips at the receiver side, not only for a single user but also for multiple users at the same time.

3.3.1 Spreading Factor, Chip Rate and Process Gain

The process of encoding a bit into several chips is called spreading. The spreading factor for this operation defines the number of chips used to encode a single bit. The speed with which the chips are transferred over the UMTS air interface is called the chip rate and is 3.84 Mchips/s, independent of the spreading factor.

As the chip rate is constant, increasing the spreading factor for a user means that his datarate decreases. Besides a higher robustness against errors, there are a number of other advantages of a higher spreading factor: The longer the code, the more codes exist that are orthogonal to each

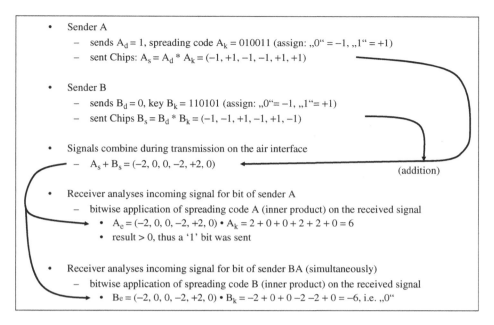

Figure 3.7 Simultaneous conversation between two users with a single base station and spreading of the data stream

other. This means that more users can simultaneously use the transmission channel compared to a system in which only shorter spreading factors are used. As more users generate more noise, it is likely that the error rate increases at the receiver side. However, as more chips are used per bit, a higher error rate can be accepted than for a smaller spreading factor. This, in turn, means that a lower signal-to-noise ratio is required for a proper reception and thus, the transmission power can be reduced if the number of users in a cell is low. As less power is required for a slower transmission, it can also be said that a higher spreading factor increases the gain of the spreading process (processing gain).

If shorter codes are used, that is, fewer chips per bit, the transmission speed per user increases. However, there are two disadvantages to this. Owing to the shorter codes, fewer people can communicate with a single base station at the same time. With a code length of eight (spreading factor 8), which corresponds to a user datarate of 384 kbit/s in the downlink direction, only eight users can communicate at this speed. With a code length of 256 on the other hand, 256 users can communicate at the same time with the base station although the transmission speed is a lot slower. Owing to the shorter spreading code, the processing gain also decreases. This means that the power level of each user has to increase to minimize transmission errors. Figure 3.8 shows these relationships in a graphical format.

3.3.2 The OVSF Code Tree

The UMTS air interface uses a constant chip rate of 3.84 Mchips/s. If the spreading factor was also constant, all users of a cell would have to communicate with the network at the same

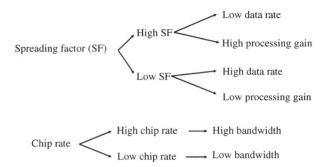

Figure 3.8 Relation between spreading factor, chip rate, processing gain and available bandwidth per user

speed. This is not desirable, as a single cell has to support many users with many different applications simultaneously. While some users may want to simply make voice calls, which require only a small bandwidth, other users might want to place video calls, watch some mobile TV (video streaming) or start a web surfing session. All these services require much higher bandwidths, and thus using the same spreading factor for all connections is not practicable.

The solution to this problem is called Orthogonal Variable Spreading Factors, or OVSF for short. While in the previous mathematical representation the spreading factors of both users were of the same length, it is possible to assign different code lengths to different users at the same time with the approach described below.

As the codes of different lengths also have to be orthogonal to each other, the codes need to fulfill the condition shown in Figure 3.9. In the simplest case (C1,1), the vector is one dimensional. On the next level, with two chips, four vectors are possible of which two are orthogonal

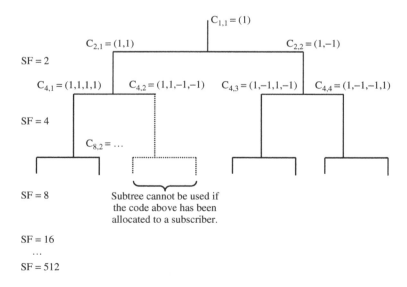

Figure 3.9 The OVSF code tree

Table 3.1   Spreading factors and data rates

Spreading Factor (Downlink)	Raw datarate (kbit/s)	User datarate (kbit/s)	Application
8	960	**384**	Packet data
16	480	**128**	Packet data
32	240	**64**	Packet data and video telephony
64	120	**32**	Packet data
128	60	**12.2**	Voice, packet data, location updates, SMS
256	30	**5.15**	Voice

to each other (C2,1 and C2,2). On the third level, with four chips, there are 16 possible vector combinations and four that are orthogonal to each other. The tree, which continues to grow for SF 8, 16, 32, 64, and so on shows that the higher the spreading factor the more the number of subscribers who can communicate with a cell at the same time.

If a mobile device, for example, uses a spreading factor of eight, all longer codes of the same branch cannot be used anymore. This is due to the codes below not being orthogonal to the code on the higher level. As the tree offers seven other SF 8 spreading factors, it is still possible for other users to have codes with higher spreading factors from one of the other vertical branches of the code tree. It is up to the network to decide how many codes are used from each level of the tree. Thus, the network has the ability to react dynamically to different usage scenarios.

Table 3.1 shows the spreading factors in the downlink direction (from the Node-B to the mobile device) as they are used in a real system. The raw datarate is the number of bits transferred per second. The user datarate results from the raw datarate after removal of the extra bits that are used for channel coding, which is necessary for error detection and correction, signaling data and channel control.

3.3.3   Scrambling in Uplink and Downlink Direction

By using OVSF codes, the datarate can be adapted for each user individually while still being able to differentiate the data streams with different speeds. Some of the OVSF codes are quite uniform. C(256,1), for example, is only comprised of chips with value '1'. This creates a problem further down the processing chain, as the result of the modulation of long sequences that never change their value would result in a very uneven spectral distribution. To counter this effect the chip stream that results from the spreading process is scrambled. This is achieved by multiplying the chip stream, as shown in Figure 3.10, with a pseudo random code called the scrambling code. The chip rate of 3.84 MChips/s is not changed by this process.

In the downlink direction the scrambling code is also used to enable the mobile device to differentiate between base stations. This is necessary as all base stations of a network transmit on the same frequency. In some cases mobile operators have bought a license for more than a single UMTS frequency. However, this was done to increase the capacity in densely populated areas and not as a means to make it easier for mobile devices to distinguish between different base stations. The use of a unique scrambling code per base station is also necessary to allow a base station to use the complete code tree instead of sharing it with the neighboring cells.

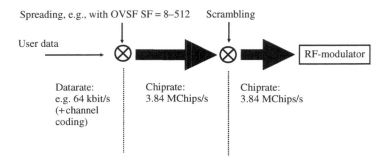

Figure 3.10   Spreading and scrambling

This means that in the downlink direction, capacity is mainly limited by the number of available codes from the code tree as well as the interference of other base stations as experienced by the mobile device.

In the uplink direction, on the other hand, each mobile device is assigned its own scrambling code. Therefore, each mobile device could theoretically use all codes of the code tree. This means that in the uplink direction the system is not limited by the number of codes but by the maximum transmitting power of the mobile device and by the interference that is created by other mobile devices in the current and neighboring cells.

Another reason for using a unique scrambling code per mobile device in the uplink direction is the signal propagation delays. As different users are at different distances from a base station the signals take a different amount of time to arrive. In the GSM radio network this was solved by controlling the timing advance. The use of a timing advance, however, is not possible in the UMTS radio network because of the soft handover state (Section 3.7.1) in which the mobile device communicates with several base stations at the same time. As the mobile device is at a different distance from each base station it communicates with simultaneously, it is not possible to synchronize the mobile device to all base stations because of the different signal propagation delays. Therefore, if no scrambling code was used, the mathematical equation shown in Figure 3.7 would not work anymore as the chips of the different senders would be out of phase with each other and the result of the equation would change (Table 3.2).

Table 3.2   Spreading and scrambling in uplink and downlink directions

	Downlink	Uplink
Spreading	• Addressing of different users • Controls the individual data rate for each user	• Controls the individual data rate for each user
Scrambling	• Ensures consistent spectral distribution • Used by the mobile device to differentiate base stations	• Ensures consistent spectral distribution • Differentiates users • Removes the need for a timing advance by preserving the orthogonal nature of the codes necessary for soft handover

3.3.4   UMTS Frequency and Cell Planning

As all cells in a UMTS radio network can use the same frequency, the frequency plan is greatly simplified compared to a GSM radio access network. While it is of paramount importance in a GSM system to ensure that neighboring cells use different frequencies, it is quite the reverse in UMTS, as all neighboring stations use the same frequency. This is possible because of the CDMA characteristics, as described in the previous paragraphs. While a thorough and dynamic frequency plan is indispensable for GSM, no frequency adaptations are necessary for new UMTS cells. If a new cell is installed to increase the bandwidth in an area that is already covered by other cells, the most important task in a UMTS network is to decrease the transmission power of the neighboring cells.

In both GSM and UMTS radio networks, it is necessary to properly define and manage the relationships between the neighboring cells. Incorrectly defined neighboring cells are not immediately noticeable but later on create difficulties for handovers (see Section 3.7.1) and cell reselections (Section 3.7.2) of moving subscribers. Properly executed cell changes and handovers also improve the overall capacity of the system as they minimize interference of mobiles that stay in cells which are no longer suitable for them.

3.3.5   The Near–Far Effect and Cell Breathing

As all users transmit on the same frequency, interference is the most limiting factor for the UMTS radio network. The following two phenomena are a direct result of the interference problem.

To keep the interference at a minimum, it is important to have a precise and fast power control. Users that are farther away from the base station have to send with more power than those closer to the base station, as the signal gets weaker the farther it has to travel. This is called the near–far effect. Even small changes in the position of the user, like moving from a free line of sight to a base station to behind a wall or tree, has a huge influence on the necessary transmission power. The importance of efficient power control for UMTS is also shown by the fact that the network can instruct each handset 1500 times per second to adapt its transmission power. A beneficial side effect of this for the mobile device is an increased operating time, which is very important for most devices as the battery capacity is quite limited.

Note: GSM also controls the transmission power of handsets. The control cycle, however, is in the order of one second as interference in GSM is less critical than in UMTS. Therefore, in a GSM network the main benefit of power control is that of increasing the operating time of the mobile device.

The dependence on low interference for each user also creates another unwanted side effect. Let us assume the following situation:

1. There are a high number of users in the coverage area of a base station and the users are dispersed at various distances from the center of the cell.
2. Because of interference the most distant user needs to transmit at the highest possible power.
3. An additional user who is located at a medium range from the center of the cell tries to establish a connection to the network for a data transfer.

In this situation the following things can happen: If the network accepts the connection request the interference level for all users will rise in the cell. All users thus have to increase

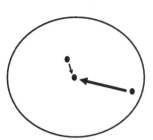

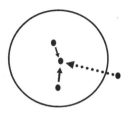

Two subscribers of a cell
one of them close
to the cell edge sending
with its maximum possible
power level.

A third subscriber would like to
communicate in the cell. This poses
a problem for the second subscriber
as he can't increase the power output
to counter the additional interference.

Figure 3.11   Cell breathing

their transmission power accordingly. The user at the border of the cell, however, is already transmitting at his maximum power and thus cannot increase the power level anymore. As a result his signal cannot be correctly decoded anymore and the connection is broken. When seen from outside the system, this means that the geographical area the cell can cover is reduced as the most distant user cannot communicate with the cell anymore. This phenomenon is called cell breathing (see Figure 3.11) as the cell expands and shrinks like a human lung, which increases and decreases its size while breathing.

To avoid this effect, the network constantly controls the signal-to-noise ratio of all active users. By actively controlling the transmission power of each user the network is aware of the impact an additional user would have on the overall situation of the cell. Therefore, the network has the possibility to reject a new user to protect the ongoing sessions.

To preserve all ongoing connections and additionally allow a new user to enter the system, it is also possible to use a different strategy. The goal of this strategy is to reduce the interference to a level that allows all users, including the prospective new one, to communicate. This can be done in a number of ways. One way is to assign longer spreading codes to already established channels. As described in Section 3.3.2, it is possible for mobile devices to reduce their transmission power by using longer spreading codes. This in turn reduces the interference for all other users. The disadvantage of using longer spreading codes is of course a reduction of the maximum transmission speed for some users. As not all connections might be impacted, there are again a number of possibilities for the selection process. Users could, for example, be assigned to different user classes. Changing spreading factors could then be done only for users of a lower user class who pay less for their subscription than others. It can also be imagined that the network could start a congestion defense mechanism at a certain load threshold before the system got into an overload situation. Once the threshold is reached, the network could then, for example, only assign short spreading factors to users with a higher priority subscription while the system load is above the threshold.

Besides cell breathing there are other interference scenarios. As already mentioned, it is necessary to increase the transmission power if the spreading factor is decreased to ensure a proper reception. Therefore, the maximum distance at which a user can be from the center of the cell also depends on the spreading factor. If a user roams between two cells it is possible that the current spreading factor would not allow data to be transferred as reliably as before because of the interference encountered at the cell edge, whereas a lower spreading factor would still allow a reliable data transfer. How this and similar scenarios at cell edges are resolved depends on the vendor's equipment and the parameter settings of the operator. As in other areas, the UMTS standard does not dictate a specific solution to these issues. Therefore, network vendors that have implemented clever solutions can gain a competitive advantage.

3.3.6 Advantages of the UMTS Radio Network Compared to GSM

While in the previous paragraphs the basic properties and methods of the UMTS Release 99 air interface have been introduced, the following paragraphs describe how it helped to overcome the limitations of GPRS.

One of the main reasons for the long delay times of early GPRS implementations was the constant reassignment of resources for bursty data traffic. UMTS Release 99 solved this issue by assigning a dedicated channel not only for voice calls but for packet data connections as well. The channel remained dedicated to the user for some time, even if there was no data transfer. A downside of this approach was that the spreading code was not available to other users. As only control information was sent over the established channel during times of inactivity, the interference level for other users decreased. As a result, some of the overall capacity of the cell was lost by keeping the spreading code assigned to a dormant user. From a user's point of view, the spreading code should only be freed up for use by someone else if the session remains dormant for a prolonged duration. Once the system decided to reassign the code to someone else, it also assigned a higher spreading factor to the dormant user, of which a greater number existed per cell. If the user resumed data transmission, there was no delay as a dedicated channel still existed. If required, the bandwidth for the user could be increased again quite quickly by assigning a code with a shorter spreading factor. The user, however, did not have to wait for this as in the meantime data transfer was possible over the existing channel.

In the uplink direction, the same methods were applied. It should be noted, however, that while the user was assigned a code, the mobile device was constantly transmitting in the uplink direction. The transmission power was lower while no user data was sent but the mobile device kept sending power control and signal quality measurement results to the network.

While this method of assigning resources was significantly superior to GPRS it soon became apparent that it had its own limitations concerning maximum data transfer rates that could be achieved and the number of simultaneous users that could be active in a cell. Later releases of the standard have therefore changed this concept and have put the logically dedicated downlink channel on a physically shared channel. This concept is referred to as HSDPA and is described in the later part of the chapter.

In both UMTS Release 99 and HSDPA, if the user remains dormant for a longer time period the network removes all resources on the air interface without cutting the logical connection. This prevents further wastage of resources and also has a positive effect on the overall operating time of a mobile device. The disadvantage of this approach is a longer reaction time once the user wants to resume the data transfer.

This is why it is beneficial to move the user into the Cell-FACH (Forward Access Channel) state after a longer period of inactivity. In this state, no control information is sent from the mobile device to the network and no dedicated channel is assigned to the connection anymore. The different connection states are described in more detail in Section 3.5.4.

The assignment of dedicated channels for both circuit- and packet-switched connections in UMTS has a big advantage for mobile users compared to GPRS. In the GPRS network, the mobile device has sole responsibility for performing a cell change. Once the cell has been changed, the mobile device first needs to listen to the broadcast channel before the connection to the network can be reestablished. Therefore, in a real network environment, a cell change interrupts an ongoing data transmission for about one to three seconds. A handover, which is controlled by the network and thus results in no or only a minimal interruption of the data transmission, is not foreseen for GPRS. Hence, GPRS users frequently experience interruptions in data transmission during cell changes while traveling in cars or trains. With UMTS there are no interruptions of an ongoing data transfer when changing cells due to a process called 'soft handover', which makes data transfers while on the move much more efficient.

Another problem with GSM is the historical dimensioning of the transmission channel for narrow band voice telephony. This limitation was overcome for GPRS by combining several timeslots for the time of the data transfer. The maximum possible datarate, however, was still limited by the overall capacity of the 200 kHz carrier. For UMTS Release 99, what were considered to be high bandwidth applications at the time were taken into consideration for the overall system design from the beginning. Owing to this, a maximum data transfer rate of 384 kbit/s could be achieved in early networks with a spreading factor of eight in the downlink direction. In the uplink direction, datarates of 64–384 kbit/s could be reached.

UMTS also enables circuit-switched 64 kbit/s data connections in the uplink and downlink directions. This speed is equal to an ISDN connection in the fixed-line network and is used for circuit-switched video telephony between UMTS users.

UMTS can also react very flexibly to the current signal quality of the user. If the user moves away from the center of the cell, the network can react by increasing the spreading factor of the connection. This reduces the maximum transmission speed of the channel, which is usually preferable to losing the connection entirely.

The UMTS network is also able to react very flexibly to changing load conditions on the air interface. If the overall interference reaches an upper limit or if a cell runs out of available codes owing to a high number of users in the cell, the network can react and assign longer spreading factors to new or ongoing connections.

3.4 UMTS Channel Structure on the Air Interface

3.4.1 User Plane and Control Plane

GSM, UMTS and other fixed and wireless communication systems differentiate between two kinds of data flows. In UMTS, these are referred to as two different planes. Data flowing in the user plane is data which is directly and transparently exchanged between the users of a connection like voice data or IP packets. The control plane is responsible for all signaling data that is exchanged between the users and the network. The control plane is thus used for signaling data to exchange messages for call establishment or messages, for example, for a

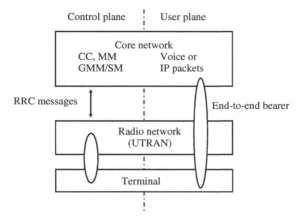

Figure 3.12   User and control planes

location update. Figure 3.12 shows the separation of user and control planes as well as some examples for protocols that are used in the different planes.

3.4.2   Common and Dedicated Channels

Both user plane data and control plane data is transferred over the UMTS air interface in so-called 'channels'. Three different kinds of channels exist:

Dedicated channels: These channels transfer data for a single user. A dedicated channel is used, for example, for a voice connection, for IP packets between the user and the network or a location update message.

Common channels: The counterpart to a dedicated channel is a common channel. Data transferred in common channels is destined for all users of a cell. An example for this type of channel is the broadcast channel, which transmits general information about the network to all users of a cell in order to inform them of, for example, the network the cell belongs to, the current state of the network, and so on. Common channels can also be used by several devices for the transfer of user data. In such a case, each device filters out its packets from the stream broadcast over the common channel and only forwards these to higher layers of the protocol stack.

Shared channels: Very similar to common channels are shared channels. These channels are not monitored by all devices but only by those that have been instructed by the network to do so. An example of such a channel is the High-Speed Downlink Shared Channel (HS-DSCH) of HSDPA (see Section 3.10).

3.4.3   Logical, Transport and Physical Channels

To separate the physical properties of the air interface from the logical data transmission, the UMTS design introduces three different channel layers. Figure 3.13 shows the channels on different layers in downlink direction while Figure 3.14 does the same for uplink channels.

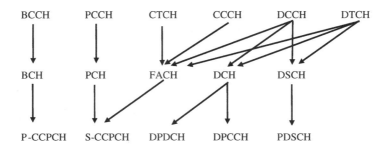

Figure 3.13   Logical, transport and physical channels in downlink direction (without HSPA)

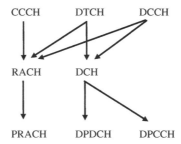

Figure 3.14   Logical, transport and physical channels in uplink direction (without HSPA)

Logical Channels

The topmost channel layer is formed by the logical channels. Logical channels are used to separate different kinds of data flows that have to be transferred over the air interface. The channels contain no information on how the data is later transmitted over the air. The UMTS standards define the following logical channels:

The Broadcast Control Channel (BCCH): This channel is monitored by all mobile devices in Idle state to receive general system information from the network. Information distributed via this channel, for example, includes how the network can be accessed, which codes are used by the neighboring cells, the LAC, the Cell-ID and many other parameters. The parameters are further grouped into System Information Block (SIB) messages to help the mobile device to decode the information and to save air interface bandwidth. A detailed description of the messages and parameters can be found in 3GPP 25.331, chapter 10.2.48.8 [3].

The Paging Control Channel (PCCH): This channel is used to inform users of incoming calls or SMS messages. Paging messages are also used for packet-switched calls if new data arrives from the network once all physical resources (channels) for a subscriber have been released owing to a long period of inactivity. If the mobile device receives a paging message it has to first report its current serving cell to the network. The network will then reestablish a logical RRC connection with the mobile device and the data waiting in the network is then delivered to the mobile device.

The Common Control Channel (CCCH): This channel is used for all messages from and to individual mobile devices (bidirectional) that want to establish a new connection with the network. This is necessary, for example, if a user wants to make a phone call, send an SMS or to establish a channel for packet-switched data transmission.

– **The Dedicated Control Channel (DCCH)**. While the three channels described above are common channels observed by many mobile devices in the cell, a DCCH only transports data for a single subscriber. A DCCH is used, for example, to transport messages for the MM and CC protocols for circuit-switched services, Packet Mobility Management (PMM) and SM messages for packet-switched services from and to the MSC and SGSN. These protocols are described in more detail in Sections 3.6 and 3.7.

– **The Dedicated Traffic Channel (DTCH)**. This channel is used for user data transfer between the network and a single user. User data can, for example, be a digitized voice signal or IP packets of a packet-switched connection. If a dedicated logical channel carries a traditional voice call, the channel is mapped to a dedicated physical channel. If the dedicated logical channel carries packet-switched data, it is also possible to map the dedicated logical connection to a common or shared physical channel as shown in Figure 3.13. In practice, a packet-switched connection is mapped to a common channel after some time of inactivity or if only little user data is transferred. The shared channel introduced with Release 99 was never used. Instead, it was replaced with the HSDSCH introduced with HSDPA in Release 5.

– **The Common Traffic Channel (CTCH)**: This channel is used for cell broadcast information that can be shown on the display of mobile devices. In practice, only few network operators make use of this. In a few countries such as the Netherlands, cell broadcast messages are used for public warning systems [4].

Transport Channels

Transport channels prepare downlink data frames for transmission over the air interface by splitting them up into smaller parts, which are encapsulated into RLC/MAC-frames that are more suitable for transmission over the air interface. The RLC/MAC header that is placed in front of each frame contains, among other things, the following information:

- length of the frame (10, 20, 40 or 80 milliseconds);
- type of integrity checking mechanism (CRC checksum);
- channel coding format for error detection and correction;
- rate matching in case the speed of the physical channel and the layers above do not match;
- control information for detection of discontinuous transmission (DTX) in case the other end has no data to send at a particular time.

All of these properties are combined into a so-called transport format. The actual channel coding, however, is only performed on the physical layer on the Node-B. This is very important as channel coding includes the addition of error detection and correction bits to the data stream, which can be a huge overhead. In Chapter 1, for example, the half-rate convolutional decoder for channel coding was introduced, which practically doubles the datarate. UMTS also makes use of this channel coder and further introduces a number of additional ones.

Logical channels are mapped to the following transport channels:

- **The Broadcast Channel (BCH)**. Transport channel variant of the logical BCCH.
- **The Dedicated Channel (DCH)**. This transport channel combines data from the logical DTCH and the logical DCCH. The channel exists in both uplink and downlink directions as data is exchanged in both directions.
- **The Paging Channel (PCH)**. Transport channel variant of the logical PCCH.
- **The Random Access Channel (RACH)**. The bidirectional logical CCCH is called RACH on the transport layer in uplink direction. This channel is used by mobile devices to send RRC Connection Request messages to the network if they wish to establish a dedicated connection with the network (e.g., to establish a voice call). Furthermore, the channel is used by mobile devices to send user packet data (in Cell-FACH state, see Section 3.5.4) if no dedicated channel exists between the mobile device and the network. It should be noted, however, that this channel is only suitable for small amounts of data.
- **The Forward Access Channel (FACH)**. This channel is used by the network to send RRC Connection Setup messages to mobile devices, which have indicated via the RACH that they wish to establish a connection with the network. The message contains information for the mobile device on how to access the network. If the network has assigned a dedicated channel, the message contains, for example, information on which spreading codes will be used in uplink and downlink directions. The FACH can also be used by the network to send user data to a mobile device in case no dedicated channel has been allocated for a data transfer. The mobile device is then in the Cell-FACH state, which is further described in Section 3.5.4. A typical channel capacity of the FACH is 32 kbit/s. In the uplink direction data is transferred via the RACH.

Physical Channels

Finally, physical channels are responsible for offering a physical transmission medium for one or more transport channels. Furthermore, physical channels are responsible for channel coding, that is, the addition of redundancy and error detection bits to the data stream.

The intermediate products between transport channels and physical channels are called Composite Coded Transport Channels (CCTrCh) and are a combination of several transport channels, which are subsequently transmitted over one or more physical channels. This intermediate step was introduced because it is not only possible to map several transport channels onto a single physical channel (e.g. the PCH and FACH on the S-CCPCH) but also to map several physical channels onto a single transport channel (e.g. the Dedicated Physical Data Channel (DPDCH) and Dedicated Physical Control Channel (DPCCH) onto the DCH).

The following physical channels are used in a cell:

- **The Primary Common Control Physical Channel (P-CCPCH)**. This channel is used for distributing broadcast information in a cell.
- **The Secondary Common Control Physical Channel (S-CCPCH)**. This channel is used to broadcast the PCH and the FACH.

- **The Physical Random Access Channel (PRACH)**. The physical implementation of the RACH.
- **The Acquisition Indication Channel (AICH)**. This channel is not shown in the channel overview figures as there is no mapping of this channel to a transport channel. The channel is used exclusively together with the PRACH during the connection establishment of a mobile device with the network. More about this channel and the process of establishing a connection can be found in Section 3.4.5.
- **The Dedicated Physical Data Channel (DPDCH)**. This channel is the physical counterpart of a dedicated channel to a single mobile device. The channel combines user data and signaling messages from (Packet) MM, CC and SM.
- **The Dedicated Physical Control Channel (DPCCH)**. This channel is used in addition to a DPDCH in both uplink and downlink directions. It contains layer 1 information like Transmit Power Control (TPC) bits for adjusting the transmission power. Furthermore, the channel is also used to transmit the so-called pilot bits. These bits always have the same value and can thus be used by the receiver to generate a channel estimation, which is used to decode the remaining bits of the DPCCH and the DPDCH. More information about the DPCCH can be found in 3GPP TS 25.211 Section 5.2.1 [5].

While the separation of channels in GSM into logical and physical channels is quite easy to understand, the UMTS concept of logical, transport and physical channels and the mappings between them is somewhat difficult to understand at first. Therefore, the following list summarizes the different kinds of channels and their main tasks:

- **Logical Channels**. These channels describe different flows of information like user data and signaling data. Logical channels contain no information about the characteristics of the transmission channel.
- **Transport Channels**. These channels prepare data packets that are received from logical channels for transmission over the air interface. Furthermore, this layer defines which channel coding schemes (e.g., error correction methods) are to be applied on the physical layer.
- **Physical Channels**. These channels describe how data from transport channels is sent over the air interface and apply channel coding and decoding to the incoming data streams.

The next paragraph gives an idea of the way the channels are used for two different procedures.

3.4.4   Example: Network Search

When a mobile device is switched on, one part of the start up procedure is the search for available networks. Once a suitable network has been found, the mobile device performs an attach procedure. Next, the mobile device is known to the network and ready to accept incoming calls, short messages, and so on. When the user switches the mobile device off, the current information about the network (e.g., the frequency, scrambling code and Cell-ID of the current cell) is saved. This enables the mobile device to skip most activities required for the network search once it is powered on again, which substantially reduces the time it takes to find and attach to the network again. In this example, it is assumed that the mobile device has no or only invalid information about the last used cell when it is powered on. This can be the case if the SIM card is used for the first time or if the cell for which information was stored on the SIM card is not found anymore.

As in all communication systems, it is also necessary in UMTS to synchronize the mobile devices with the network. Without correct synchronization it is not possible to send an RRC Connection Request message at the correct time or to detect the beginning of an incoming data frame. Therefore, the mobile device's first task after it is switched on is to synchronize to the cells of the networks around it. This is done by searching all frequency bands assigned to UMTS for Primary Synchronization Channels (P-SCH). As can be seen in Figure 3.15, a UMTS data frame consists of 15 slots in which 2560 chips per slot are usually transported. On the P-SCH only the first 256 chips per slot are sent and all base stations use the same code. If several signals (originating from several base stations) are detected by the mobile at different times owing to the different distances of the mobile device to the various cells, the mobile device synchronizes to the timing of the burst with the best signal quality.

Once a P-SCH is found, the mobile device is synchronized to the beginning of a slot. In the next step, the mobile device then has to synchronize itself with the beginning of a frame. To do this, the mobile device will search for the Secondary Synchronization Channel (S-SCH). Again only 256 chips per slot are sent on this channel. However, on this channel each slot has a different chip pattern. As the patterns and the order of the patterns are known, the mobile device is able to determine the slot that contains the beginning of a frame.

If an operator only has a license for a single channel, all cells of the network operator send on the same frequency. The only way to distinguish them from each other is by using a different

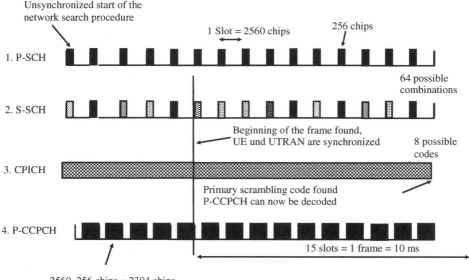

Figure 3.15   Network search after the mobile device is switched on

scrambling code for each cell. The scrambling code is used to encode all downlink channels of a cell including the P-CCPCH, which contains the system broadcast information. The next step of the process is therefore to determine the primary scrambling code of the selected cell. The first part of this process was already started with the correct identification of the S-SCH and the chip pattern. Altogether, 64 different S-SCH chip patterns are specified in the standard. This means that in theory the mobile device could distinguish up to 64 individual cells at its current location. In an operational network, however, it is very unlikely that the mobile device would receive more than a few cells at a time. To determine the primary scrambling code, the mobile device then decodes the Common Pilot Channel (CPICH), which broadcasts another known chip pattern. Eight possible primary scrambling codes are assigned to each of the 64 chip patterns that are found on the S-SCH. To find out which code is used by the cell out of the eight scrambling codes for all other channels, the mobile device now applies each of the eight possible codes on the scrambled chip sequence and compares the result to the chip pattern that is expected to be broadcasted on the CPICH. As only one of the scrambling codes will yield the correct chip pattern the mobile device can stop the procedure as soon as it has found the correct one.

Once the primary scrambling code has been found by using the CPICH, the mobile device can now read the system information of the cell which is broadcast via the P-CCPCH. The P-CCPCH is always encoded with spreading code C256,1 with a spreading factor of 256, which is easy to find by the mobile device even under difficult radio conditions. Only after having deciphered the information that is broadcast on this channel is the mobile aware of which network the cell belongs to. The following list shows some parameters that are broadcast on the P-CCPCH:

- The identity of the network the cell belongs to (MCC/MNC), location area (LAC) and Cell-ID.
- Cell access restrictions. This suggests which groups of subscribers are allowed to communicate with the cell. Usually all subscribers are allowed to communicate with a cell. Only under certain conditions will the network operator choose to temporarily restrict access to parts of the network for some subscribers. This can help during catastrophic events to allow important users of the network like the police and doctors to communicate with facilities like hospitals. Without access restrictions, cells quickly overload during such events as the number of call attempts of normal users increase dramatically and can thus delay important calls.
- Primary Scrambling Codes and frequencies of neighboring cells. As described above, the frequencies of the other cells in the area are usually the same as the frequency of the current cell. Only in areas of very high usage might operators deploy cells in other frequency bands to increase the overall available bandwidth. Both scrambling codes and frequencies of neighboring cells are needed by the mobile device to be able to easily find and measure the reception quality of other cells while they are in idle mode for cell reselection purposes.
- Frequency information of neighboring GSM cells. This information is used by the mobile to be able to reselect a GSM cell in case the signal quality of the current cell deteriorates and no suitable neighboring UMTS cell can be received.
- Parameters that influence the Cell Reselection Algorithm. This way the network is able to instruct the mobile device to prefer some cells over others.
- Maximum transmission power the mobile is allowed to use when sending a message on the RACH.

- Information about the configuration of the PRACH and S-CCPCH which transport the RACH and FACH, respectively. This is necessary because some parameters, like the spreading factor are variable to allow the operator to control the bandwidth of these channels. This is quite important as they do not only transport signaling information but also transport user data, as described below.

If the cell belongs to the network the mobile device wants to attach to, the next step in the process is to connect to the network by performing a circuit-switched location update and a packet-switched attach procedure. These procedures use the higher protocol layers of the MM and PMM, respectively, which are also used in GSM and GPRS. For UMTS, both protocol stacks were only slightly adapted. Further information on these procedures is described in Sections 1.8.1 and 2.8.1.

3.4.5   Example: Initial Network Access Procedure

If the mobile device is in Idle state and wants to establish a connection with the network, it has to perform an initial network access procedure. This may be done for the following reasons:

- to perform a location update;
- for a mobile-originated call;
- to react to a paging message;
- to start a data session (Packet Data Protocol (PDP) context activation);
- to access the network during an ongoing data session for which the physical air interface connection was released by the network owing to long inactivity.

For all scenarios above, the mobile device needs to access the network to request a connection over which further signaling messages can be exchanged. As can be seen in Figure 3.16, the mobile device starts the initial network access procedure by sending several preambles with a length of 4096 chips. The time required to transmit the 4096 chips is exactly 1 millisecond. If the mobile device receives no answer from the network, it increases the transmission power and repeats the request. The mobile device keeps increasing the transmission power for the preambles until a response is received or the maximum transmission power and number of retries have been reached without a response. This is necessary as the mobile device does not know which transmission power level is sufficient to access the network. Thus the power level is very low at the beginning, which on the one hand creates only low interference for other subscribers in the cell but on the other hand does not guarantee success. To allow the network to answer, the preambles are spaced three slots apart. Once the preamble is received correctly, the network then answers on the AICH. If the mobile device receives the message correctly, it is then aware of the transmission power to use and proceeds by sending a 10- or 20-millisecond frame on the PRACH, which contains an RRC Connection Request message. As the spreading factor of the PRACHs is variable, the message can contain between 9 and 75 bytes of information.

To avoid collisions of different mobile devices, the PRACH is divided into 15 slots. Furthermore, there are 16 different codes for the preamble. Thus, it is very unlikely that two mobile devices use the same slot with the same code at the same time. Nevertheless, in case this happens, the connection request will fail and has to be repeated by the mobile devices as their

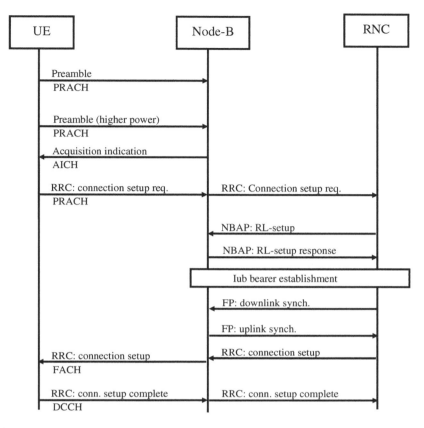

Figure 3.16 Initial Network Access Procedure (RRC Connection Setup) as described in 3GPP TS 25.931 [6]

requests cancel out each other. Once the RNC has received the RRC Connection Request message as shown in Figure 3.16, it will allocate the required channels in the radio network and on the air interface. There are two possibilities the RNC can choose from:

- The RNC can use a DCH for the connection as shown in Figure 3.16. Thus, the mobile device changes into the Cell-DCH RRC state (see also Section 3.5.4). Using a DCH is a good choice for the network in case the Connection Request Message indicates that the mobile device wishes to establish a user data connection (voice or packet-switched data).
- The RNC can also decide to continue to use the RACH and FACH for the subsequent exchange of messages. The mobile device will thus change into the Cell-FACH state. The decision to use a shared channel instead of a DCH can be made, for example, if the Connection Request message indicates to the network that the channel is only required for signaling purposes, like for performing a location update procedure.

After choosing a suitable channel and reserving the necessary resources, the RNC replies with a RRC Connection Setup message to the mobile device via the FACH. If a DCH was established for the connection, the mobile device will switch to the new channel and return an RRC Connection Setup complete message. This message confirms the correct establishment of the connection, and is also used by the mobile device to inform the network about all optional features such as HSDPA parameters, HSUPA parameters, Dual-Carrier operation support and many other features that it supports. This has become very important as practically all features that were specified in the subsequent 3GPP Release versions are optional and hence the network cannot assume support for them on the mobile device's side. The network stores this information and makes use of it later on when it needs to decide on how to configure the channels to the mobile device.

After this basic procedure, the mobile device can then proceed to establish a higher layer connection between itself and the core network to request the establishment of a phone call or data connection (PDP context). A number of these scenarios are shown in Section 3.8.

3.4.6 The Uu Protocol Stack

UMTS uses channels on the air interface as was shown in the previous paragraph. As on any interface, several protocol layers are used for encapsulating the data and sending it to the correct recipient. In UMTS Release 99 the RNC is responsible for all protocol layers except for the physical layer which is handled in the Node-B. The only exception to this rule is the BCCH, which is under the control of the Node-B. This is due to the fact that the BCCH only broadcasts static information that does not have to be repeatedly sent from the RNC to the Node-B.

As is shown in Figure 3.17, higher layer Packet Data Units (PDUs) are delivered to the RNC from the core network. This can be user data like IP packets or voice frames, as well as Control Plane messages of the MM, CM, PMM or SM subsystems.

If the PDUs contain IP user data frames, the Packet Data Convergence Protocol (PDCP) can optionally compress the IP header. The compression algorithm used by UMTS is described in RFC 2507 [6]. Depending on the size of the transmitted IP frames, header compression

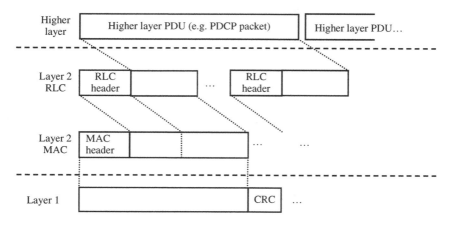

Figure 3.17   Preparation of user data frames for air interface (Uu) transmission

can substantially increase the transmission speed. Small frames in particular benefit from this as the IP header requires a proportionally oversized part of the frame. In practice, it can be observed that this functionality is not yet widely used.

The RLC layer is aware of the physical properties of the air interface and splits the packets it receives from higher layers for transmission over the air interface. This procedure is called segmentation and is required as PDCP frames that contain IP frames can be of variable size and can even be over 1000 bytes long. Frames on the air interface, however, are usually much smaller and are always of the same length. The length of those frames is determined by the spreading factor, the Transmission Time Interval (TTI, 10–80 milliseconds) and the applied coding scheme.

Just like GSM and GPRS, the UMTS radio network has been designed to send only small frames over the air interface. This has the advantage that in case of packet loss or corruption only a few bytes have to be retransmitted. Depending on the spreading factor and thus the speed of the connection, the frame sizes vary. For a 384 kbit/s bearer with a TTI of 10 milliseconds, for example, each data frame contains 480 bytes of user data. For a 64 kbit/s bearer with a TTI of 20 milliseconds, a frame contains only 160 bytes. For a voice call with a TTI of 20 milliseconds and a datarate of 12.2 kbit/s, a frame contains only 30 bytes.

If RLC frames are smaller than a frame on the air interface, it is also possible to concatenate several RLC frames for a single TTI. In the event that there is not enough data arriving from higher layers to fill an air interface frame, padding is used to fill the frame. Instead of padding the frame, it is also possible to use the remaining bits for RLC control messages.

Depending on the kind of user data one of three different RLC modes is used:

- The RLC transparent mode is used primarily for the transmission of circuit-switched voice channels and for the information that is broadcast on the BCCH and the PCCH. As the length of voice frames does not vary and as they are sent in a predefined format every 20 milliseconds, padding is also not necessary. Therefore, no adaptation or control functionality is required on the RLC layer, hence the use of the RLC transparent mode.
- The RLC non-acknowledged mode offers segmentation and concatenation of higher layer frames as described above. Furthermore, this mode allows marking of the beginning and end of layer 3 user data frames. Thus, it is possible to always completely fill an air interface frame regardless of the higher layer frames. As no acknowledgement for RLC frames is required in this mode, frames that are not received correctly or lost cannot be recovered on this layer.
- The third mode is the RLC acknowledged mode (AM), which is mostly used to transfer IP frames. In addition to the services offered by the non-acknowledged mode, this mode offers flow control and automatic retransmission of erroneous or missing blocks. Similar to Transmission Control Protocol (TCP), a window scheme is used to acknowledge the correct reception of a block. By using an acknowledgement window it is not necessary to wait for a reply for every transmitted block. Instead, further blocks can be transmitted up to the maximum window size. Up to this time, the receiver has the possibility to acknowledge frames, which in turn advances the window. If a block was lost, the acknowledgement bit in the window will not be set, which automatically triggers a retransmission. The advantage of this method is that the data flow, in general, is not interrupted by a transmission error. The RLC window size can be set between 1 and 2^{12} frames and is negotiated between the mobile device and RNC. This flexibility is the result of the experience gained with GPRS.

There, the window size was static; it offered only enough acknowledgement bits for 64 frames. In GPRS, this proved to be problematic, especially for coding schemes three and four during phases of increased block error rates (BLER), which lead to interrupted data flows as frames cannot be retransmitted quickly enough to advance the acknowledgement window.

Once the RLC layer has segmented the frames for transmission over the air interface and has added any necessary control information, the MAC layer performs the following operations:

- Selection of a suitable transport channel: As was shown in Figure 3.13, logical channels can be mapped onto different transport channels. User data of a DTCH can, for example, be transferred either on a DCH or on the FACH. The selection of the transport channel can be changed by the network at any time during the connection to increase or decrease the speed of the connection.
- Multiplexing of data on common and shared channels: The FACH can not only be used to transport RRC messages for different users but can also carry user data frames. The MAC layer is responsible for mapping all logical channels selected on a single transport channel and for adding a MAC header. The header describes, among other things, the subscriber for whom the MAC-frame is intended. This part of the MAC layer is called MAC c/sh (common/shared).
- For DCHs, the MAC layer is also responsible for multiplexing several data streams on a single transport channel. As can be seen in Figure 3.13, several logical user data channels (DTCH) and the logical signaling channel (DCCH) of a user are mapped onto a single transport channel. This permits the system to send user data and signaling information of the MM, PMM, CC and SM subsystems in parallel. This part of the MAC layer is called the MAC-d (dedicated).

Before the frames are forwarded to the physical layer, the MAC layer includes additional information in the header to inform the physical layer of the transport format it should select for transmission of the frames over the air interface. This so-called Transport Format Set (TFS) describes the combination of datarate, the TTI of the frame and the channel coding and puncturing scheme to be used.

For most channels, all layers described before are implemented in the RNC. The only exception is the physical layer, which is implemented in the Node-B. The Node-B, therefore, is responsible for the following tasks:

In order not to send the required overhead for error detection and correction over the Iub interface, channel coding is performed in the Node-B. This is possible as the header of each frame contains a TFS field that describes which channel encoder and puncturing scheme is to be used. UMTS uses the half-rate convolutional decoder already known from GSM as well as a new 1/3 rate and Turbocode coder for very robust error correction. These coders double or even triple the number of bits. It should be noted that puncturing is used to remove some of the redundancy again before the transmission to adapt the data to the fixed frame sizes of the air interface. Later, the physical layer performs the spreading of the original data stream by converting the bits into chips, which are then transferred over the air interface.

Finally, the modulator converts the digital information into an analog signal which is sent over the air interface. QPSK modulation is used for the UMTS Release 99 air interface, which

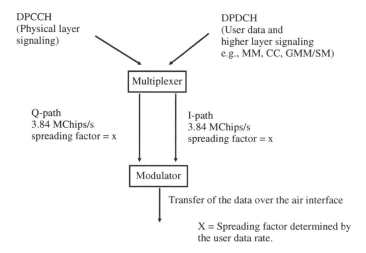

Figure 3.18   User data transmission in downlink direction via the complex I- and Q-path

transmits two chips per transmission step. This is done in the Node-B in downlink direction by sending one chip over the complex I-path and a second chip over the complex Q-path. As each path uses a fixed transmission rate of 3.84 MChips/s, the total datarate of the transmission is 2×3.84 MChips/s. The DPDCH and the DPCCH, which only use a small percentage of the frames, especially for low spreading factors, are thus time multiplexed in the downlink direction as shown in Figure 3.18.

For the uplink direction, which is the direction from the mobile device to the network, a slightly different approach was chosen. As in the downlink direction, QPSK modulation is used. Instead of multiplexing user and signaling data over both the I- and Q-path, user data is only sent on the I-path in the uplink. The Q-path is used exclusively for the transmission of the DPCCH, which carries layer 1 messages for power control (see 3GPP 25.211, 5.2.1 [5]). Thus, only one path is used for the transmission of user data in the uplink direction. This means that for an equal bandwidth in uplink and downlink direction, the spreading factor in uplink direction is only half that of the downlink direction.

Note: DPCCH is used only to transmit layer 1 signaling for power control. Control and signaling information of the MM, PMM, CC and SM subsystems that are exchanged between the mobile device and the MSC or SGSN are not transferred over the DPCCH but use the logical DCCH. This channel is sent together with the logical DTCH (user data) in the DPDCH transport channel (see Figures 3.13, 3.18 and 3.19).

The decision to use only the I-path for user data in the uplink direction was made for the following reason: While a dedicated channel has been assigned, there will be times in which no user data has to be transferred in the uplink direction. This is the case during a voice call, for example, while the user is not talking. During packet calls, it also happens quite frequently that no IP packets have to be sent in the uplink direction for some time. Thus, switching off the transmitter during that time could save battery capacity. The disadvantage of completely switching off the uplink transmission path, however, is that the interference caused by this can be heard, for example, in radio receivers which are close by. This can be observed with

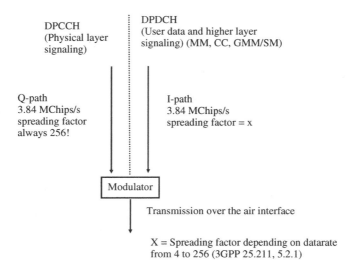

Figure 3.19   User data transmission via the I-path only

GSM mobile devices, for example, which use only some timeslots on the air interface and thus have to frequently activate and deactivate the transmitter. In UMTS, only the transmission on the I-path is stopped while the DPCCH on the Q-path continues to be transmitted. This is necessary, as power control and signal quality information need to be sent even if no user data is transferred to maintain the channel. The transmission power is thus only reduced and not completely switched off. The typical interference of GSM mobile devices in radio receivers which are close to the device, can thus, not be observed anymore with a UMTS mobile device.

3.5   The UMTS Terrestrial Radio Access Network (UTRAN)

3.5.1   Node-B, Iub Interface, NBAP and FP

The base station, called Node-B in the 3GPP standards, is responsible for all functions required for sending and receiving data over the air interface. This includes, as shown in Section 3.3, channel coding, spreading and despreading of outgoing and incoming frames as well as modulation. Furthermore, the Node-B is also responsible for the power control of all connections. The Node-B just receives a transmission quality target from the RNC for each connection and then decides on its own if it is necessary to increase or decrease the transmission power in both uplink and downlink directions to meet the target.

Size and capacity of a Node-B are variable. Typically, the Node-B is used in a sectorized configuration. This means that the 360 degrees coverage area of a Node-B is divided into several independent cells, each covering a certain area. Each cell has its own Cell-ID, scrambling code and OVSF tree. Each cell also uses its own directional antennas, which cover either 180 degrees (2 sector configuration) or 120 degrees (3 sector configuration). The capacity of the Iub interface, which connects the Node-B to an RNC, depends mainly on the number of sectors of the Node-B.

While GSM only uses some of the 64 kbit/s timeslots on an E-1 link to the base station, UMTS base stations require a much higher bandwidth. To deliver high datarates, Node-Bs were initially connected to the RNC with at least one E-1 connection (2 Mbit/s). If a Node-B served several sectors, multiple E-1 links were required. Owing to the rising datarates enabled by HSPA, even the aggregation of several E-1 lines has become insufficient. E-1 lines are also quite expensive, which is another limiting factor. In the meantime, however, high-speed DSL connections, fiber and microwave Ethernet lines have become available. Therefore, in most networks, E-1 connections to most Node-Bs have been replaced by links based on these technologies, with the IP protocol replacing the ATM transport protocol that was used over E-1 lines.

For regions with low voice and data traffic, a Node-B may only have a single cell. To decrease costs, transmission power is increased to cover a larger omnidirectional geographical area. From the outside, it is not always possible to distinguish between different configurations, as even a single cell can use a sectorized antenna installation. In the downlink direction, the signal is sent over all antennas and thus the total capacity of the base station is identical to a configuration that uses an omnidirectional antenna. In the uplink direction, a sectorized antenna installation has a big advantage, as the signal coming from the subscriber can be more easily received owing to the much higher gain of a sectorized antenna.

For very dense traffic areas, like streets in a downtown area, a Node-B microcell can be an alternative to a sectorized configuration. A microcell is usually equipped only with a single transceiver that covers only a very small area, for example, several dozens of meters of a street. As the necessary transmission power for such a small coverage area is very low, most network vendors have specialized micro Node-Bs with very compact dimensions. Usually, these micro Node-Bs are not much bigger than a PC workstation.

For the exchange of control and configuration messages on the Iub interface, the Node-B Application Part (NBAP) is used between the RNC and the Node-B. It has the following tasks:

- cell configuration;
- common channel management;
- dedicated channel management such as the establishment of a new connection to a subscriber;
- forwarding of signal and interference measurement values of common and dedicated channels to the RNC;
- control of the compressed mode, which is further explained in Section 3.7.1.

User data is exchanged between the RNC and the Node-Bs via the Frame Protocol (FP), which has been standardized for dedicated channels in 3GPP 25.427 [7]. The FP is responsible or the correct transmission and reception of user data over the Iub interface and transports user data frames in a format that the Node-B can directly transform into a Uu (air interface) frame. This is done by evaluating the Traffic Format Identifier (TFI), which is part of every FP frame. The TFI, among other things, instructs the Node-B to use a certain frame length (e.g., 10 milliseconds) and which channel coding algorithm to apply.

The FP is also used for the synchronization of the user data connection between the RNC and the Node-B. This is especially important for the data transfer in the downlink direction, as the Node-B has to send an air interface frame every 10, 20, 40 or 80 milliseconds to the mobile device. In order not to waste resources on the air interface and to minimize the delay it is necessary that all Iub frames arrive at the Node-B in time. To ensure this, the RNC and Node-B exchange synchronization information at the setup of each connection and also when the synchronization of a channel has been lost.

Finally, FP frames are also used to forward quality estimates from the Node-B to the RNC. These help the RNC during the soft handover state of a dedicated connection to decide which Node-B has delivered the best data frame for the connection. This topic is further discussed in Section 3.7.1.

3.5.2   The RNC, Iu, Iub and Iur Interfaces, RANAP and RNSAP

The heart of the UMTS radio network is the RNC. As can be seen in Figures 3.20 and 3.21, all interfaces of the radio network are terminated by the RNC.

In the direction of the mobile subscriber the Iub interface is used to connect several hundred Node-Bs to an RNC. During the first years after the initial deployment of UMTS networks, most Node-Bs were connected to the RNC via 2 Mbit/s E-1 connections either via fixed-line or microwave links. The number of links used per Node-B mainly depended on the number of sectors and number of frequencies used. Today, most sites are connected to the RNC via high-speed IP-based links over DSL, fiber or Ethernet microwave, with backhaul speeds of several hundred megabits, if required.

An initial disadvantage of IP-based links was that the transport protocol used on them, for example, Ethernet, was not synchronous and therefore could not be used by the base stations to synchronize themselves with the rest of the network. As a consequence, protocol extensions had to be developed to be able to recover a very precise clock signal from such links before they could be relied on as the only means to connect Node-Bs to the RNCs.

The RNC is connected to the core network via the Iu interface. As shown in Figure 3.1, UMTS continues to use independent circuit-switched and packet-switched networks for the following services:

For voice and video telephony services, the circuit-switched core network, already known from GSM, continues to be used. The MSC therefore remains the bridge between the core and

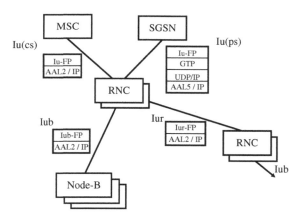

Figure 3.20   RNC protocols and interfaces for user data (user plane)

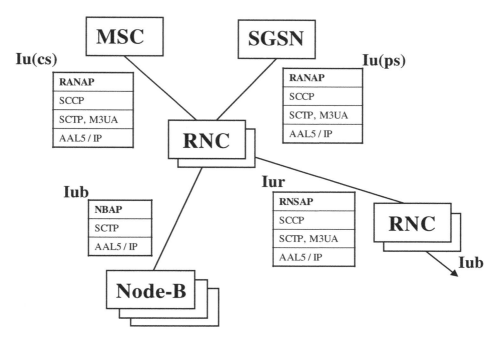

Figure 3.21   RNC protocols and interfaces used for signaling (control plane)

access network. Owing to the new functionalities offered by UMTS like, for example, video telephony, a number of adaptations were necessary on the interface that connects the MSC to the radio network. While in GSM, the Transcoding and Rate Adaptation Unit (TRAU) was logically part of the radio network, it was decided to put this functionality into the core network for UMTS. This was done because even in GSM, the TRAU is physically located near the MSC to save transmission resources as described in Section 1.5. In the Release 4 BICN network architecture, the UMTS TRAU is part of the MG. The interface between the MSC/TRAU and RNC has been named Iu(cs), which indicates that this interface connects the radio network to the circuit-switched part of the core network. The Iu(cs) interface therefore corresponds to the GSM A-interface and reuses many functionalities on the higher layers for MM and CC.

The BSSMAP protocol, which is used on the GSM A-interface, has been enhanced and modified for UMTS and renamed Radio Access Network Application Part (RANAP). In the standards, RANAP is described in 3GPP TS 25.413 [8] and forms the basis for MM, CC and the SM. Furthermore, RANAP is used by the MSC and SGSN for requesting the establishment and clearing of radio bearers (RABs) by the RNC.

In practice, the same MSC can be used with both the UTRAN (via the Iu(cs) interface) and the GSM radio network (via the A-interface). With GSM and the A-interface, the MSC can only handle 12.2 kbit/s circuit-switched connections for voice calls and 9.6 or 14.4 kbit/s channels for data calls. With UMTS and the Iu(cs) interface, the MSC is also able to establish 64 kbit/s circuit-switched connections to the RNC, which equals the speed of an ISDN B-channel. This functionality is mainly used for circuit-switched video telephony. By using optimized video and speech compression algorithms, which are part of the H.323 M (mobile) standard, this bearer ideally fulfills the tough requirements of this service concerning guaranteed bandwidth, delay time and optimized handling of handovers in the UTRAN. The Iu(cs) interface is, like all other interfaces in the UTRAN, based on ATM or IP and can thus use a number of different transmission technologies.

All packet-switched services, which in most cases require a connection to the Internet, are routed to and from the core network via the Iu(ps) interface. The functionality of this interface corresponds to the GSM/GPRS Gb interface that has been described in Chapter 2. SGSNs usually support both the Gb and Iu(ps) interface in a single node, which allows the use of only a single SGSN in a region to connect both types of radio network to the packet-switched core network.

Similar to the Iu(cs) interface, the higher layer GSM/GPRS signaling protocols were reused for UMTS and only slightly enhanced for the new capabilities of the radio network. For the lower layers, however, ATM or IP are used instead of the old Frame Relay Protocol.

The handling of user data has changed significantly for the SGSN with UMTS. In the GSM/GPRS system, the SGSN is responsible for processing incoming GTP packets from the GGSN and converting them into a BSSGP frame for transmission to the correct PCU and vice versa. In UMTS, this is no longer necessary as the SGSN can forward the GTP packets arriving from the GGSN directly to the RNC via an IP connection and can send GTP packets it receives from the RNC to the GGSN. The UMTS SGSN is thus no longer aware of the cell in which a subscriber is currently located. This change was made mainly for the following two reasons:

- The SGSN has been logically separated from the radio network and its cell-based architecture. It merely needs to forward GTP packets to the RNC, which then processes the packets and decides to which cell(s) to forward them. This change is especially important for the implementation of the soft handover mechanism, which is further described in Section 3.7.1, as the packet can be sent to a subscriber via several Node-Bs simultaneously. This complexity, however, is concealed from the SGSN as it is a pure radio network issue that is outside of the scope of a core network node. As a consequence, a UMTS SGSN is only aware of the current Serving RNC (S-RNC) of a subscriber.
- Using GTP and IP on the Iu(ps) interface on top of the ATM or an Ethernet transport layer significantly simplifies the protocol stack when compared to GSM/GPRS. The use of GTP and IP via the ATM Adaptation Layer (AAL) 5 or directly over Ethernet is also shown in Figure 3.20.

The SGSN is still responsible for the Mobility and Session Management (GMM/SM) of the subscribers as described in Chapter 2. Only a few changes were made to the protocol to address the specific needs of UMTS. One of those was made to allow the SGSN to request the setup of a radio bearer when a PDP context is established. This concept is not known in GSM/GPRS as 2G subscribers do not have any dedicated resources on the air interface. As described in Chapter 2, GPRS users are only assigned a certain number of timeslots for a short time, which are shared or immediately reused for other subscribers once there is no more data to transmit.

In Release 99 networks a different concept was used at first: Here, the RNC assigned a dedicated radio bearer (RAB) for a packet-switched connection in a very similar manner to circuit-switched voice calls. On the physical layer, this meant that the user got his own PDTCH and PDCCH for the packet connection. The bandwidth of the channel remained assigned to the subscriber even if not fully used for some time. When no data had to be sent in the downlink direction, DTX was used as described in Section 3.5.4. This reduced interference in the cell and helped the mobile device to save energy. The RNC could then select from different spreading factors during the setup of the connection to establish bearers with a guaranteed bandwidth of 8, 32, 64, 128 or 384 kbit/s. Later on, the RNC could then change the bandwidth at any time by assigning a different spreading factor to the connection, which was useful, for example, if the provided bandwidth was not sufficient or not fully used by the subscriber for some time. As the standard is very flexible in this regard, different network vendors had implemented different strategies for the radio resource management.

With 3GPP Release 5, the introduction of HSDPA has fundamentally changed this behavior for packet-switched connections as dedicated channels on the air interface proved to be inflexible in practice. Instead, data packets for several users can be sent over the air interface on a very fast shared channel. Mobile devices assigned to the shared channel continuously listen to shared control channels and when they receive the information that packets will be scheduled for them on the high-speed channel, they retrieve them from there.

Packet-switched data on a shared or dedicated physical channel and a circuit-switched voice or video call can be transmitted simultaneously over the air interface. Hence, one of the big limitations of GSM/GPRS in most deployed networks today has been resolved. To transmit several data streams for a user simultaneously, the RNC has to be able to modify a radio bearer at any time. If a circuit-switched voice call is added to an already existing packet-switched data connection, the RNC modifies the RAB to accommodate both streams. It is of course also possible to add a packet-switched data session to an ongoing circuit-switched voice call.

Another option for packet-switched data transfer over the air interface is to send data to the user via the FACH. For data in the uplink direction, the RACH is used. This is an interesting concept as the primary role of those channels is to carry signaling information for radio bearer establishments. As the capacity of those channels is quite limited and also has to be shared, the use of these common channels only makes sense for small amounts of data or as a fallback in case a user has not transmitted any data over a dedicated connection or the high-speed shared channel for some time. Another disadvantage of using common channels is that the mobile device is responsible for the mobility management and therefore, no seamless handover to other cells is possible (see Section 3.7.1). Therefore whenever the network detects that the amount of data transferred to or from a mobile device increases again, a dedicated or high-speed shared connection is quickly reestablished.

Independent of whether a dedicated, common or shared channel is assigned at the request of the SGSN during the PDP context activation, the bandwidth of the established connection depends on a number of factors. Important factors are, for example, the current load of a cell and the reception conditions of the mobile device at its current location. Furthermore, the number of available spreading codes and distance of the mobile device from the Node-B are also important factors.

The mobile device can also influence the assignment of radio resources during establishment of a PDP context. By using optional parameters of the 'at + cgdcont' command (see Section 2.9) the application can ask the network to establish a connection for a certain Quality of Service (QoS) level. The QoS describes properties for a new connection like the minimal acceptable datarate or the maximum delay time allowed, which the network has to guarantee throughout the duration of the connection. It is also possible to use different APNs to let the

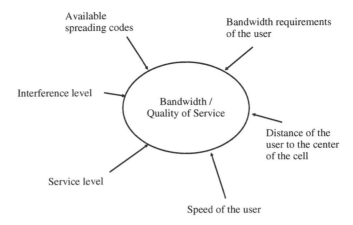

Figure 3.22   Factors influencing the Quality of Service and the maximum bandwidth of a connection

network automatically assign the correct QoS settings to a connection. The HLR therefore stores a QoS profile for each user that defines which APNs a user is allowed to use and which QoS level a user is allowed to request for a new connection (see Figure 3.22).

The assignment of resources on the air interface can also be influenced by the service level assigned to a user. By this process network operators can, for example, allocate higher maximum datarates to users who pay more for their subscription.

The Iur interface completes the overview of the UTRAN interfaces for this chapter. This interface connects RNCs with each other to support the soft handover procedure between Node-Bs that are connected to different RNCs. Further information about this topic can be found in Section 3.7.1. Furthermore, the Iur interface allows keeping up a packet-switched connection that is currently in the Cell-FACH, Cell-PCH or URA-PCH state if the functionality is used in the network. More information about the different connection states can be found in Section 3.5.4. The protocol responsible for these tasks is called the Radio Network Subsystem Application Part (RNSAP).

3.5.3   Adaptive Multirate (AMR) NB and WB Codecs for Voice Calls

For UMTS, it was decided to use the AMR codec for voice encoding. AMR was already introduced as an optional voice codec for GSM, as described in Chapter 1; however, in UMTS, AMR is mandatory. With AMR, the codec is no longer only negotiated at the establishment of a voice call but the system can change the codec every 20 milliseconds. As the name, AMR, already suggests, this functionality is quite useful in adapting to a number of changes that can occur during the lifetime of a call.

If the reception quality deteriorates during a call, the network can decide to use a voice codec with a lower bit rate. If the spreading factor of the connection is not changed, more bits of the bearer can then be used to add additional redundancy. A lower bit rate codec naturally lowers the quality of the voice transmission, which is however still better than a reduction in voice quality owing to an increased error rate. If the reception quality increases again during the connection, AMR returns to a higher bit rate codec and decreases the number of redundancy bits again.

Table 3.3 AMR codecs and bit rates

Codec mode	Bit rate (kbit/s)
AMR_12.20	12.20 (GSM EFR)
AMR_10.20	10.20
AMR_7.95	7.95
AMR_7.40	7.40 (IS-641)
AMR_6.70	6.70 (PDC-EFR)
AMR_5.90	5.90
AMR_5.15	5.15
AMR_4.75	4.75

Another application of AMR is to increase the number of simultaneous calls in a cell during cell congestion. In this case, a higher spreading factor is used for a connection, which only allows lower bit rate AMR codes to be used. This somewhat reduces the voice quality for the subscriber but increases the number of possible simultaneous voice calls.

Table 3.3 gives an overview of the different AMR codecs that have been standardized in 3GPP TS 26.071 [9]. While UMTS mobile devices have to support all bit rates, network support is optional.

A further significant voice quality improvement in mobile networks has been achieved by introducing the AMR-Wideband (AMR-WB) codec. Although previous AMR codecs, now also referred to as AMR-Narrowband (AMR-NB) codecs, digitized the voice signal up to an audible frequency of 3400 Hz, AMR-WB codecs have an upper limit of 7000 Hz. Therefore, much more of the audible spectrum is captured in the digitization process that results in a much improved sound quality at the receiver side. This, however, comes at the expense of the resulting data stream not being compatible anymore with the PCM G.711 codec used in fixed-line networks. In mobile networks, AMR-WB uses the G.722.2 codec with a datarate of 12.65 kbit/s over the air interface, which is about the same as is required for NB-AMR voice calls.

WB-AMR requires a Release 4 MSC in the network as the voice data stream can no longer be converted to PCM without reverting to a narrowband voice signal. For a connection between two 3G AMR-WB-compatible devices, no transcoding is necessary anymore. This is also referred to as Transcoding Free Operation (TrFO). On the radio network side, only software changes are required as the requirements for an AMR-WB bearer are very similar to that of AMR-NB.

If a call that is established from an AMR-WB-capable device terminates to a device that is only AMR-NB-capable, there are several possibilities for handling the connection. One way to establish the channel is to use AMR-WB from the originator to the MG of the MSC where the channel is transcoded into AMR-NB for the terminator. Another implementation possibility is for the MSC to wait with the bearer establishment of the originator until the capabilities of the terminator are known and then decide whether to establish a narrowband or a wideband connection to the originator. And finally, it is also possible to establish a wideband connection to the originator during call establishment and to modify the bearer if the MSC determines afterward that the terminating side supports only narrowband AMR.

While an AMR-WB connection is established, it can also become necessary to introduce a transcoder temporarily or to even permanently change from a wideband to a narrowband codec.

If the user uses the keypad to type DTMF (Dual-Tone Multi Frequency) tones, for example, to enter a password for the voice mail system, the corresponding tones are generated in the MSC. Therefore, the MSC interrupts the transparent end-to-end connection during this time to play the tone.

Although there are quite a number of AMR-WB-capable UMTS networks in practice today, the support in GSM networks is more limited. If a wideband voice call is handed over from UMTS to a GSM radio network not supporting the codec, it is therefore necessary to switch the connection from AMR-WB to AMR-NB during the handover. Unfortunately, this has a negative audible effect on the voice quality. As many networks are configured defensively and thus move a voice call from UMTS to GSM long before it would be necessary from a signal strength point of view, it can be advantageous to lock a device to 'UMTS-only' mode in areas well covered by a UMTS network to prevent such an inter-system handover from taking place.

Another situation in which a transcoder has to be deactivated during an ongoing connection is when an end-to-end connection is extended into a conference call with several parties. The conference call is established in the MG and therefore requires a transcoder, which is implemented today with the AMR-NB codec.

As can be seen from these scenarios, the introduction of AMR-WB and overcoming the frequency limit of PCM of 3400 Hz required much more than just the support of a new codec in the UTRAN radio network.

In practice, there are currently a number of additional limitations as to when AMR-WB can be used. In case several network operators in a country have an AMR-WB-capable network, calls between the two networks are still established with a narrowband codec. This is because the interconnection between the two networks is still based on E-1 links, which are only PCM capable over which the AMR-WB codec cannot be transported. Over the coming years, it is expected that network operators will replace their PCM interconnectivity with IP connections between their MGs, thus removing this limitation.

Fixed-line networks that have been migrated to using the Internet Protocol in recent years are also often wideband speech codec compatible. Unfortunately, fixed-line networks use the G.722 codec with a higher bitrate than the G.722.2 codec that is used in wireless networks. As a consequence, a transcoder would be needed between networks to convert between the two wideband codecs. Such transcoders are usually not yet in place today, which is why speech calls between fixed and wireless networks have to be transcoded to a narrowband PCM signal at the border between the two networks.

3.5.4 Radio Resource Control (RRC) States

The activity of a subscriber determines in which way data is transferred over the air interface between the mobile device and the network. In UMTS, a mobile device can therefore be in one of five RRC states as shown in Figure 3.23.

Idle State

In this state, a mobile device is attached to the network but does not have a physical or logical connection with the radio network. This means that the user is involved neither in a voice call nor in a data transfer. From the packet-switched core network point of view, the subscriber might still have an active PDP context (i.e., an IP address) even if no radio resources are

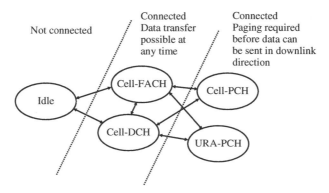

Figure 3.23   Radio Resource Control (RRC) states

assigned at the moment. Owing to the user's period of inactivity, the radio network has decided to release the radio connection. This means that if the user wants to send some data again (e.g., request a new web page), the mobile device needs to request the establishment of a new radio bearer.

Cell-DCH State

The Cell-DCH RRC state is used similarly to the GSM dedicated mode for circuit-switched voice calls. While in this state, a physical connection is established between the mobile device and the network. In the UTRAN this means that the mobile device has been assigned its own spreading code in the downlink direction and its own spreading and scrambling codes in the uplink direction.

The Cell-DCH state is also used for packet-switched connections. At first, the term contradicts the packet-switched approach. The advantage of packet-switched connections is that resources are shared and used only while they are needed to transfer data. In Release 99, however, air interface resources are not immediately freed once there is no more data to be transferred. If a subscriber does not send or receive data for some time, only control information is sent over the established channel. Other subscribers benefit indirectly from this owing to the reduced overall interference level of the cell during such periods as shown in Figure 3.24. If new data arrives, which has to be sent over the air interface, no new resources have to be assigned as the dedicated channel is still established. Once data is sent again the interference level increases again for other subscribers in the cell. This effect can be observed, for example, during web surfing sessions. For this application, the page requests of different subscribers are statistically multiplexed, which helps to counter the effect of the dedicated bearer approach without the negative effects of resource assignments as is the case in a GSM/GPRS radio network.

A mobile device that is assigned resources on High-Speed Downlink Shared Channels is also considered to be in Cell-DCH state, even though no dedicated resources for user data transfer exist on the air interface. Instead, the device has to share the high-speed channel with other users. The dedicated control of the connection, however, is very similar to that for subscribers who have dedicated air interface resources for circuit-switched voice and video

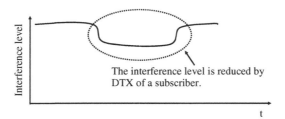

Figure 3.24 Discontinuous Transmission (DTX) on a dedicated channel reduces the interference for other subscribers

calls and packet-switched data transfers. Hence, the following section on mobility management in dedicated mode is valid for both setups.

By using signal measurements of the mobile device and the Node-B, it is possible to control the power level of each mobile device in a cell, which is a task that is shared between the Node-B and the RNC. By using the downlink of the PDCCH the network is able to instruct the mobile device to adapt its transmission power to the current conditions, that is, 1500 times a second. The rate at which power control is performed shows the importance of this factor for UMTS, interference being the major limiting factor in the number of connections that can be established simultaneously in a cell.

While in the Cell-DCH state, the mobile continuously measures the reception quality of the current and neighboring cells and reports the results to the network. On the basis of these values the RNC can then decide to start a handover procedure when required. While the GSM radio network uses a static reporting interval, a much more flexible approach was selected for UMTS. On the one hand, the RNC can instruct the mobile device similar to the GSM approach, to send periodic measurement reports. The measurement interval is now flexible and can be set by the network between 0.25 and 64 seconds. On the other hand, the network can also instruct the mobile device to send measurement reports only if certain conditions are met. Measurement reports are then sent only to the network if the measurement values reach a certain threshold. This removes some signaling overhead. Another advantage of this method for the RNC is that it has to process fewer messages for each connection compared to periodical measurement reports. In practice, both periodic and event-based measurement reports are used depending on the network vendor.

Cell-FACH State

The Cell-FACH state is mainly used when only a small amount of data needs to be transferred to or from a subscriber. In this mode, the subscriber does not get a dedicated channel but uses the FACH to receive data. As described in Section 3.4.5, the FACH is also used for carrying signaling data such as RRC connection setup messages for devices that have requested access to the network via the RACH. The FACH is a 'Common Channel' as it is not exclusively assigned to a single user. Therefore, the MAC header of each FACH data frame has to contain a destination ID consisting of the S-RNTI (Serving – Radio Network Temporary ID), which was assigned to a mobile device during connection establishment, and the ID of the S-RNC. The mobile devices have to inspect the header of each FACH

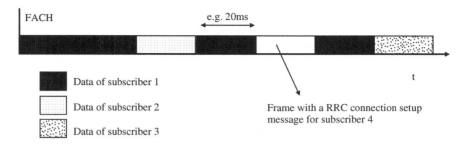

Figure 3.25 Data of different subscribers is time multiplexed on the FACH

data frame and only forward those frames to higher layers of the protocol stack that contain the mobile device's ID (see Figure 3.25). The approach of Cell-FACH RRC state is thus similar to Ethernet (802.11) and GSM/GPRS for packet-switched data transmission. If data is received in the downlink direction, no resources have to be assigned and the data can be sent to the subscriber more or less quickly depending on the current traffic load of the FACH. As several subscribers share the same channel, the network cannot ensure a certain datarate and constant delay time for any mobile device in the Cell-FACH state. Furthermore, it should be noted that the FACH usually uses a high spreading factor, which limits the total available bandwidth for subscribers on this channel. Typically, the FACH is configured as a 32 kbit/s channel.

Compared to the Cell-DCH state in which the mobility of the subscriber is controlled by the network, no such control has been foreseen for the Cell-FACH state. In the Cell-FACH state the mobile device itself is responsible for changing cells and this is called Cell Update instead of Handover. As the network does not control the Cell Update it is also not possible to ensure an uninterrupted data transfer during the procedure. Owing to these reasons the Cell-FACH RRC state is not suitable for real-time or streaming applications. In practice, it can be observed that a dedicated connection is established, even for small screen web browsing that requires only a little data, but is released again very quickly after the data transfer. More about the use of the different RRC states in operational networks can be found in Section 3.13.2.

The Cell-FACH state is also suitable for the transmission of MM and PMM signaling messages between the mobile device and the MSC or SGSN. As the mobile device already indicates the reason for initiating the connection to the network in the RRC Connection Setup message, the network can flexibly decide if a DCH is to be used for the requested connection or not. In practice, it can be observed that this flexibility is not always used as some networks always establish a DCH independently of the reason for the connection setup.

If the mobile device is in Cell-FACH state, uplink data frames are sent via the RACH whose primary task is to forward RRC Connection Setup Request messages. As described in Section 3.4.5, access to the RACH is a time-intensive procedure that causes some delay before the actual data frame can be sent.

There are two possibilities for a mobile device to change to the Cell-FACH state. As already discussed, the network can decide during the RRC Connection setup phase to use the FACH for MM/PMM signaling or user data traffic. In addition, it is possible to enter the Cell-FACH state from the Cell-DCH state. The RNC can decide to modify the radio bearer this way if, for example, no data has been sent or received by the mobile device for some time. This reduces the power consumption of the mobile device. As long as only small amounts of data are exchanged, the Cell-FACH state is usually maintained. If the data volume increases again, the network can immediately establish a new dedicated bearer and instruct the mobile device to enter Cell-DCH state to be able to transfer data more quickly.

Cell-PCH and URA-PCH States

The optional Cell-Paging Channel (Cell-PCH) RRC state and the UTRAN Registration Area – Paging Channel (URA-PCH) RRC state can be used to reduce the power consumption of the mobile device during extended times of inactivity. Similar to the Idle state, no resources are assigned to the mobile device. If data arrives for a subscriber from the network, the mobile device needs to be paged first. The mobile device then responds and implicitly changes back to Cell-FACH state.

As the name Cell-PCH already indicates, the subscriber is only paged in a single cell if new data from the core network arrives. This means that the mobile device has to send a Cell Update message to the RNC whenever it selects a new cell. In the URA-PCH state, the mobile only informs the RNC whenever it enters a new URA. Consequently, the paging message needs to be sent to all cells of the URA in case of incoming data (see Section 3.7.3).

The difference between the Cell-PCH and URA-PCH state compared to the Idle state is that the network and mobile device still maintain a logical connection and can restart data transfers in the uplink direction much quicker as no reestablishment of the core network connection, no new authentication procedure and no reactivation of ciphering is necessary. As the RRC states are managed by the RNC, the SGSN, as a core network component, has no information on the RRC state of the mobile device. Therefore, the SGSN simply forwards all incoming data packets from the GGSN to the RNC regardless of the current state of the mobile. If the mobile is currently in either the Cell-PCH or the URA-PCH state the RNC needs to buffer the packets, page the mobile device, wait for an answer and then establish a physical connection to the mobile device again. If the mobile device is in Cell-DCH or Cell-FACH state, on the other hand, the RNC can directly forward any incoming packets. The distinction between a logical and a physical connection has been made to separate the connection between the mobile device and core network (SGSN and MSC) on the one hand and the connection between the mobile device and the RNC on the other. The advantage of this concept is the decoupling of the MSC and SGSN from the properties and functionality of the radio network. Hence, it is possible to evolve the radio network and core network independently from each other.

Although early networks mainly used the Cell-DCH, Cell-FACH and Idle states for data connectivity, it can be observed today that many networks now also use the Cell-PCH and URA-PCH states to reduce battery consumption and signaling traffic.

As described in Chapter 2, the GSM/GPRS SGSN is aware of the state of a mobile device as the Idle, Ready and Standby states as well as the Ready timer is administered by the SGSN. Thus, a core network component performs radio network tasks like Cell Updates. This has the advantage that the SGSN is aware of the cell in which a subscriber is currently located, which can be used for supplementary location-dependent functionalities. The advantage of implementing the UMTS state management in the RNC is the distribution of this task over several RNCs and thus a reduction of the signaling load of the SGSN, as well as a clear separation between core network and radio access network responsibilities (Table 3.4).

Table 3.4 RNC and SGSN states

	RNC state	SGSN state
Idle	Not connected	Not connected
Cell-DCH	Connected, data is sent via the DCH or HS-DSCH	Connected
Cell-FACH	Connected, incoming data is sent immediately via the FACH (Common Channel)	Connected
Cell-PCH	Connected, but subscriber has to be paged and needs to reply before data can be forwarded. Once the answer to the paging has been received, the subscriber is put in either Cell-FACH or Cell-DCH state	Connected
URA-PCH	Same as Cell-PCH. Furthermore, the network only needs to be informed of a cell change in case the mobile device is moved into a cell, which is part of a different UTRAN Registration Area	Connected

3.6 Core Network Mobility Management

From the point of view of the MSC and the SGSN, the mobile device can be in any of the MM or PMM states described below. The MSC knows the following MM states:

- **MM Detached**. The mobile device is switched off and the current location of the subscriber is unknown. Incoming calls for the subscriber cannot be forwarded to the subscriber and are either rejected or forwarded to another destination if the Call Forward Unreachable (CFU) supplementary service is activated.
- **MM Idle**. The mobile device is powered on and has successfully attached to the MSC (see Attach procedure). The subscriber can at any time start an outgoing call. For incoming calls, the mobile device is paged in its current Location Area.
- **MM Connected**. The mobile device and MSC have an active signaling and communication connection. Furthermore, the connection is used for a voice or a video call. From the point of view of the RNC, the subscriber is in the Cell-DCH RRC state as this is the only bearer that supports circuit-switched connections.

The SGSN implements the following PMM states:

- **PMM Detached**. The mobile device is switched off and the location of the subscriber is unknown to the SGSN. Furthermore, the mobile device cannot have an active PDP context, that is, no IP address is currently assigned to the subscriber.
- **PMM Connected**. The mobile device and the SGSN have an active signaling and communication connection. The PMM connected state is only maintained while the subscriber has an active PDP context, which effectively means that the GGSN has assigned an IP address for the connection. In this state, the SGSN simply forwards all incoming data packets to the S-RNC. In contrast to GSM/GPRS, the UMTS SGSN is aware only of the S-RNC for the subscriber and not of the current cell. This is due not only to the desired separation of radio network and core network functionality, but also due to the soft handover mechanism (see Section 3.7). The SGSN is also not aware of the current RRC state of the mobile device. Depending on the QoS profile, the network load, the current data transfer activity and the

required bandwidth, the mobile device can be either in Cell-DCH, Cell-FACH, Cell-PCH or URA-PCH state.

• **PMM Idle**. In this state, the mobile device is attached to the network but no logical signaling connection is established with the SGSN. This can be the case, for example, if no PDP context is active for the subscriber. If a PDP context is established, the RNC has the possibility to modify the RRC state of a connection at any time. This means, that the RNC, for example, can decide after a period of inactivity of the connection to set the mobile device into the RRC Idle state. Subsequently, as the RNC no longer controls the mobility of the subscriber, it requests the SGSN to set the connection into PMM Idle State as well. Therefore, even though the subscriber no longer has a logical connection to either the RNC or the SGSN, the PDP context remains active and the subscriber can keep the assigned IP address. For the SGSN, this means that if new data arrives for the subscriber from the GGSN, a new signaling and user data connection has to be established before the data can be forwarded to the mobile device.

3.7 Radio Network Mobility Management

Depending on the MM state of the core network, the radio network can be in a number of different RRC states. How the mobility management is handled in the radio network depends on the respective state. Table 3.5 gives an overview of the MM and PMM states in the core network and the corresponding RRC states in the radio network.

3.7.1 Mobility Management in the Cell-DCH State

For services like voice or video communication it is very important that little or no interruption of the data stream occurs during a cell change. For these services, only the Cell-DCH state can be used. In this state, the network constantly controls the quality of the connection and is able to redirect the connection to other cells if the subscriber is moving. This procedure is called handover or handoff.

A handover is controlled by the RNC and triggered based on measurement values of the quality of the uplink signal measured by the base station and measurement reports on downlink quality sent by the mobile device. Measurement reports can be periodic or event triggered. Different radio network vendors use different strategies for measurement reporting. Unlike in GSM where only the signal strength, referred to as Received Signal Strength Indication (RSSI),

Table 3.5 Core network and radio network states

MM states and possible RRC states	MM Idle	MM connected	PMM Idle	PMM connected
Idle	×	–	×	–
Cell-DCH	–	×	–	×
Cell-FACH	–	–	–	×
Cell-PCH	–	–	–	×
URA-PCH	–	–	–	×

is used for the decision, UMTS needs additional criteria as neighboring base stations transmit on the same frequency. A mobile device thus not only receives the signal of the current serving base station but also the signals of neighboring base stations, which, from its point of view, are considered to be noise. In UMTS, the following values are used:

- **RSSI**. To describe the total signal power received in milliwatts. The value is usually expressed in dBm (logarithmic scale) and typical values are −100 dBm for a low signal level to −60 dBm for a very strong signal level.
- Received Signal Code Power (**RSCP**). The power the pilot channel of a base station is received with. The RSCP can be used, for example, to detect UMTS cell edge scenarios where no neighboring UMTS cell is available to maintain the connection. In this case, the network takes action when the RSCP level falls below a network-operator-defined threshold. If the network is aware of neighboring GSM cells, it can activate the compressed mode so that the mobile device can search and report neighboring GSM cells to which the connection could be handed over.
- **EcNo**. The received energy per chip (Ec) of the pilot channel divided by the total noise power density (No). In other words, the EcNo is the RSCP divided by the RSSI. The better this value the better the signal can be distinguished from the noise. The EcNo is usually expressed in decibels as it is a relative value. The value is negative as a logarithmic scale is used and the RSCP is smaller than the total received power. The EcNo can be used to compare the relative signal quality of different cells on the same frequency. Their relative difference to each other, independent of their absolute signal strengths, can then be used, for example, to decide which of them should be the serving cell.

In UMTS a number of different handover variants have been defined.

Hard Handover

This is as shown in Figure 3.26. This kind of handover is very similar to the GSM handover. By receiving measurement results from the mobile device of the active connection and measurement results of the signal strength of the broadcast channel of the neighboring cells, the RNC

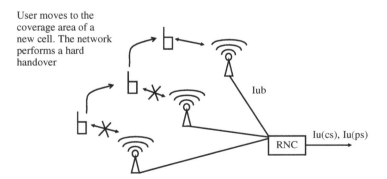

Figure 3.26   UMTS hard handover

is able to recognize if a neighboring cell is more suitable for the connection. To redirect the call into the new cell, a number of preparatory measures have to be performed in the network before the handover is executed. This includes, for example, the reservation of resources on the Iub interface and, if necessary, also on the Iur interface. The procedure is similar to the resource reservation of a new connection.

Once the new connection is in place, the mobile device receives a command over the current connection to change into the new cell. The handover command contains, among other parameters, the frequency of the new cell and the new channelization and scrambling code to be used. The mobile device then suspends the current connection and attempts to establish a connection in the new cell. The interruption of the data stream during this operation is usually quite short and takes about 100 milliseconds on average, as the network is already prepared for the new connection. Once the mobile device is connected to the new cell the user data traffic can resume immediately. This kind of handover is called UMTS hard handover, as the connection is briefly interrupted during the process.

Soft Handover

With this kind of handover, a voice call is not interrupted at any time during the procedure. On the basis of signal quality measurements of the current and neighboring cells, the RNC can decide to set the mobile device into soft handover state. All data from and to the mobile device will then be sent and received not only over a single cell but also over two or even more cells simultaneously. All cells that are part of the communication are put into the so-called Active Set of the connection. If a radio connection of a cell in the Active Set deteriorates, it is removed from the connection. Thus it is ensured that despite the cell change, the mobile device never losses contact with the network. The Active Set can contain up to six cells at the same time, although in operational networks no more than two or three cells are used at a time. Figure 3.27 shows a soft handover situation with three cells.

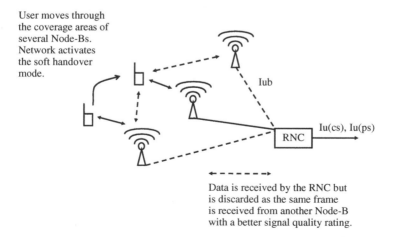

Figure 3.27 Connections to a mobile device during a soft handover procedure with three cells

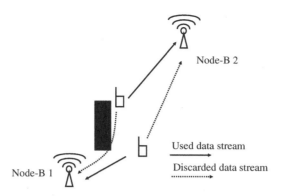

Figure 3.28   Soft handover reduces the energy consumption of the mobile due to lower transmission power

The soft handover procedure has a number of advantages over the hard handover described before. As no interruption of the user data traffic occurs during the soft handover procedure the overall connection quality increases. As the soft handover procedure can be initiated while the signal quality of the current cell is still acceptable the possibility of a sudden loss of the connection is reduced.

Furthermore, the transmission power and hence the energy consumption of the mobile device can be reduced in some situations as shown in Figure 3.28. In this scenario, the subscriber first roams into an area in which it has a good coverage by cell 1. As the subscriber moves, there are times when buildings or other obstacles are in the way of the optimal transmission path to cell 1. As a consequence, the mobile device needs to increase its transmission power. If the mobile device is in soft handover state however, cell 2 still receives a good signal from the mobile device and can thus compensate for the deterioration of the transmission path to cell 1. As a consequence, the mobile device is not instructed to increase the transmission power. This does not mean, however, that the connection to cell 1 is released immediately, as the network speculates on an improvement of the signal conditions.

As the radio path to cell 1 is not released, the RNC receives the subscriber's data frames from both cell 1 and cell 2 and can decide, on the basis of the signal quality information included in both frames, that the frame received from cell 2 is to be forwarded into the core network. This decision is made for each frame, that is, the RNC has to make a decision for every connection in handover state every 10, 20, 40 or 80 milliseconds, depending on the size of the radio frame.

In the downlink direction, the mobile device receives identical frames from cell 1 and cell 2. As the cells use different channelization and scrambling codes the mobile device is able to separate the two data streams on the physical layer (see Figure 3.29). This means that the mobile device has to decode the data stream twice, which of course slightly increases the power consumption as more processing power is required.

From the network point of view, the soft handover procedure has an advantage because the mobile device uses less transmission power compared to a single cell scenario in order to reach at least one of the cells in the Active Set, and so interference is reduced in the uplink direction. This increases the capacity of the overall system, which in turn increases the number of subscribers that can be handled by a cell.

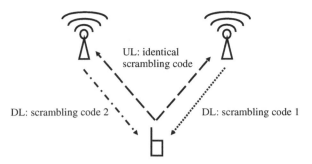

Figure 3.29   Use of scrambling codes while a mobile device is in soft handover state

On the other hand, there are some disadvantages for the network, as in the downlink direction, data has to be duplicated so that it can be sent over two or even more cells. In the reverse direction, the RNC receives a copy of each frame from all cells of the Active Set. Thus, the capacity that has to be reserved for the subscriber on the different interfaces of the radio network is much higher than that for a subscriber who only communicates with a single cell. Therefore, good network planning tries to ensure that there are no areas of the network in which more than three cells need to be used for the soft handover state.

A soft handover gets even more complicated if cells need to be involved that are not controlled by the S-RNC. In this case, a soft handover is only possible if the S-RNC is connected to the RNC that controls the cell in question. RNCs in that role are called the Drift RNCs (D-RNC). Figure 3.30 shows a scenario that includes an S-RNC and D-RNC. If a foreign cell needs to be included in the Active Set, the S-RNC has to establish a link to the D-RNC via the Iur interface. The D-RNC then reserves the necessary resources to its cell on the Iub interface and acknowledges the request. The S-RNC in turn informs the mobile device to include the

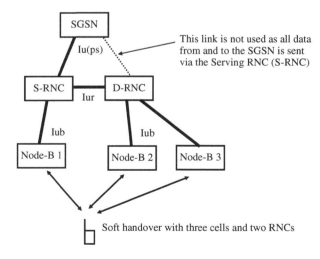

Figure 3.30   Soft handover with S-RNC and D-RNC

new cell in its Active Set via an 'Update Active Set' message. From this point onward, all data arriving at the S-RNC from the core network will be forwarded via the Iub interface to the cells that are directly connected to the S-RNC and also via the Iur interface to all D-RNCs, which control a cell in the Active Set. These in turn forward the data packets to the cells under their control. In the reverse direction, the S-RNC is the point of concentration for all uplink packets as the D-RNCs forward all incoming data packets for the connection to the S-RNC. It is then the task of the S-RNC to decide which of the packets to use on the basis of the signal quality indications embedded in each frame.

A variation of the soft handover is the so-called softer handover, which is used when two or more cells of the same Node-B are part of the Active Set. For the network, the softer handover has the advantage that no additional resources are necessary on the Iub interface as the Node-B decides which of the frames, received from the mobile device via the different cells, are to be forwarded to the RNC. In the downlink direction, the point of distribution for the data frames is also the Node-B, that is, it duplicates the frames it receives from the RNC for all cells that are part of the Active Set of a connection.

One of the most important parameters of the GSM air interface is the timing advance. Mobile devices that are farther away from the base station have to start sending their frames earlier compared to mobile devices that are closer to the base station, owing to the time it takes the signal to reach the base station. This is called timing advance control. In UMTS controlling the timing advance is not possible. This is because while a mobile device is in soft handover state, all Node-Bs of the Active Set receive the same data stream from the mobile device. The distance of the mobile device to each Node-B is different though, and thus each Node-B receives the data stream at a slightly different time. For the mobile device, it is not possible to control this by starting to send data earlier, as it only sends one data stream in the uplink direction for all Node-Bs. Fortunately, it is not necessary to control the timing advance in UMTS as all active subscribers transmit simultaneously. As no timeslots are used, no collisions can occur between the different subscribers. To ensure the orthogonal nature of the channelization codes of the different subscribers it would be necessary, however, to receive the data streams of all mobile devices synchronously. As this is not possible, an additional scrambling code is used for each subscriber that is multiplied by the data that has already been treated with the channelization code. This decouples the different subscribers and thus a time difference in the arrival of the different signals can be tolerated.

The time difference of the multiple copies of a user's signal is very small compared to the length of a frame. While the transmission time of a frame is 10, 20, 40 or 80 milliseconds, the delay experienced on the air interface of several Node-Bs is less than 0.1 milliseconds even if the distances vary by 30 km. Thus, the timing difference of the frames on the Iub interface is negligible.

If a subscriber continues to move away from the cell in which the radio bearer was initially established, there will be a point at which not a single Node-B of the S-RNC is part of the transmission chain anymore. Figure 3.31 shows such a scenario. As this state is a waste of radio network resources, the S-RNC can request a routing change from the MSC and the SGSN on the Iu(cs)/Iu(ps) interface. This procedure is called a Serving Radio Network Subsystem (SRNS) Relocation Request. If the core network components agree to perform the change, the D-RNC becomes the new S-RNC and the resources on the Iur Interface can be released.

An SRNS relocation is also necessary if a handover needs to be performed due to degrading radio conditions and no Iur connection is available between two RNCs. In this case, it is not the

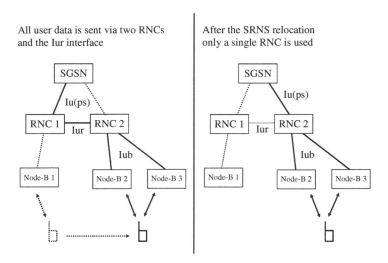

Figure 3.31   SRNS relocation procedure

optimization of radio network resources that triggers the procedure but the need to maintain the radio bearer. Along with the SRNS relocation it is necessary to perform a hard handover into the new cell, as a soft handover is not possible due to the missing Iur interface.

When the first GSM networks were built at the beginning of the 1990s, many earlier generation networks already covered most parts of the country. The number of users was very small though, so it was not immediately necessary to reach the same coverage area with GSM as well. When the first UMTS networks became operational, the situation had changed completely. Owing to the enormous success of GSM, most people in Europe already possessed a mobile phone. As network deployment is a lengthy and costly process it was not possible to provide countrywide coverage for UMTS right from the start. Therefore, it was necessary to ensure a seamless integration of UMTS into the already existing GSM infrastructure. This meant that right from the beginning the design of UMTS mobile devices had to incorporate GSM and GPRS. Thus, while a user roams in an area covered by UMTS, both voice calls and packet data are handled by the UMTS network. If the user roams into an area that is only covered by a 2G network, the mobile device automatically switches over to GSM, and packet-switched connections use the GPRS network. In order not to interrupt ongoing voice or data calls, the UMTS standards also include procedures to handover an active connection to a 2G network. This handover procedure is called intersystem handover (see Figure 3.32).

In UMTS there are a number of different possibilities to perform an intersystem handover. The first intersystem handover method is the blind intersystem handover. In this scenario, the RNC is aware of GSM neighboring cells for certain UMTS cells. In the event of severe signal quality degradation, the RNC reports to the MSC or SGSN that a handover into a 2G cell is necessary. The procedure is called a 'blind handover' because no measurement reports of the GSM cell are available for the handover decision.

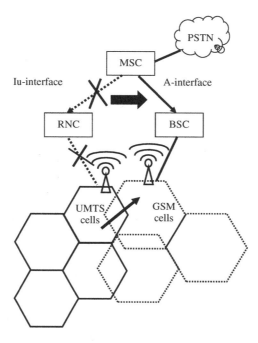

Figure 3.32   3G to 2G handover

The advantage of this procedure is the simple implementation in the network and in the mobile devices. However, there are a number of downsides as well:

- The network does not know if the GSM cell can be found by the mobile device.
- The mobile device and the target GSM cell are not synchronized. This considerably increases the time it takes for the mobile device to contact the new cell once the handover command has been issued by the network. For the user, this means that during a voice call he might notice a short interruption of the voice path.
- If a UMTS cell has several GSM neighboring cells, as shown in Figure 3.33, the RNC has no means to distinguish which would be the best one for the handover. Thus, such a network layout should be avoided. In practice, however, this is often not possible.

To improve the success rate and quality of intersystem handovers, the UMTS standards also contain a controlled intersystem handover procedure that is usually used in practice today. To perform a controlled handover, UMTS cells at the border of the coverage area inform mobile devices about both UMTS and GSM neighboring cells. A mobile device can thus measure the signal quality of neighboring cells of both systems during an active connection. As described before, there are several ways to report the measurement values to the RNC. The RNC, in turn, can then decide to request an intersystem handover from the core network on the basis of current signal conditions rather than purely guessing that a certain GSM cell is suitable for the handover.

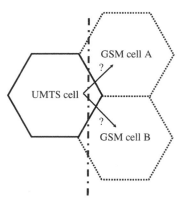

Figure 3.33   A UMTS cell with several GSM neighboring cells presents a problem for blind intersystem handovers

Performing neighboring cell signal strength measurements is quite easy for UMTS cells as they usually use the same frequency as the current serving cell. The mobile device thus merely applies the primary codes of neighboring cells on the received signal to get signal strength indications for them. For the mobile device this means that it has to perform some additional computing tasks during an ongoing session. For neighboring GSM cells, the process is somewhat more complicated as they transmit on different frequencies and thus cannot be received simultaneously with the UMTS cells of the Active Set. The same problem occurs when signal quality measurements need to be made for UMTS cells that operate on a different frequency to increase the capacity of the radio network. The only way for the mobile device to perform measurements for such cells, therefore, is to stop transmitting and receiving frames in a predefined pattern to perform measurements on other frequencies. This mode of operation is referred to as compressed mode and is activated by the RNC, if necessary, in the mobile device and all cells of the Active Set of a connection. The standard defines three possibilities for implementing compressed mode. While network vendors can choose which of the options described below they want to implement, the support of all options is required in the mobile device:

- **Reduction of the spreading factor**. For this option, the spreading factor is reduced for some frames. Thus, more data can be transmitted during these periods, which increases the speed of the connection. This allows the insertion of short transmission gaps for interfrequency measurement purposes without reducing the overall speed of the connection. As the spreading factor changes, the transmission power has to be increased to ensure an acceptable error rate.
- **Puncturing**. After the channel coder has added error correction and error detection bits to the original data stream, some of them are removed again to have time for interfrequency measurements. To keep the error rate of the radio bearer within acceptable limits, the transmission power has to be increased.
- **Reduction of the number of user data bits per frame**. As fewer bits are sent per frame, the transmission power does not have to be increased. The disadvantage is the reduced user datarate while operating in compressed mode.

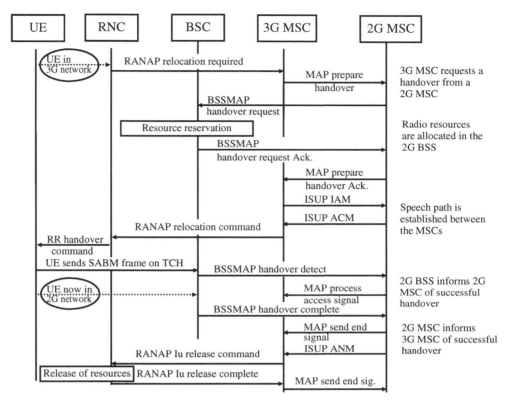

Figure 3.34   3G–2G intersystem hard handover message flow

The goal of the measurements while in compressed mode is to be able to successfully decode the Frequency Correction Channel (FCCH) and the Synch Channel (SCH) of the surrounding GSM cells. For further information on these channels see Section 1.7.3.

Figure 3.34 shows how an intersystem handover from UMTS to GSM is performed. The procedure starts on the UTRAN side just like a normal inter-MSC handover by the RNC sending an SRNS relocation request. As the SRNS relocation is not known in GSM, the 3G MSC uses a standard 2G prepare handover message to initiate the communication with the 2G MSC. Thus, for the 2G MSC, the handover looks like a normal GSM to GSM handover and is treated accordingly.

3.7.2   Mobility Management in Idle State

While in Idle state, the mobile device is passive, that is, no data is sent or received. Nevertheless, there are a number of tasks that have to be performed periodically by the mobile device.

To be able to respond to incoming voice calls, short messages, MMS messages, etc., the PCH is monitored. If a paging message is received that contains the subscriber's International Mobile Subscriber Identity (IMSI) or Temporary Mobile Subscriber Identity (TMSI), the mobile device reacts and establishes a connection with the network. As the monitoring of the paging channel consumes some power, subscribers are split into a number of groups based on their IMSI (paging group). Paging messages for a subscriber of each group are then broadcast at certain intervals. Thus, a mobile device does not have to listen for incoming paging messages all the time but only at a certain interval. At all other times, the receiver can be deactivated and thus battery capacity can be saved. A slight disadvantage of this approach is, however, that the paging procedure takes a little bit longer than if the PCH was constantly monitored by the mobile device.

In the event that the subscriber has an active PDP context while the mobile device is in Idle state, the network will also need to send a paging message in case of an incoming IP frame. Such a frame could, for example, originate from a messaging application. When the mobile device receives a paging message for such an event, it has to reestablish a logical connection with the network before the IP frame can be forwarded.

In Idle state, the mobile device is responsible for mobility management; that is, changing to a more suitable cell when the user is moving. As the network is not involved in the decision-making process, the procedure is called cell reselection.

While in Idle state, no physical or logical connection exists between the radio network and the mobile device. Thus, it is necessary to reestablish a physical connection over the air interface if data needs to be transported again. For the circuit-switched part of the network the RRC Idle state, therefore, implies that no voice connection is established. For the SGSN, on the other hand the situation is different. A PDP context can still be established in Idle state, even though no data can be sent or received. To transfer data again, the mobile device needs to reestablish the connection, and the network then either establishes a DCH or uses the FACH for the data exchange. In practice, it can be observed that the time it takes to reestablish a channel is about 2.5–3 seconds. Therefore, the mobile device should only be put into Idle state after a prolonged period of inactivity as this delay has a negative impact on the Quality of Experience of the user, for example, during a web-browsing session. Instead of an instantaneous reaction to the user clicking on a link, the user notices an undesirably long delay before the new page is presented.

While a mobile device is in Idle state, the core network is not aware of the current location of the subscriber. The MSC is only aware of the subscriber's current location area. A location area usually consists of several dozen cells and therefore, it is necessary to page the subscriber for incoming calls. This is done via a paging message that is broadcast on the PCH in all cells of the location area. This concept has been adopted from GSM without modifications and is described in more detail in Section 1.8.1.

From the point of view of the SGSN, the same concept is used in case an IP packet has to be delivered while the mobile device is in Idle state. For the packet-switched part of the network, the cells are divided into routing areas (RA). An RA is a subset of a location area but most operators use only a single routing area per location area. Similar to the location area, the routing area concept was adopted from the 2G network concept without modification.

In the event that the mobile device moves to a new cell that is part of a different location or routing area, a location or a routing area update has to be performed. This is done by establishing a signaling connection, which prompts the RNC to set the state of the mobile device to Cell-DCH or Cell-FACH. Subsequently, the location or routing area update is performed transparently over the established connection with the MSC and the SGSN. Once the updates are performed, the mobile device returns to Idle state.

3.7.3   Mobility Management in Other States

In Cell-FACH, Cell-PCH or URA-PCH state, the mobile device is responsible for mobility management and thus for cell changes. The big difference between these states and the Idle state is that a logical connection exists between the mobile device and the radio network when a packet session is active. Depending on the state, the mobile device has to perform certain tasks after a cell change.

In Cell-FACH state the mobile device can exchange data with the network at any time. If the mobile device performs a cell change it has to inform the network straight away via a cell update message. Subsequently, all data is exchanged via the new cell. If the new cell is connected to a different RNC, the cell update message will be forwarded to the S-RNC of the subscriber via the Iur interface. As the mobile device has a logical connection to the network, no Location or Routing Area Update is necessary if the new cell is in a different area. This means that the core network is not informed that the subscriber has moved to a new Location or Routing Area. This is, however, not necessary as the S-RNC will forward any incoming data over the Iur interface via the D-RNC to the subscriber. In practice, changing the cell in Cell-FACH state results in a short interruption of the connection, which is tolerable as this state is not used for real-time or streaming services.

If the new serving cell is connected to an RNC that does not have an Iur interface to the S-RNC of the subscriber, the cell update will fail. As the new RNC cannot inform the S-RNC of the new location of the subscriber, it will reset the connection and the mobile device automatically defaults to Idle state. To resume data transmission, the mobile device then performs a location update with the MSC and SGSN as shown in Figure 3.35.

As the SGSN detects during the location and routing area update that there is still a logical connection to a different RNC, it sends a message to the previous RNC that the subscriber is no longer under its control. Thus, it is ensured that all resources that are no longer needed to maintain the connection are released.

From the MM point of view, the Cell-PCH is almost identical to the Cell-FACH state. The only difference is that no data can be transmitted to the mobile device in Cell-PCH state. If data is received for the mobile device while being in Cell-PCH state, the RNC needs to page

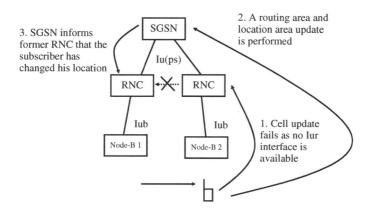

Figure 3.35   Cell change in PMM connected state to a cell that cannot communicate with the S-RNC

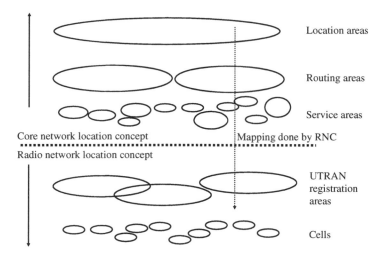

Figure 3.36   Location concepts of radio- and core network

the mobile device first. Once the mobile device responds, the network can then put the mobile device in Cell-DCH or Cell-FACH state and the data transfer can resume.

And finally, there is the URA-PCH state, which only requires a cell update message to be sent to the network if the subscriber roams into a new URA. The URA is a new concept that has been introduced with UMTS. It refines a location area as shown in Figure 3.36.

The core network is not aware of UTRAN Registration Areas. Furthermore, even single cells have been abstracted into so-called Service Areas. This is in contrast to a GSM/GPRS network, where the MSC and SGSN are aware of the location area and even the cell-ID in which the mobile device is located during an active connection. In UMTS, the location area does not contain single cells but one or more service areas. It is possible to assign only a single cell to a service area to be able to better pinpoint the location of a mobile device in the core network. By this abstraction it was possible to clearly separate the location principles of the core network, which is aware of Location Areas, Routing Areas and Services Areas and the radio network that deals with UTRAN registration areas and single cells. Core network and radio network are thus logically decoupled. The mapping between the location principles of core and radio network is done at the interface between the two networks, the RNC.

3.8   UMTS CS and PS Call Establishment

To establish a circuit-switched or packet-switched connection, the mobile device has to contact the network and request the establishment of a session. The establishment of the user data bearer is then performed in several phases.

As a first step, the mobile device needs to perform an RRC Connection Setup procedure as shown in Figure 3.37, to establish a signaling connection. The procedure itself was introduced in Figure 3.16. The goal of the RRC connection setup is to establish a temporary radio channel that can be used for signaling between the mobile device, the RNC and a core network node.

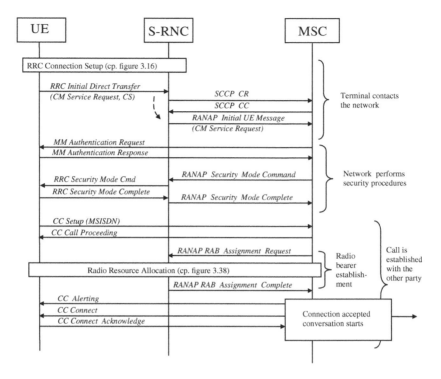

Figure 3.37   Messaging for a mobile-originated voice call (MOC)

The RNC can choose either to assign a dedicated channel (Cell-DCH state) or to use the FACH (Cell-FACH state) for the subsequent exchange of messages.

If a circuit-switched connection is established as shown in Figure 3.37, the mobile device sends a CM Service Request DTAP message (see Section 1.4) over the established signaling connection to the RNC that transparently forwards the message to the MSC. DTAP messages are exchanged between the RNC and the MSC via the connection-oriented Signaling Connection and Control Part (SCCP) protocol. Therefore, the RNC has to establish a new SCCP connection before the message can be forwarded.

Once the MSC has received the CM Service Request message, it verifies the identity of the subscriber via the attached TMSI or IMSI. This is done in a challenge and response procedure similar to GSM. In addition to the mobile device authentication already known from GSM, a UMTS network has to authenticate itself to the user to protect against air interface eavesdropping with a false base station. Once the authentication procedure has been performed, the MSC activates the ciphering of the radio channel by issuing a Security Mode command. Optionally, the MSC afterward assigns a new TMSI to the subscriber, which, however, is not shown in Figure 3.37, for clarity. Details of the UMTS authentication and ciphering process are described in Section 3.9.

After successful authentication and activation of the encrypted radio channel, the mobile device then proceeds to inform the MSC of the exact reason of the connection request. The CC setup message contains, among other things, the telephone number (MSISDN) of the destination. If the MSC approves the request, it returns a Call Proceeding message to the mobile device and starts two additional procedures simultaneously.

At this point, only a signaling connection exists between the mobile device and the radio network, which is not suitable for a voice call. Thus, the MSC requests the establishment of a speech path from the RNC via an RAB Assignment Request message. The RNC proceeds by reserving the required bandwidth on the Iub interface and instructs the Node-B to allocate the necessary resources on the air interface. Furthermore, the RNC also establishes a bearer for the speech path on the Iu(cs) interface to the MSC. As a dedicated radio connection was already established for the signaling in our example, it is only modified by the Radio Resource Allocation procedure (Radio Link Reconfiguration). The reconfiguration includes, for example, the allocation of a new spreading code as the voice bearer requires a higher bandwidth connection than a slow signaling connection. If the RNC has performed the signaling via the FACH (Cell-FACH state), it is necessary at this point to establish a DCH and to move the mobile device over to a dedicated connection. Figure 3.38 shows the necessary messages for this step of the call establishment.

Simultaneous with the establishment of the resources for the traffic channel in the radio network, the MSC tries to establish the connection to the called party. This is done, for example, via ISUP signaling to the Gateway MSC for a fixed-line destination as described in Section 1.4. If the destination is reachable, the MSC informs the caller by sending Call Control 'Alerting' and 'Connect' messages.

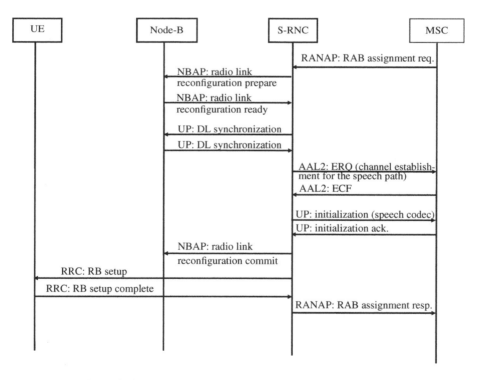

Figure 3.38 Radio resource allocation for a voice traffic channel

The establishment of a packet-switched connection is referred to in the standards as PDP context activation. From the user's point of view, the activation of a PDP context means getting an IP address to be able to communicate with the Internet or another IP network. Further background information on the PDP context activation can be found in Chapter 2. As shown for a voice call in the previous example, the establishment of a packet-switched connection also starts with an RRC connection setup procedure.

Once the signaling connection has been established successfully, the mobile device continues the process by sending an 'Activate PDP Context Request' message via the RNC to the SGSN as shown in Figure 3.39. As shown in the previous example, this triggers the authentication of the subscriber and activation of the air interface encryption. Once encryption is in place, the SGSN continues the process by establishing a tunnel to the GGSN, which in turn assigns an IP address to the user. Furthermore, the SGSN requests the establishment of a suitable bearer from the RNC taking QoS parameters (e.g., minimal bandwidth, latency) into account that are sent to the SGSN at the beginning of the procedure in the 'Activate PDP Context request' message. For normal Internet connectivity, no special QoS settings are usually requested by the mobile devices but if requested, these values can be modified by the SGSN or GGSN based on information contained in the user's profile in the HLR in the event that the user has not subscribed to the requested QoS or in case the connection requires a different QoS setting. The establishment of the RAB is done in the same way as shown in Figure 3.38 for a circuit-switched channel. However, as the bearer for a packet-switched connection uses other QoS attributes, the parameters inside the messages will be different.

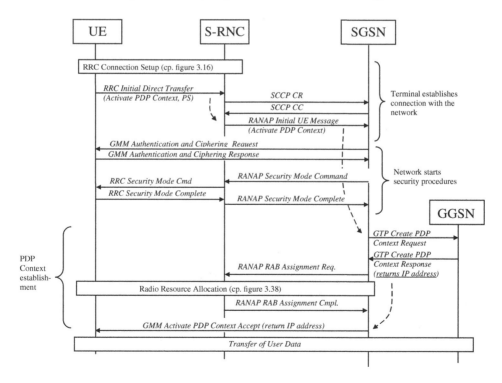

Figure 3.39   PDP context activation

3.9 UMTS Security

Like GSM, UMTS has strong security measures to prevent unauthorized use and eavesdropping on user data traffic and conversations. UMTS also includes enhancements to overcome a number of weaknesses that have been found, over the years, in the way GSM protects networks and users. The following are the main weaknesses:

- The GSM circuit-switched part does not protect the link between the base station and the BSC. In many cases microwave links are used, which are vulnerable to third party monitoring.
- GSM allows man-in-the-middle attacks with equipment that masquerades as a GSM base station.
- The CK length used in GSM is 64 bits. Although secure when GSM was first developed in the early 1990s, the length is considered insufficient today.
- A number of weaknesses with the A5/1 stream cipher have been detected, as described in Chapter 1, which allow decryption of a voice conversation with the appropriate equipment.

UMTS addresses these weaknesses in a number of ways. As in GSM, a one-pass authentication and key agreement (AKA) procedure is used with immediate activation of ciphering after successful authentication. The general principle is described in Chapter 1. When a mobile device attaches to the network after power-up, it tries to register with the network by initiating location and routing area update procedures. At the beginning of the message exchange the mobile device transmits its identity (IMSI or TMSI and PTMSI), which it retrieved from the SIM. If the subscriber is not known by the MSC/VLR and/or the SGSN, authentication information has to be requested from the authentication center, which is part of the HLR. In addition to the random number (RAND), the expected response (XRES) and the CK, which are also used in GSM, two additional values are returned. These are the integrity key (IK) and the authentication token (AUTN). Together, these five values form an authentication vector.

The AUTN serves two purposes. The AuC generates the AUTN from a RAND and the secret key of the subscriber. It is then forwarded together with the RAND to the mobile device in an MM authentication request message. The mobile device then uses the AUTN to verify that the authentication procedure was initiated by an authorized network. The AUTN additionally includes a sequence number, which is increased in both the network and the mobile device after every successful authentication. This prevents attackers from using intercepted authentication vectors for fake authentications later on.

Like in GSM, a UMTS device has to generate a response value that it returns to the network in the MM authentication response message. The MSC/VLR or SGSN then compares the response value to the XRES value, which they received as part of the authentication vector from the HLR/AuC. If both values match, the subscriber is authenticated.

In a further step, ciphering between the mobile device and the network is activated when the network sends a RANAP Security Mode Command message to the RNC. This message contains the 128-bit CK. While in GSM, ciphering for circuit-switched calls is a functionality of the base station, UMTS calls are ciphered by the RNC. This prevents eavesdropping on the Iub interface between the RNC and the base station. An RRC security mode command message informs the mobile device that ciphering is to be activated. Like in GSM, the CK is not sent to the mobile as this would compromise security. Instead, the mobile calculates the CK itself by using, among other values, its secret key and the RAND.

Security mode command messages activate not only ciphering but also integrity checking for signaling messages, which was not performed in GSM. While ciphering is optional, it is mandatory for integrity checking to be activated after authentication. Integrity checking is performed for RRC, CC, SM, MM and GMM messages between the mobile device and the network. User data, on the other hand, has to be verified by the application layer, if required. To allow the receiver to check the validity of a message, an integrity stamp field is added to the signaling messages. The most important parameters for the RNC to calculate the stamp are the content of the signaling message and the IK, which is part of the authentication vector. Integrity checking is done for both uplink and downlink signaling messages. To perform integrity checking for incoming messages and to be able to append the stamp for outgoing messages, the mobile device calculates the IK itself after the authentication procedure. The calculation of the key is performed by the SIM card, using the secret key and the RAND which were part of the authentication request message. This way, the IK is also never exchanged between the mobile device and the network.

Keys for ciphering and integrity checking have a limited lifetime to prevent attempts to break the cipher or integrity protection by brute force during long duration monitoring attacks. The values of the expiry timers are variable and are sent to the mobile device during connection establishment. Upon expiry, a new set of ciphering and IKs are generated with a reauthentication between the mobile device and the network.

Authentication, ciphering and integrity checking are performed independently for circuit-switched and packet-switched connections. This is because the MSC handles circuit-switched calls while the SGSN is responsible for packet sessions. As these devices are independent they have to use different sets of authentication vectors and sequence numbers.

UMTS also introduces new algorithms to calculate the different parameters used for authentication, ciphering and integrity checking. These are referred to as f0–f9. Details on the purpose and use of these algorithms can be found in 3GPP TS 33.102 [10].

On the user side all actions that require the secret key are performed on the SIM card to protect the secret key. As older GSM SIM cards cannot perform the new UMTS authentication procedures, a backward compatibility mode has been specified to enable UMTS-capable mobile devices to use UMTS networks with an old GSM SIM card. The UMTS ciphering and IKs are then computed by the mobile device based on the GSM CK (note: not the secret key!) that is returned by the SIM card. A drawback of this fallback method is, however, that although the network can still properly authenticate the mobile device, the reverse is not possible as the validation of the AUTN on the user's device requires the secret key, which is only stored in the SIM card. On the network side, the HLR is aware that the subscriber uses a non-UMTS-capable SIM card and generates an adapted authentication vector without an AUTN.

3.10   High-Speed Downlink Packet Access (HSDPA) and HSPA+

UMTS networks today no longer use DCHs for high-speed packet transfer. With 3GPP Release 5, HSDPA was introduced to be more flexible for bursty data transmissions and to deliver much higher datarates per cell and per user than before. Data rates that could be achieved in practice at first ranged between 1 and 8 Mbit/s depending on the radio signal conditions and distance to the base station. With further extensions to HSDPA in the subsequent versions of the 3GPP specifications that are described in this chapter and used in practice today, even higher data rates are possible, exceeding 30 Mbit/s under ideal radio signal conditions.

Important standards documents that were created or enhanced for HSDPA are the overall system description Stage 2 in 3GPP TS 25.308 [11], the physical layer description TR 25.858 [12], physical layer procedures in TS 25.214 [13], Iub and Iur interface enhancements in TR 25.877 [14], RRC extensions in TS 25.331 [3] and signaling procedure examples in TS 25.931 [15].

3.10.1 HSDPA Channels

Figures 3.40 and 3.41 show how HSDPA combines the concepts of dedicated and shared channels. For user data in the downlink direction one or more High-Speed Physical Downlink Shared Channels (HS-PDSCH) are used. These can be shared between several users. Hence, it is possible to send data to several subscribers simultaneously or to increase the transmission speed for a single subscriber by bundling several HS-PDSCH where each uses a different code.

Each HS-PDSCH uses a spreading factor length of 16, which means that in theory, up to 15 simultaneous HS-PDSCH channels can be used in a single cell. When reception conditions permit, higher order modulation and coding can be used to increase the transmission speed. In operational networks, the number of HS-PDSCH channels available per cell depends on the number of voice and circuit-switched video calls that are handled by the cell for other users in parallel, which require a DCH. In practice, many network operators use at least two 5-MHz channels per sector in areas of heavy usage so that voice calls have less impact on the capacity available for high-speed Internet access. Parameters like bandwidth, delay and lossless handovers are not guaranteed for an HSDPA connection as the bandwidth available to a user depends, among other factors, on the current signal quality and the number of simultaneous users of the current cell. HSDPA thus sacrifices the concept of a DCH with a guaranteed

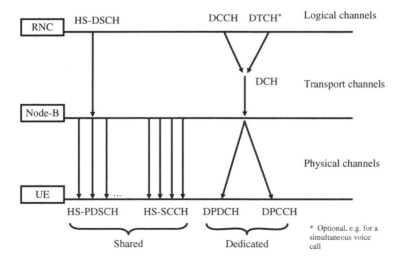

Figure 3.40 Simplified HSDPA channel overview in downlink direction

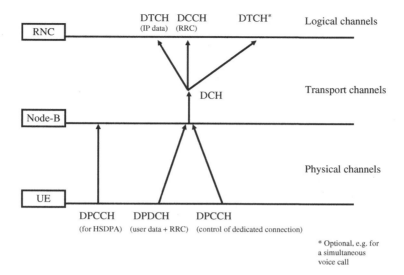

Figure 3.41 Simplified HSDPA channel overview in uplink direction

bandwidth for a significantly increased bandwidth. For many applications like web surfing or the transfer of big files or e-mails with file attachments, this is very beneficial.

The assignment of timeslots on HS-DSCH channels to a user is done via several simultaneous broadcast High-Speed Shared Control Channels (HS-SCCHs), which use a spreading factor length of 128. A mobile device has to be able to receive and decode at least four of those channels simultaneously. Thus, it is possible to inform many users at the same time about which of the HS-PDSCH channel's data is sent to them in the next timeslot.

In addition to the shared channels, an HSDPA connection requires a number of dedicated channels per subscriber:

- A High Speed Dedicated Physical Control Channel (HS-DPCCH) in the uplink direction with a spreading factor of 256 for HSDPA controls information like acknowledgments and retransmission requests for bad frames as well as for transmitting signal quality information. This channel uses its own channelization code and is not transmitted with other channels by using time or IQ-multiplexing.
- A Dedicated Control Channel (DCCH) for RRC messages in the uplink and downlink directions between the RNC and the mobile device which is used for tasks like mobility management that is, for example, necessary for cell changes.
- A Dedicated Physical Control Channel (DPCCH) for transmit power control information.
- A DTCH for IP user data in the uplink direction, as HSDPA only uses shared channels in the downlink direction. The uplink bearer can have a bandwidth of 64, 128 or 384 kbit/s if a Release 99 DCH is used or more if HSUPA is supported by the network and the mobile device.
- Optionally, an additional DTCH is used in the uplink and downlink directions in case a circuit-switched connection (e.g. for a voice call) is established during a HSDPA connection. The channel can have a bandwidth of up to 64 kbit/s.

3.10.2   Shorter Delay Times and Hybrid ARQ (HARQ)

Apart from offering increased bandwidth to individual users and increasing the capacity of a cell in general, another goal of HSDPA was to reduce the round-trip delay (RTD) time for both stationary and mobile users. HSDPA further reduces the RTD times experienced with Release 99 dedicated channels of 160–200 milliseconds to about 100 milliseconds. This is important for applications like web browsing, as described in Section 3.13, and also for EDGE, as described in Section 2.12.1, which require several frame round trips for the DNS query and establishment of the TCP connections before the content of the web page is sent to the user. To reduce the round-trip time, the air interface block size has been reduced to 2 milliseconds. This is quite small compared to the block sizes of dedicated channels of at least 10 milliseconds.

Owing to the frequently changing signal conditions experienced when the user is, for example, in a car or train, or even while walking in the street, transmission errors will frequently occur. Owing to error detection mechanisms and retransmission of faulty blocks, no packet loss is experienced on higher layers. However, every retransmission increases the overall delay of the connection. Higher layer protocols like TCP, for example, react very sensitively to changing delay times and interpret them as congestion. To minimize this effect, HSDPA adds an error detection and correction mechanism on the MAC layer in addition to the mechanisms which already exist on the RLC layer. This mechanism is directly implemented in the Node-B and is called Hybrid Automatic Retransmission Request (HARQ). Together with a block size of 2 milliseconds instead of at least 10 milliseconds for DCHs, an incorrect or missing block can be retransmitted by the Node-B in less than 10 milliseconds. This is a significant enhancement over Release 99 dedicated channels as they only use a retransmission scheme on the RLC layer, which needs at least 80–100 milliseconds for the detection and retransmission of a faulty RLC frame.

Compared to other error detection and correction schemes that are used, for example, on the TCP layer, HARQ does not use an acknowledgement scheme based on a sliding window mechanism but sends an acknowledgement or error indication for every single frame. This mechanism is called Stop and Wait (SAW). Figure 3.42 shows how a frame is transmitted in the downlink direction, which the receiver cannot decode correctly. The receiver therefore sends an error indication to the Node-B, which in turn retransmits the frame. The details of how the process works are given below.

Before the transmission of a frame, the Node-B informs the mobile device of the pending transmission on the HS-SCCH. Each HS-SCCH frame contains the following information:

- ID of the mobile device for which a frame is sent in one or more HS-PDSCH channels in the next frame;
- channelization codes of the HS-PDSCH channels that are assigned to a mobile device in the next frame;
- transport Format and Resource indicator (channel coding information);
- modulation format (QPSK, 16-QAM, 64-QAM, MIMO);
- HARQ process number (see below);
- if the block contains new data or is used for retransmission and which redundancy version (RV) is used (see below).

Each frame on the HS-SCCH is split into three slots. The information in the control frame is arranged in such a way that the mobile device has all information necessary to receive the

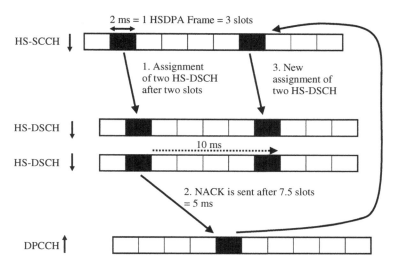

Figure 3.42 Detection and report of a missing frame with immediate retransmission within 10 milliseconds

frame once it has received the first two of the three slots. Thus, the network does not wait till the complete control frame is sent but already starts sending the user data on the HS-PDSCH once the mobile device has received the first two slots of the control frame. This means that the Shared Control Channel and the Downlink Shared Channels are sent with a time shift of one slot. After the reception of a user data frame, the mobile device has exactly 5 milliseconds to decode the frame and to check if it was received correctly. If the frame was sent correctly, the mobile device sends an Acknowledge (ACK) message in the uplink direction on the High Speed Dedicated Physical Control Channel (HS-DPCCH). If the mobile device is not able to decode the packet correctly, a Not Acknowledge (NACK) message is sent. To save additional time, the uplink control channel is also slightly time-shifted against the downlink shared channel. This allows the network to quickly retransmit a frame.

As HARQ can only transmit a frame once the previous frame has been acknowledged, the mobile device must be able to handle up to eight simultaneous HARQ processes. Thus, it is ensured that the data flow is not interrupted by a problem with a single frame. As higher layers of the protocol stack expect the data in the right order, the data stream can only be forwarded once a frame has been received correctly. Therefore, the mobile device has to have a buffer to store frames of other HARQ processes that need to be reassembled with other frames that have not yet been received correctly.

For the network, there are two options for retransmitting a frame. If the Incremental Redundancy method is used, the network uses error correction information that was punctured out after channel coding to make data fit into the MAC-frame. Puncturing is a method that is already used in UMTS Release 99, GPRS and EDGE, and further information can be obtained in Section 2.3.3. If a frame needs to be retransmitted, the network sends different redundancy bits and the frame is thus said to have a different RV of the data. By combining the two frames, the overall redundancy is increased on the receiving side and the chance that the frame can be decoded correctly increases. If the frame still cannot be decoded, there is still enough redundancy information left that has not yet been sent to assemble a third version of the frame.

The second retransmission method is called Chase Combining and it retransmits a frame with the same RV as before. Instead of combining the two frames on the MAC layer, this method combines the signal energy of the two frames on the physical layer before attempting to decode the frame again. The method that is used for retransmission is controlled by the network. However, the mobile device can indicate to the network during bearer establishment which of the two methods it supports.

3.10.3   Node-B Scheduling

The HS-DSCH channels have been designed in a way that different channels can be assigned to different users at the same time. The network then decides for each frame which channels to assign to which users. As shown before, the HS-SCCH channels are used to inform the mobile devices on which channels to listen for their data. This task is called scheduling. To quickly react to changing radio conditions of each subscriber, the scheduling for HSDPA has not been implemented on the RNC as for other channels but directly on the Node-B. This can also be seen in Figure 3.40 as the HS-SCCHs originate from the Node-B. This means that for HSDPA, yet another task that was previously located in the RNC for DCHs has been moved to the border node of the network. This way, the scheduler can, for example, react very quickly to deteriorating radio conditions (fading) of a mobile device. Rather than sending frames to a mobile device while it is in a deep fading situation and thus most likely unable to receive the frame correctly, the scheduler can use the frames during this time for other mobile devices. This helps to increase the total bandwidth available in the cell as less frames have to be used for retransmission of bad or missing blocks. Studies like [16] and [17] have shown that a scheduler that takes channel conditions into consideration can increase the overall cell capacity by about 30% for stationary users. Apart from the signal quality of the radio link to the user, the scheduling is also influenced by other factors like the priority of the user. As with many other functionalities the standard does not say which factors should influence the scheduling in which way, and thus a good scheduling implementation of a vendor can lead to an advantage.

As the RNC has no direct influence on the resource assignment for a subscriber, it is also not aware how quickly data can be sent. Hence, a flow control mechanism is required on the Iub interface between the RNC and the Node-B. For this reason, the Node-B has a data buffer for each user priority from which the scheduler then takes the data to be transmitted over the air interface. In order for the RNC to find out how much space is left in those buffers, it can send a Capacity Request message to the Node-B, which reports to the RNC the available buffer sizes using a Capacity Allocation message. It should be noted that a Node-B does not administer a data buffer per user but only one data buffer per user priority.

3.10.4   Adaptive Modulation and Coding, Transmission Rates and Multicarrier Operation

To reach the highest possible datarate during favorable transmission conditions, several new modulation schemes have been introduced with HSDPA over several 3GPP releases, in addition to the already existing QPSK modulation that transfers 2 bits per transmission step:

- 16-QAM, 4 bits per step. The name is derived from the 16 values that can be encoded in 4 bits (2^4).
- 64-QAM, 6 bits per step.
- Two simultaneous data streams transmitted on the same frequency with MIMO.

To further increase the single-user peak datarate, dual-carrier HSDPA (also referred to as dual-cell HSDPA) was specified to bundle two adjacent 5-MHz carriers. At the time of publication, many networks have deployed this functionality. In the subsequent versions of the standard, aggregation of more than two carriers was specified, as well as combining 5-MHz carriers in different bands. However, because of the quick adoption of LTE, it is unlikely that these features will be seen in practice in the future.

In addition to changing the modulation scheme, the network can also alter the coding scheme and the number of simultaneously used HS-DSCH channels for a mobile device on a per frame basis. This behavior is influenced by the Channel Quality Index (CQI), which is frequently reported by the mobile device. The CQI has a range from 1 (very bad) to 31 (very good) and tells the network how many redundancy bits are required to keep the Block Error Rate (BLER) below 10%. For a real network, this means that under less favorable conditions more bits per frame are used for error detection and correction. This reduces the transmission speed but ensures that a stable connection between network and mobile device is maintained. As modulation and coding is controlled on a per user basis, bad radio conditions for one user have no negative effects for other users in the cell to which the same HS-DSCHs are assigned for data transmission.

By adapting the modulation and coding schemes, it is also possible to keep the power needed for the HSDPA channels at a constant level or to only vary it when the DCH load of the cell changes. This is different from the strategy of Release 99 dedicated channels. Here, the bandwidth of a connection is stable while the transmission power is adapted depending on the user's changing signal quality. Only if the power level cannot be increased any more to ensure a stable connection does the network take action and increase the spreading factor to reduce the bandwidth of the connection.

The capabilities of the mobile device and of the network limit the theoretical maximum data rate. The standard defines a number of different device categories, which are listed in 3GPP TS 25.306 [18]. Table 3.6 shows some of these categories and their properties. Not listed in the table is category 12, which was used by early HSDPA devices that are no longer available. Such devices could support five simultaneous high-speed channels and QPSK modulation only. The resulting datarate was 1.8 Mbit/s.

With a Category 24 mobile device, found in practice today, that supports QPSK, 16-QAM, 64-QAM and Dual Carrier operation, the following maximum transmission speed can be reached: 42,192 bits per TTI (which is distributed over 2×15 HS-PDSCH channels) every 2 milliseconds $= (1/0.002) \times 42,192 = 42.2$ Mbit/s. This corresponds to a speed of 1.4 Mbit/s per channel with a spreading factor of 16. Compared to a Release 99 DCH with 384 kbit/s which uses a spreading factor of 8, the transmission is around eight times faster. This is achieved by using the 64-QAM modulation instead of QPSK, which increases the maximum speed threefold, and by reducing the number of error detection and correction bits while signal conditions are favorable.

Table 3.6 A selection of HSDPA mobile device categories

HS-DSCH category	Maximum number of simultaneous HS-PDSCH	Best modulation	MIMO/Dual carrier	Code rate	Maximum datarate (Mbit/s)
6	5	16-QAM	–	0.76	3.6
8	10	16-QAM	–	0.76	7.2
9	15	16-QAM	–	0.7	10.1
10	15	16-QAM	–	0.97	14.4
14	15	64-QAM	–	0.98	21.1
16	15	16-QAM	MIMO	0.97	27.9
20	15	64-QAM	MIMO	0.98	42.2
24	15	64-QAM	DC	0.98	42.2
28	15	64-QAM	DC + MIMO	0.98	84.4

In practice, many factors influence how fast data can be sent to a mobile device. The following list summarizes, once again, the main factors:

- signal quality;
- number of active HSDPA users in a cell;
- number of established channels for voice and video telephony in the cell;
- number of users that use a dedicated channel for data transmission in a cell;
- mobile device category;
- antenna and transceiver design in the mobile device;
- sophistication of interference cancellation algorithms in the mobile device;
- bandwidth of the connection of the Node-B to the RNC;
- interference generated by neighboring cells;
- achievable throughput in other parts of the network, as high datarates cannot be sustained by all web servers or other end points.

Transmission speeds that can be reached with HSDPA also have an impact on other parts of the mobile device. Apart from increased processing power required in the mobile device, the interface to an external device, such as a notebook, needs to be capable of handling data at these speeds. The maximum transmission rate of Bluetooth, for example, does not exceed 2 Mbit/s. This is not sufficient for the HSDPA datarates achieved in practice today. As a consequence, the only suitable technologies today are USB 2.0 and Wi-Fi (Wireless LAN). While USB 2.0 is mostly used by external 3G Universal Serial Bus (USB) sticks today, 3G to Wi-Fi routers or the Wi-Fi tethering functionality built into most smartphones today are ideally suited to connect several devices over UMTS to the Internet. In principle, these devices work in the same way as the multipurpose WLAN access points typically found in private households today for DSL connectivity (see Figure 6 in Chapter 5). The advantage of using Wi-Fi for connectivity is that no special configuration beyond the Wi-Fi password is required in client devices.

3.10.5   Establishment and Release of an HSDPA Connection

To establish an HSDPA connection, an additional DCH is required to be able to send data in the uplink direction as well. If the network detects that the mobile device is HSDPA capable during the establishment of an RRC connection, it automatically allocates the necessary resources during the setup of the connection as shown in Figure 3.43.

To establish an HSDPA connection, the S-RNC informs the Node-B that a new connection is required and the Node-B will configure the HS-PDSCH accordingly. In a further step, the RNC then reserves the necessary resources on the Iub interface between itself and the Node-B. Once this is done, the network is ready for the high-speed data transfer and informs the mobile device via a RRC Radio Bearer Reconfiguration message that data will now be sent on the HS-DSCH. Once data is received by the RNC from the SGSN, flow control information is exchanged between the Node-B and the RNC. This ensures that the data buffer of the Node-B is not flooded as the RNC has no direct information on or control of how fast the incoming data can be sent to the mobile device. When the Node-B receives data for a user, it is then the task of the HSDPA scheduler in the Node-B to allocate resources on the air interface and to inform the user's mobile device via the shared control channels whenever it sends data on one or more HS-PDSCHs.

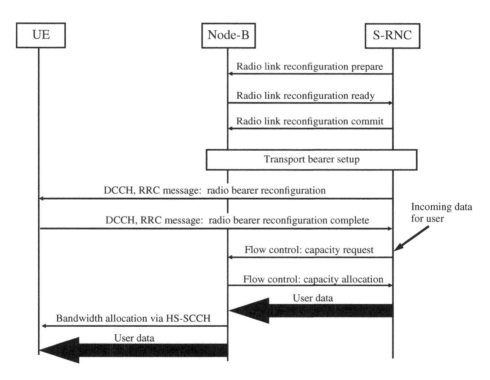

Figure 3.43   Establishment of an HSDPA connection

While the mobile device is in HSDPA reception mode, it has to constantly monitor all assigned HS-SCCH channels and also maintain the necessary DCHs. This of course results in higher power consumption, which is acceptable while data is transferred. If no data is transferred for some time, this state is quite unfavorable as the power consumption remains high and thus the runtime of the mobile device decreases. This state is also not ideal for the network as bandwidth on the air interface is wasted for the dedicated channel of the HSDPA connection. Thus, the network can decide to release the HSDPA connection after a period of time and put the subscriber into the Cell-FACH state (see Section 3.5.4). In this state, the mobile device can still send and receive data, but the bandwidth is very small. Nevertheless, this is quite acceptable as an HSDPA connection can be reestablished again very quickly, when required.

3.10.6   HSDPA Mobility Management

HSDPA has been designed for both stationary and mobile users. Therefore, it is necessary to maintain the connection while the user is moving from cell to cell. For this reason, the mobile device keeps a so-called Active Set for the DCH of the HSDPA connection, which is required for the soft handover mechanism, as described in Section 3.7.1. In contrast to a pure dedicated connection, the mobile device only receives its data over one of the Node-Bs of the Active Set. On the basis of the configuration of the network, the mobile device then reports to the RNC if a different cell of the Active or Candidate Sets would provide better signal quality than the current cell. The RNC can then decide to redirect the data stream to a different cell. As the concept is different from the UMTS soft handover, the standards refer to this operation as cell change procedure.

Compared to the Cell Update procedure of (E)GPRS, the cell change procedure of HSDPA is controlled by the network and not by the mobile device. As the mobile device is already synchronized with the new cell, a cell change only leads to a short interruption of the data transfer on the HS-PDSCHs.

Depending on the relationship between the old and the new cell, there are several different kinds of cell changes:

- **Intra Node-B cell change**. Old and new cell are controlled by the same Node-B. This is the simplest version of the operation as data that is still available in the buffer of the Node-B can simply be sent over the new cell.
- **Inter Node-B cell change**. Old and new cells belong to different Node-Bs. In this scenario, the RNC has to instruct the new Node-B to allocate resources for the HSDPA connection. This is done in a similar way to establishing a new connection as shown in Figure 3.43. User data that is still buffered in the old Node-B is lost and has to be retransmitted by the RLC layer which is controlled in the RNC.
- **Cell change with Iur interface**. If the old and new cells are under the control of different RNCs, the HSDPA connection has to be established over the Iur interface.
- **Cell change without Iur interface**. If the old and new cells are under the control of different RNCs which are not connected via the Iur interface, an SRNS relocation has to be performed, which also involves core network components (SGSN and possibly also the MSC).
- **Old and new cells use different frequencies (interfrequency cell change)**. In this scenario additional steps are required in the mobile device to find cells on different frequencies and to synchronize them before data transmission can resume.

- **Inter-RAT cell change**. If the subscriber leaves the UMTS coverage area completely, a cell change procedure from UMTS/HSDPA to GSM has also been specified. Similar to the interfrequency cell change described above, HSDPA connections can use a compressed mode similar to that of dedicated channels to allow the mobile device to search for cells on other frequencies.

During all scenarios it is, of course, also possible that an additional voice or video call is established. This further complicates the cell change/handover as this connection also has to be maintained next to the data connection and handed over into a new cell.

3.11   High-Speed Uplink Packet Access (HSUPA)

Owing to the emergence of peer-to-peer applications like multimedia calls, video conferencing and social networking applications, the demand for uplink bandwidth is continually increasing. Other applications like sending e-mails with large file attachments or large MMS messages sent by a user also benefit from higher uplink datarates. UMTS uplink speeds have not been enhanced until 3GPP Release 6. Hence, for a long time the uplink was still limited to $64-128$ kbit/s and to 384 kbit/s in some networks under ideal conditions despite the introduction of HSDPA in 3GPP Release 5. The solution to satisfy the increasing demand in the uplink direction is referred to as Enhanced Uplink (EUL) in 3GPP. In the public network situation, the feature is referred to as HSUPA. HSUPA increases theoretical uplink user datarates to up to 5.76 Mbit/s in 3GPP Release 6 and 11.5 Mbit/s in 3GPP Release 7. Further enhancements have been made in subsequent 3GPP releases, but it is unlikely that these enhancements will be implemented in practice due to the emergence of LTE. When taking realistic radio conditions into account, the number of simultaneous users, mobile device capabilities, etc. user speeds of $1-4$ Mbit/s are reached in practice today.

For the network, HSUPA has a number of benefits as well. For HSDPA, an uplink DCH is required for all mobile devices that receive data via the HSDSCHs for TCP acknowledgements and other user data. This is problematic for bursty applications as a DCH in the uplink direction wastes uplink resources of the cell despite the mobile device reducing its power output during periods when no user data is sent. Nevertheless, HSUPA continues to use the dedicated concept of UMTS for the uplink by introducing an Enhanced Dedicated Channel (E-DCH) functionality for the uplink only. However, the E-DCH concept includes a number of enhancements to decrease the impact of bursty applications on the DCH concept. To have both high-speed uplink and downlink performance using an E-DCH introduced with HSUPA only makes sense when combined with HSDSCHs that were introduced with HSDPA.

While a Release 99 DCH ensures a constant bandwidth and delay time for data packets with all its advantages and disadvantages discussed in previous chapters, the E-DCH trades in this concept for higher datarates. Thus, while still being a dedicated channel, an E-DCH does not necessarily guarantee a certain bandwidth to a user in the uplink direction anymore. For many applications, this is quite acceptable and allows the increase of the number of simultaneous users that can share the uplink resources of a cell. This is because the network can control the uplink noise in a much more efficient way by dynamically adjusting the uplink bandwidth on a per subscriber basis in a cell to react to changing radio conditions and traffic load. This reduces the overall cost of the network by requiring less base stations for the same number of users, which, in turn, can result in cheaper subscription costs.

The E-DCH concept also ensures full mobility for subscribers. However, the radio algorithms are clearly optimized to ensure the highest throughput for low speed or stationary use.

The main purpose of the E-DCH concept is to support streaming (e.g., Mobile TV), inter-active (e.g., web browsing) and background services (e.g., FTP). To ensure good performance for real-time applications like IMS video conferencing, the E-DCH enhancements also contain optional mechanisms to ensure a minimal bandwidth to a user. As these methods are optional and packet-based real-time applications are not yet widely used, current implementations focus on the basic E-DCH concept. Despite this, Voice and Video over IP services can still be used over a nonoptimized E-DCH without any problems as long as the cell's bandwidth is suffi-cient to satisfy the demand of all users currently transmitting data in a cell regardless of the application.

As the uplink bandwidth increases and fast retransmissions are introduced, the E-DCH approach also further reduces the RTD times for applications like web surfing and interactive gaming to around 60 milliseconds.

Finally, it is important to note that the E-DCH concept is backward compatible. Thus, a cell can support Release 99 mobile devices that were only designed for DCHs, HSDPA mobile devices that require a DCH in the uplink direction, and mobile devices that support a combi-nation of HSDPA in the downlink and HSUPA in uplink.

As the E-DCH concept is an evolution of existing standards, it has triggered the creation of a number of new documents as well as the update of a number of existing specifications. Most notably, 3GPP TR 25.896 [19] was created to discuss the different options that were analyzed for HSUPA. Once consensus on the high level architecture was reached, 3GPP TS 25.309 [20] was created to give a high level overview of the selected solution. Among the specification documents that were extended are 3GPP TS 25.211 [5], which describes physical and transport channels, and 3GPP TS 25.213 [21], which was extended to contain information about E-DCH spreading and modulation.

3.11.1 E-DCH Channel Structure

For the E-DCH concept a number of additional channels were introduced in both uplink and downlink directions as shown in Figure 3.44 and 3.45. These are used in addition to existing channels, which are also shown in the figure. For further explanation of these channels, see Section 3.4.3 for Release 99 channels and Section 3.10.1 for HSDPA.

As shown on the left side in Figure 3.44, HSUPA introduces a new transport channel which is called the E-DCH. While still being a DCH for a single user, the dedicated concept was adapted to use a number of features that were already introduced with HSDPA for the downlink direction. Therefore, the following overview just gives a short introduction to the feature and the changes required to address the needs of a dedicated channel:

- **Node-B scheduling**. While standard DCHs are managed by the RNC, E-DCHs are managed by the Node-B. This allows a much quicker reaction to transmission errors, which in turn decreases the overall RTD time of the connection. Furthermore, the Node-B is able to react much more quickly to changing conditions of the radio environment and variations of user demands for uplink resources, which help to better utilize the limited bandwidth of the air interface.

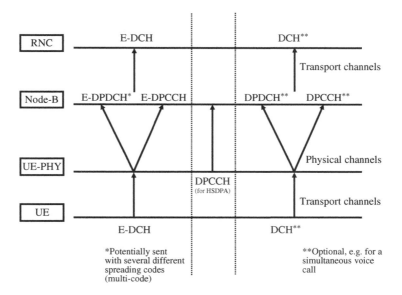

Figure 3.44   Transport and Physical Channels used for HSUPA

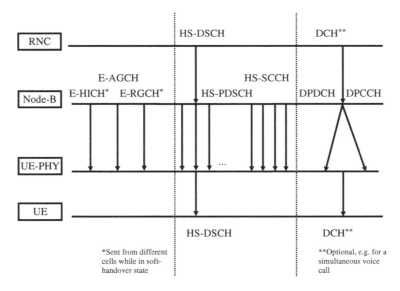

Figure 3.45   Simultaneous downlink channels for simultaneous HSUPA, HSDPA and DCH use

- **HARQ**. Instead of leaving the error detection and correction to the RLC layer alone, the E-DCH concept uses the HARQ scheme, which is also used by HSDPA in the downlink direction. This way, errors can be detected on a per MAC-frame basis by the Node-B. For further details see Section 3.10.2, which describe the HARQ functionality of HSDPA in the downlink direction. While the principle of HARQ in the uplink direction is generally the same, it should be noted that the signaling of acknowledgements is done in a slightly different way due to the nature of the DCH approach.
- **Chase Combining and Incremental Redundancy**. These are used in a similar way for E-DCH as described in Section 3.10.5 for HSDPA to retransmit a frame when the HARQ mechanism reports a transmission error.

On the physical layer, the E-DCH is split into two channels: The Enhanced Dedicated Physical Data Channel (E-DPDCH) is the main transport channel and is used for user data (IP frames carried over RLC/MAC-frames) and layer 3 RRC signaling between the mobile device on the one side and the RNC on the other. As described below, the spreading factor used for this channel is quite flexible and can be dynamically adapted from 64 to 2 depending on the current signal conditions and the amount of data the mobile device wants to send. It is even possible to use several channelization codes at the same time to increase the overall speed. This concept is called a multicode channel and is similar to the HSDPA concept of assigning frames on several downlink shared channels to a single mobile device. As described in more detail below, the maximum number of simultaneous code channels has been limited to four per mobile device, with two channels being used with SF = 2 and the other two with SF = 4. In terms of frame length, 10 milliseconds are used for the E-DPDCH by default with 2 millisecond frames being standardized as optional.

The Enhanced Dedicated Physical Control Channel (E-DPCCH) is used for physical layer control information. For each E-DPDCH frame, a control frame is sent on the E-DPCCH to the Node-B, which most importantly contains the 7-bit Traffic Format Combination ID (TFCI). Only by analyzing the TFCI is the Node-B able to decode the MAC-frame on the E-DPDCH as the mobile device can choose the spreading factor and coding of the frame from a set given to it by the Node-B to adapt to the current signal conditions and uplink user data buffer state. Furthermore, each frame on the E-DPCCH contains a 2-bit Retransmission Sequence Number (RSN) to signal HARQ retransmissions and the RV (see Section 3.10.2) of the frame. Finally, the control frame contains a so-called 'Happy' bit to indicate to the network if the maximum bandwidth currently allocated to the mobile device is sufficient or if the mobile device would like the network to increase it. While the spreading factor of the physical data channel is variable, a constant spreading factor of 256 is used for the E-DPCCH.

A number of existing channels, which might also be used together with an E-DCH, are shown in the middle and on the right of Figure 3.44. Usually, an E-DCH is used together with HSDPA HSDSCHs, which require a separate DPCCH to send control information for downlink HARQ processes. To enable applications like voice and video telephony during an E-DCH session a mobile must also support simultaneous Release 99 dedicated data and control channels in the uplink as well. This is necessary as these applications require a fixed and constant bandwidth of 12.2 and 64 kbit/s, respectively. In total, an E-DCH-capable mobile device must therefore be able to simultaneously encode the data streams of at least five uplink channels. If multicode operation for the E-DPDCH is used, up to eight code channels are used in the uplink direction at once.

In the downlink direction, HSUPA additionally introduces two mandatory and one optional channel to the other already numerous channels that have to be monitored in the downlink direction. Figure 3.45 shows all channels that a mobile device has to decode while having an E-DCH assigned in the uplink direction, HSDPA channels in the downlink direction and an additional DCH for a simultaneous voice or video session via a circuit-switched bearer.

While HSUPA only carries user data in the uplink direction, a number of control channels in the downlink direction are nevertheless necessary. For the network to be able to return acknowledgements for received uplink data frames to the mobile device, the Enhanced HARQ Information Channel (E-HICH) is introduced. The E-HICH is a dedicated channel, which means that the network needs to assign a separate E-HICH to each mobile device currently in E-DCH state.

To dynamically assign and remove bandwidth to and from individual users quickly, a shared channel called the Enhanced Access Grant Channel (E-AGCH) is used by the network that must be monitored by all mobile devices in a cell. A fixed spreading factor of 256 is used for this channel. Further details about how this channel is used to issue grants (bandwidth) to the individual mobile devices are given in Section 3.11.3.

Finally, the network can also assign an Enhanced Relative Grant Channel (E-RGCH) to individual mobile devices to increase or decrease an initial grant that was given on the E-AGCH. The E-RGCH is again a dedicated channel, which means that the network has to assign a separate E-RGCH to every active E-DCH mobile device. The E-RGCH is optional, however, and depending on the solutions of the different network vendors there might be networks in which this channel is not used. If not used, only the E-AGCH is used to control uplink access to the network. Note that although all channels are called 'enhanced', none of these channels have a Release 99 predecessor.

Besides these three control channels, an E-DCH mobile device must also be able to decode a number of additional downlink channels simultaneously. As HSUPA is used together with HSDPA, the mobile device also needs to be able to simultaneously decode the HS-DSCHs as well as up to four HS-SCCH. If a voice or video call is established besides the high-speed packet session, the network will add another two channels in the downlink direction as shown in Figure 3.45 on the right hand side. In total, an E-DCH mobile must, therefore, be capable of decoding 10–15 downlink channels at the same time. If the mobile device is put into soft handover state by the network (see Section 3.7.1) the number of simultaneous channels increases even further as some of these channels are then broadcast via different cells of the mobile device's Active Set.

3.11.2   The E-DCH Protocol Stack and Functionality

To reduce the complexity of the overall solution, the E-DCH concept introduces two new layers called the MAC-e and MAC-es. Both layers are below the existing MAC-d layer. As shown in Figure 3.46, higher layers are not affected by the enhancements and thus the required changes and enhancements for HSUPA in both the network and the mobile devices are minimized.

While on the mobile device the MAC-e/es layers are combined, the functionality is split on the network side between the Node-B and the RNC. The lower layer MAC-e functionality is implemented on the Node-B in the network. It is responsible for scheduling, which is further described below, and the retransmission (HARQ) of faulty frames.

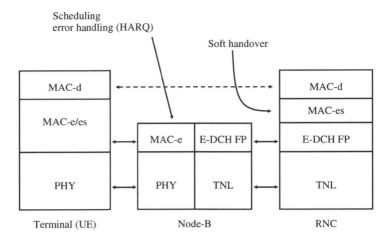

Figure 3.46 E-DCH protocol stack

The MAC-es layer in the RNC on the other hand is responsible for recombining frames received from different Node-Bs in case an E-DCH connection is in soft handover state. Furthermore, the RNC is also responsible for setting up the E-DCH connection with the mobile device at the beginning. This is not part of the MAC-es layer but part of the RRC algorithm, which has to be enhanced for HSUPA as well. As the RNC treats an E-DCH channel like a DCH, the mobile device is in Cell-DCH state while an E-DCH is assigned. While scheduling of the data is part of the Node-B's job, overall control of the connection rests with the RNC. Thus, the RNC can decide to release the E-DCH to a mobile device after some period of inactivity and put the mobile device into Cell-FACH state. Therefore, HSUPA becomes part of the Cell-DCH state and thus part of the overall Radio Resource Management as described in Section 3.5.4.

One of the reasons for enhancing the dedicated connection principle to increase uplink speeds instead of using a shared channel approach is that it enables the soft handover principle to be used in the uplink direction. This is not possible with a shared channel approach, which is used by HSDPA in the downlink direction because cells would have to be synchronized to assign the same timeslots to a user. In practice, this would create a high signaling overhead in the network. By using DCHs the timing between the different mobile devices that use the same cells in soft handover state is no longer critical as they can send at the same time without being synchronized. The only issue arising from sending at the same time is the increased noise level in the cells. However, neighboring cells can minimize this by instructing mobiles in soft handover state to decrease their transmission power via the Relative Grant Channel (E-RGCH) as further described below. Using soft handover in the uplink direction might prove to be very beneficial as the mobile device's transmit power is much lesser than that of the Node-B. Furthermore, there is a higher probability that one of the cells can pick up the frame correctly and thus the mobile device has to only retransmit a frame if all cells of the Active Set send a negative acknowledge message for a frame. This in turn reduces the necessary transmission power on the mobile device side and increases the overall capacity of the air interface.

Another advantage of the dedicated approach is that mobile devices also do not have to be synchronized within a single cell and thus do not have to wait for their turn to send data. This further reduces the RTD times.

3.11.3   E-DCH Scheduling

If the decision is made by the RNC to assign an E-DCH to the mobile device, the bearer establishment or modification messaging is very similar to establishing a standard DCH. During the E-DCH establishment procedure, the RNC informs the mobile device of the Transport Format Combination Set (TFCS) that can be used for the E-DCH. A TFCS is a list (set) of datarate combinations, coding schemes and puncturing patterns for different transport channels that can be mapped on to the physical channel. In practice, at least two channels, a DTCH for user data and a DCCH for RRC messages, are multiplexed over the same physical channel (E-DPDCH). This is done in the same way as for a standard dedicated channel. By using this list, the mobile device can later select a suitable TFC for each frame depending on how much data is currently waiting in the transmission buffer and the current signal conditions. By allowing the RNC to flexibly assign a TFC set to each connection, it is possible to restrict the maximum speed on a per subscriber basis based on the subscription parameters. During the E-DCH setup procedure, the mobile device is also informed of which of the cells of the Active Set will be the serving E-DCH cell. The serving cell is defined as being the cell over which the network later controls the bandwidth allocations to the mobile device.

Once the E-DCH has been successfully established, the mobile device has to request a bandwidth allocation from the Node-B. This is done by sending a message via the E-DCH, even though no bandwidth has so far been allocated. The bandwidth request contains the following information for the Node-B:

- UE estimation of the available transmit power after subtracting the transmit power already necessary for the DPCCH and other currently active DCHs;
- indication of the priority level of the highest priority logical channel currently established with the network for use via the E-DCH;
- buffer status for the highest priority logical channel;
- total buffer status (taking into account buffers for lower priority logical channels).

Once the Node-B receives the bandwidth request, it takes the mobile device's information into account together with its own information about the current noise level, bandwidth requirements of other mobile devices in the cell and the priority information for the subscriber it has received from the RNC when the E-DCH was initially established. The Node-B then issues an absolute grant, also called a scheduling grant, via the E-AGCH, which contains information about the maximum power ratio the mobile can use between the E-DPDCH and the E-DPCCH. As the mobile has to send the E-DPCCH with enough power to be correctly received at the Node-B, the maximum power ratio between the two channels implicitly limits the maximum power that can be used for the E-DPDCH. This in turn limits the number of choices the mobile device can make from the TFC set that was initially assigned by the RNC. Therefore, as some TFCs cannot be selected anymore, the overall speed in the uplink direction is implicitly limited.

Furthermore, an absolute grant can be addressed to a single mobile device only or to several mobile devices simultaneously. If the network wants to address several mobile devices at once, it has to issue the same Enhanced Radio Network Temporary ID (E-RNTI) to all group members when their E-DCH is established. This approach minimizes signaling when the network wants to schedule mobile devices in the code domain.

Another way to dynamically increase or decrease a grant given to a mobile device or a group of mobile devices is the use of relative grants, which are issued via the optional relative grant channel (E-RGCH). These grants are called relative grants because they can increase or decrease the current power level of the mobile step by step with an interval of one TTI or slower. Thus, the network is quickly able to control the power level and, therefore, implicitly the speed of the connection every 2 or 10 milliseconds. Relative grants can also be used by all cells of the Active Set. This allows cells to influence the noise level of E-DCH connections currently controlled by another cell to protect their selves from too much noise being generated in neighboring cells. This means that the mobile device needs to be able to decode the E-RGCH of all cells of the Active Set. As shown in Figure 3.47, each cell of the Active Set can assume one of three roles:

- One of the cells of the Active Set is the serving E-DCH cell from which the mobile receives absolute grants via the E-AGCH (cell 4 in Figure 3.47). The serving E-DCH cell can, in addition, instruct the mobile device to increase, hold or decrease its power via commands on the E-RGCH.
- The serving E-DCH cell and all other cells of the Node-B that are part of the Active Set of a connection (cell 3 and 4 in Figure 3.47) are part of the serving radio link set. The commands

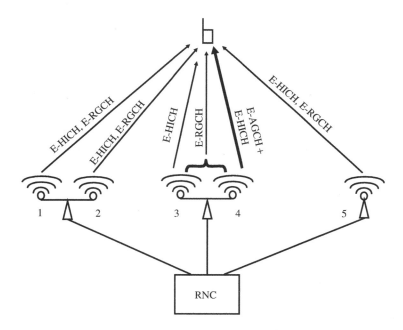

Figure 3.47   Serving E-DCH cell, serving RLS and non-serving RLS

sent over the E-RGCH of these cells are identical and thus the mobile device can combine the signals for decoding.
- All other cells of the Active Set are part of the non-serving radio link set (cell 1, 2 and 5 in Figure 3.47). The mobile device has to decode all E-RGCHs of these cells separately. Cells in the non-serving RLS can only follow send, hold or down commands.

If an 'up' command is received from the serving RLS, the mobile device is allowed to increase its transmission power only if at the same time a no 'down' command is received by one or more cells of the non-serving RLS. In other words, if a 'down' command is received by the mobile device from any of the cells, the mobile device has to immediately decrease its power output. This way, only the serving E-DCH is able to increase or decrease the power output of the mobile via the relative grant channels while all other cells of the non-serving RLS are only permitted to decrease the power level.

It should be noted that in a real environment, it is unlikely that five cells as shown in Figure 3.47 are part of the Active Set of a connection, as the benefit of the soft handover would be compromised by the excessive use of air interface and Iub link resources. Thus, in a normal environment, it is the goal of radio engineering to have two or at most three cells in the Active Set of a connection in soft handover state.

As has been shown, the Node-B has quite a number of different pieces of information to base its scheduling decision on. The standard, however, does not describe how these pieces of information are used to ensure a certain QoS level for the different connections and leaves it to the network vendors to implement their own algorithms for this purpose. Again, the standards encourage competition between different vendors, which unfortunately increases the overall complexity of the solution.

To enable the use of the E-DCH concept for real-time applications like voice and video over IP in the future, the standard contains an optional scheduling method, which is called a nonscheduled grant. If the RNC decides that a certain constant bandwidth and delay time is required for an uplink connection, it can instruct the Node-B to reserve a sufficiently large power margin for the required bandwidth. The mobile device is then free to send data at this speed to the Node-B without prior bandwidth requests. If such E-DCH connections are used, which is again implementation dependent, the Node-B has to ensure that even peaks of scheduled E-DCH connections do not endanger the correct reception of the nonscheduled transmissions.

3.11.4 E-DCH Mobility

Very high E-DCH datarates can only be achieved for stationary or low-mobility scenarios owing to the use of low spreading factors and few redundancy bits. Nevertheless, the E-DCH concept uses a number of features to enable high datarates also in high-speed mobility scenarios. To this end, macro diversity (soft handover) can be used as shown in Figure 3.47. This means that the uplink data is received by several cells, which forward the received frames to the RNC. Each cell can then indicate to the mobile device if the frame has been received correctly and thus, only the frame has to be repeated if none of the cells were able to decode the frame correctly. This is especially beneficial for mobility scenarios in which reception levels change quickly because of obstacles suddenly appearing inbetween the mobile device and one of the cells of the Active Set as shown earlier. Furthermore, the use of soft handover ensures that no interruptions in the uplink occur while the user is moving through the network with the mobile device.

For capacity reasons, network operators usually use several 5-MHz carriers in a cell today. If a device moves to a different location that is served by a cell with only a single carrier, a soft handover procedure is used if the previous carrier and the new carrier are on the same frequency. If the device is served on a carrier frequency that is not present in the new cell, an inter-frequency hard-handover is required as the carrier of the new cell cannot be decoded at the same time as the carriers in the current Active Set that transmit on a different frequency.

3.11.5 E-DCH-Capable Devices

E-DCH-capable devices once again require increased processing power and memory capabilities compared to HSDPA devices to sustain the high datarates offered by the system in both downlink (HSDPA) and uplink (HSUPA) directions. To benefit from the evolution of mobile device hardware, the standard defines a number of mobile device categories that limit the maximum number of spreading codes that can be used for an E-DCH and their maximum length. This limits the maximum speed that can be achieved with the mobile device in the uplink direction. Table 3.7 shows a number of typical E-DCH mobile device categories and their maximum transmission speeds under ideal transmission conditions. The highest number of simultaneous spreading codes an E-DCH mobile device can use is four, with two codes having a spreading factor of two and two codes having a spreading factor of four. The maximum user data rates are slightly lower than the listed transmission speeds as the transport block also includes the frame headers of different protocol layers. Under less ideal conditions, the mobile device might not have enough power to transmit using the maximum number of codes allowed and might also use a more robust channel coding method that uses smaller transport block sizes, as more bits are used for redundancy purposes. In addition, the Node-B can also restrict the maximum power to be used by the mobile device, as described above, to distribute the available uplink capacity of the cell among the different active users.

In practice, most devices on the market today are E-DCH category 6 capable of a theoretical maximum uplink data throughput of 5.76 Mbit/s. Under good radio conditions and when close to a base station, uplink speeds of 3–4 Mbit/s can be reached.

Table 3.7 Spreading code sets and maximum resulting speed of different E-DCH categories

Category	Modulation/Dual cell	Maximum E-DPDCH set of the mobile device category	Maximum transport block size for a TTI	Maximum speed (Mbit/s)
2	QPSK	2 × SF-4	14.592 bits (10 ms)	1.5
6	QPSK	2 × SF-2 + 2 × SF-2	20.000 bits (10 ms)	2.0
6	QPSK	2 × SF-4 + 2 × SF-2	11.484 bits (2 ms)	5.7
7	16-QAM	2 × SF-2 + 2 × SF-2	22.996 bits (2 ms)	11.5

3.12   Radio and Core Network Enhancements: CPC and One Tunnel

While the evolution of wireless networks was mainly focused on increasing datarates for quite some time, other factors such as reducing power consumption and increasing efficiency of the core network architecture are also very important to keep the overall system viable. Starting with 3GPP Release 7, a number of enhancements were specified in that direction, CPC and One Tunnel.

CPC is a package of features to improve the handling of mobile subscribers while they have a packet connection established, that is, while they have an IP address assigned. Taken together, they have the following benefits:

- Reduction of power consumption;
- Reduction of the number of state changes;
- Minimization of delays between state changes;
- Reduction of signaling overhead;
- An increase in the number of mobile devices per cell that can be served simultaneously.

CPC does not introduce new revolutionary features. Instead, already existing features are modified to achieve the desired results. 3GPP TR 25.903 [22] gives an overview of the proposed changes and the following descriptions refer to the chapters in the document that have been selected for implementation.

3.12.1   A New Uplink Control Channel Slot Format

While a connection is established between the network and a mobile device, several channels are used simultaneously. This is because it is not only user data which is being sent but also control information to keep the link established, to control transmit power, and so on. Currently, the Uplink Dedicated Control Channel (UL DPCCH) is transmitted continuously, even during times of inactivity to remain synchronized. This way, the mobile device can resume uplink transmissions without delay whenever required.

The control channel carries four parameters:

1. Transmit power control (TPC).
2. Pilot (used for channel estimation of the receiver).
3. TFCI.
4. Feedback indicator (FBI).

The pilot bits are always the same and allow the receiver to get a channel estimate before decoding user data frames. While no user data frames are received, however, the pilot bits are of little importance. What remains important is the TPC. The idea behind the new slot format is to increase the number of bits to encode the TPC and decrease the number of pilot bits while the uplink channel is idle. This way, additional redundancy is added to the TPC field. As a consequence, the transmission power for the control channel can be lowered without risking corruption of the information contained in the TPC. Once user data transmission resumes, the standard slot format is used again and the transmission power used for the control channel is increased again.

3.12.2   CQI Reporting Reduction and DTX and DRX

CQI Reporting Reduction

To make the best use of the current signal conditions in the downlink direction, the mobile has to report to the network how well its transmissions are received. The quality of the signal is reported to the network with the CQI alongside the user data in the uplink direction. To reduce the transmit power of the mobile device while data is being transferred in the uplink direction but not in the downlink direction, this feature reduces the number of CQI reports.

UL HS-DPCCH Gating (Gating = Switch-Off)

When no data is being transmitted in either the uplink or the downlink direction, the uplink control channel (UL DPCCH) for HSDPA is switched off. Periodically, it is switched on for a short time to transmit bursts to the network to maintain synchronization. This improves battery life for applications such as web browsing, lowers battery consumption for VoIP and reduces the noise level in the network (i.e., allowing more simultaneous VoIP users). Figure 3.48 shows the benefits of this approach.

F-DPCH Gating

Mobile devices in HSDPA active mode always receive a Dedicated Physical Channel (DPCH) in the downlink direction, in addition to high-speed shared channels, which carries power control information and Layer 3 radio resource (RRC) messages, for example, for handovers, channel modifications, and so on. The Fractional-DPCH feature puts the RRC messages on the HSDPA shared channels and the mobile thus only has to decode the power control information from the DPCH. At all other times the DPCH is not used by the mobile (thus it is fractional). During these times, power control information is transmitted for other mobiles using the same spreading code. This way, up to 10 mobile devices use the same spreading code for the dedicated physical channel but listen to it at different times. This means that fewer spreading

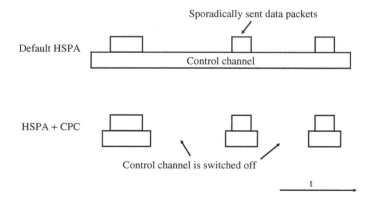

Figure 3.48   Control channel switch-off during times with little activity

codes are used by the system for this purpose, which in turn leaves more resources for the high-speed downlink channels or allows a significantly higher number of users to be kept in HSPA Cell-DCH state simultaneously.

3.12.3   HS-SCCH Discontinuous Reception

While a mobile is in HSPA mode, it has to monitor one or more HS-SCCHs to see when packets are delivered to it on the high-speed shared channels. This monitoring is continuous, that is, the receiver can never be switched off. For situations when no data is transmitted or the average data transfer rate is much lower than that which could be delivered over the high-speed shared channels, the Node-B can instruct the mobile device to only listen to selected slots of the shared control channel. The slots that the mobile does not have to observe are aligned as much as possible with the uplink control channel gating (switch-off) times. Therefore, there are times when the mobile device can power down its receiver to conserve energy. Should more data were to arrive again from the network than what can be delivered with the selected DRX cycle at some point, the DRX mode is switched off and the network can once again schedule data in the downlink continuously.

3.12.4   HS-SCCH-less Operation

This feature is not intended to improve battery performance but to increase the number of simultaneous real-time IMS VoIP users in the network. VoIP requires relatively little bandwidth per user and hence the number of simultaneous users can be high. On the radio link, however, each connection has a certain signaling overhead. Therefore, more users mean more signaling overhead, which decreases the overall available bandwidth for user data. In the case of HSPA, the main signaling resources are the HS-SCCHs. The more the number of active users, the more will be their proportional requirement of the available bandwidth.

HS-SCCH-less operation aims at reducing this overhead. For real-time users that require only limited bandwidth, the network can schedule data on high-speed downlink channels without prior announcements on a shared control channel. This is done as follows: The network instructs the mobile to listen to not only the HS-SCCH but, in addition, to all packets being transmitted on one of the HSDSCHs. The mobile device then attempts to blindly decode all packets received on that shared channel. To make blind decoding easier, packets which are not announced on a shared control channel can only have one of four transmission formats (number of data bits) and are always modulated using QPSK. These restrictions are not an issue for performance, since HS-SCCH-less operation is only intended for low bandwidth real-time services.

The checksum of a packet is additionally used to identify the device for which the packet is intended. This is done by using the mobile device's MAC address as an input parameter for the checksum algorithm in addition to the data bits. If the device can decode a packet correctly and if it can reconstruct the checksum, it is the intended recipient. If the checksum does not match then either the packet is intended for a different mobile device or a transmission error has occurred. In both cases, the packet is discarded.

In case of a transmission error, the packet is automatically retransmitted since the mobile device did not send an acknowledgement (HARQ ACK). Retransmissions are announced on the shared control channel, which requires additional resources, but should not happen frequently as most packets should be delivered properly on the first attempt.

It should be noted at this point that at the time of publication, HS-SCCH-less operation is not used in networks, as IMS VoIP has not yet been deployed in 3G networks.

3.12.5   Enhanced Cell-FACH and Cell-/URA-PCH States

The CPC features described above aim to reduce power consumption and signaling overhead in HSPA Cell-DCH state. The CPC measures therefore increase the number of mobile devices that can be in Cell-DCH state simultaneously and allow a mobile device to remain in this state for a longer period of time even if there is little or no data being transferred. Eventually, however, there is so little data transferred that it no longer makes sense to keep the mobile in Cell-DCH state; that is, it does not justify even the reduced signaling overhead and power consumption. In this case, the network can put the connection into Cell-FACH or even into Cell-PCH or URA-PCH state to reduce energy consumption even further. The downside of this is that a state change back into Cell-DCH state takes much longer and that little or no data can be transferred during the state change. In Release 7 and 8, the 3GPP standards were thus extended to also use the HSDSCHs for these states as described in 3GPP TR 25.903 [22]. In practice, this is done as follows:

- **Enhanced Cell-FACH**. In the standard Cell-FACH state, the mobile device listens to the secondary common control physical channel in the downlink direction for incoming RRC messages from the RNC and for user data (IP packets). With the Enhanced Cell-FACH feature, the network can instruct a mobile device to observe a high-speed downlink control channel or the shared data channel directly for incoming RRC messages from the RNC and for user data. The advantage of this approach is that in the downlink direction, information can be sent much faster. This reduces latency and speeds up the Cell-FACH to Cell-DCH state change procedure. Unlike in Cell-DCH state, no other uplink or downlink control channels are used. In the uplink direction, data packets can be sent in two ways. The first method is to use the RACH as before to respond to RRC messages from the RNC and to send its IP packets. An additional method was introduced with 3GPP Release 8, which foresees a special E-DCH for faster data transmission. Both methods limit the use of adaptive modulation and coding since the mobile cannot send frequent measurement reports to the base station to indicate the downlink reception quality. Furthermore, it is also not possible to acknowledge proper receipt of frames. Instead, the RNC informs the base station when it receives measurement information in radio resource messages from the mobile device.
- **Enhanced Cell-/URA-PCH states**. In these two states, the mobile device is in a deep sleep state and only observes the paging information channel so as to be alerted of an incoming paging message that is transmitted on the PCH. To transfer data, the mobile device moves back to Cell-FACH state. If the mobile device and the network support Enhanced Cell-/URA-PCH states, the network can instruct the mobile device not to use the slow PCH to receive paging information but to use a HSDSCH instead. The high-speed downlink channel is then also used for subsequent RRC commands, which are required to move the device back into a more active state. Like the measure above, this significantly decreases the wakeup time.

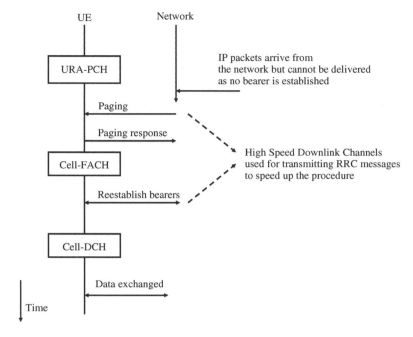

Figure 3.49 Message exchange to move a mobile device from URA-PCH state back to Cell-DCH state when IP packets arrive from the network

Figure 3.49 shows how this works in practice. While the message exchange to notify the mobile device of incoming data and to move it to another activity state remains the same, using the HSDSCHs for the purpose speeds up the procedure by several hundred milliseconds.

Although the CPC features, such as the various DRX and DTX enhancements as well as the fractional DPCH functionality, have recently seen an uptake in life networks to increase the number of simultaneous users, it can be observed that the enhanced cell states are not in use at the time of publication of this current edition. At this point in time, it is uncertain whether network operators are interested in deploying those features due to the quick uptake of LTE. As a consequence, operators might not see a high value in adding those features in the relatively limited bandwidth used by UMTS compared to LTE. Also, operators are cautious when introducing such new features, as they significantly increase the complexity of the air interface as both old and new mobile devices have to be supported simultaneously. This rising complexity is especially challenging for the development of devices and networks, as it creates additional interaction scenarios that become more and more difficult to test and debug. Today, devices are already tested with network equipment of several vendors and different software versions. Adding yet another layer of features makes this even more complex.

3.12.6 Radio Network Enhancement: One Tunnel

In the core network, one particular feature has been standardized to decrease latency and processing load on network nodes that do not necessarily have to be in the transmission path for user data.

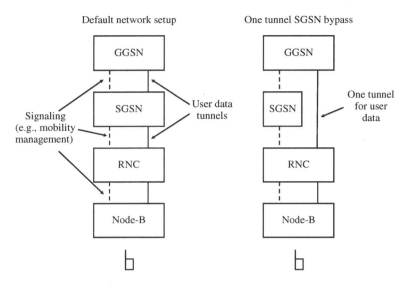

Figure 3.50   Current network architecture vs. the one-tunnel enhancement

Figure 3.50 shows the default path of user data between a mobile device and the Internet through the UMTS network. In the default architecture, the packet is sent through the GGSN, the SGSN, RNC and the base station. All user data packets are tunneled through the network as described above, since the user's location can change at any time. The current architecture uses a tunnel between the GGSN and the SGSN and a second tunnel between the SGSN and the RNC. All data packets therefore have to pass through the SGSN, which terminates one tunnel, extracts the packets and puts them into another tunnel. This requires both time and processing power.

Since both the RNC and the GGSN are IP routers, this process is not required in most cases. The one-tunnel approach, now standardized in 3GPP (see [23] and [24]), allows the SGSN to create a direct tunnel between the RNC and the GGSN. This way it removes itself from the transmission path. Mobility and Session Management, however, remain on the SGSN, which means, for example, that it continues to be responsible for mobility management and tunnel modifications in case the mobile device moves to an area served by another RNC. For the user, this approach has the advantage that the packet delay is reduced. From a network point of view, the advantage is that the SGSN requires less processing resources per active user, which helps to reduce equipment costs. This is especially important as the amount of data traversing the packet-switched core network is rising significantly.

A scenario where the one-tunnel option is not applicable is international roaming. Here, the SGSN has to be in the loop to count the traffic for interoperator billing purposes. Another case where the one-tunnel option cannot be used is when the SGSN is asked by a prepaid system to monitor the traffic flow. This is only a small limitation, however, since in practice it is also possible to perform prepaid billing via the GGSN.

Proprietary enhancements even aim to terminate the user data tunnel at the Node-B, bypassing the RNC as well. However, this has not found the widespread support in 3GPP.

3.13   HSPA Performance in Practice

After the introduction of how UMTS works from a technical point of view, this section takes a look at which features are used in practice and how. Looking back at early networks, voice calls are still handled the same way now as then. The way packet data is transferred, however, has completely changed with HSPA. This section is therefore a snapshot of how the system performs in practice at the time of publication.

3.13.1   Throughput in Practice

In practice, the experienced throughput depends on a variety of different factors:

- The maximum throughput capability of the mobile device.
- The sophistication of the receiver and antenna of the mobile device.
- The capability of the network.
- The radio conditions at the place of use. This includes the signal level received from a base station and the interference of neighbor cell transmissions on the same frequency.
- The bandwidth of the backhaul link between the base station and the rest of the network.
- The number of other users in the cell that actively exchange data at the same time.

With a HSDPA category 24 and HSUPA category 6 device, downlink speeds of 30 Mbit/s and uplink speeds of around 3–4 Mbit/s can be reached in practice under very good signal conditions. But even under less favorable conditions, speeds of several megabits per second can still be achieved. It can be observed that even small changes in the position of the device or rotation of the antenna can have a significant impact on the overall transmission speed. For example, by moving the device only by a few centimeters, throughput can easily change by several megabits per second. When using a 3G USB stick indoors in stationary situations, it is often advisable to use a USB extension cable and put the USB stick closer to a window to increase data transfer speeds. Also, increasing the height of the location of the USB stick improves reception, especially under very weak signal conditions. This way it is possible to reach speeds well over 1 Mbit/s, even if the device receives the network with a very low signal. Over time, higher and higher theoretical peak data rates have been specified in the standards. This comes at the expense of the size of the areas in which they can be reached. A study published in Bergman [25] shows that transmission modes requiring MIMO and 64-QAM modulation can only be used close to the base stations. In the majority of the coverage area the same lower order modulation and coding schemes are used as before. However, subscribers in those areas can still benefit from higher data rates close to the base station. As subscribers close to the base stations can transmit and receive data faster, more time is available to send data to other subscribers. In other words, the overall throughput that can be reached in the cell increases, which in turn benefits all subscribers in the cell.

Another significant impact on individual speeds and overall cell throughput is the development of advanced receivers in the mobile devices. Advanced HSPA devices use reception (RX) diversity, that is, two antennas and two receiver chains, significantly enhancing the device's ability to receive a data stream. Even more advanced HSPA receivers that have found their way into today's devices use interference cancellation algorithms that are able to distinguish between noise created by neighboring base stations transmitting on the same frequency and the signal received from the serving base stations. This is especially beneficial at the cell edge.

A detailed analysis of current and future network capacity and throughput for HSPA and other technologies can be found in Sauter [26].

3.13.2   Radio Resource State Management

From a throughput point of view an ideal mobile device would always be instantly reachable and always ready to transfer data whenever something is put into the transmission buffer from higher protocol layers. The downside of this is a high power consumption that, as will be shown in the next section, would drain the battery within only a few hours. Therefore, a compromise is necessary.

In practice today, mobile devices are either in the RRC Idle state when not transferring any data or in the Cell-PCH or URA-PCH state. In these states, the mobile only listens to the PCH and only reestablishes a physical connection when it receives a paging message or when new data packets arrive from applications running on the device. The time it takes to switch from RRC Idle to Cell-DCH on the high-speed shared channel is around 2.5 seconds and is the major source of delay noticeable to a user when surfing the web. In Cell-PCH and URA-PCH states, the return to the fast Cell-DCH state takes only around 0.7 seconds as the logical connection to the RNC remains in place despite the physical bearer no longer being present. As a consequence, no complex connection request, authentication and activation of ciphering procedures are required. In practice, it can be observed that many networks have adopted this approach, which is especially beneficial for users due to the noticeably shorter delay when clicking on a link on a web page before a new page is loaded. While in the Cell-DCH state, the mobile device can transmit and receive at any time and round-trip packet delay is in the order of 60–85 milliseconds with a category 24 HSDPA and category 6 HSUPA device. As the Cell-DCH state requires a significant amount of power on the mobile device's side even if no data is transferred, most UMTS networks move the connection to a more power-conserving state after an inactivity time of only a few seconds. Typical inactivity timer values range between 5 and 10 seconds. After the inactivity timer has expired, many networks then assign the Cell-FACH state to the device. In this state, only a narrow downlink channel is observed and no control and channel feedback information is sent or received from the network. As the channel is relatively narrow, RTD times are around 300 milliseconds. In case there is renewed data traffic beyond a network-configurable threshold, the connection is set into Cell-DCH state again.

If no further data traffic occurs, the network will set the connection to an even more power-conserving state after a further 10–15 seconds. In practice, this is either the Idle state or the Cell-PCH or the URA-PCH state. Some networks even skip the Cell-FACH state entirely and move the connection from Cell-DCH directly to the Cell-PCH or URA-PCH state. In practice, it can be observed today that most mobile devices do not wait for the network to act but request the release of the physical connection themselves if they come to the conclusion that it is unlikely that further data will be sent. This mechanism is described in the next section.

3.13.3   Power Consumption

Figure 3.51 shows the power consumption in the different states. On the left, the mobile device is in Cell-DCH state on high-speed channels. Power consumption is very high and ranges

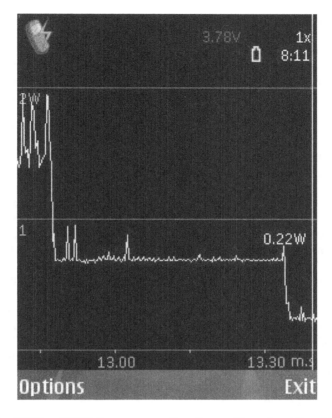

Figure 3.51   FACH and DCH Power consumption. Source: Reproduced from Nokia © 2010

between 1.5 and 2 W. Typical mobile device batteries have a capacity of around 5–8 watt hours. Therefore, being in Cell-DCH state continuously would drain the battery in a few hours.

In the middle of Figure 3.51, the mobile device is in Cell-FACH state and power consumption is reduced to about 0.8 W. In the example, the idle timer in the network is set to 45 seconds and after expiry the mobile is set into Idle state. Here, the power requirements go down to 0.3 W. Most of it, however, is used for the background light of the display. Once the light is turned off, power requirements are reduced to less than 0.1 W.

The autonomy time of a mobile device is mainly influenced by the type of applications running on a device and the timer settings in the network. If, for example, an e-mail client is used on a device that checks the inbox in the network once every 10 minutes or requires a keep-alive message to be sent in this interval to keep the logical connection to the server open, a radio connection has to be established six times an hour. If the Cell-DCH timer is set to 15 seconds, the mobile device remains in this highly power-consuming state for 1.5 minutes per hour in addition to the time it takes to communicate with the server, usually a few seconds per attempt. Further, typical networks have a Cell-FACH timer of 10–30 seconds. If the lower value is used for this example, the mobile device will be in this state for an additional minute per hour. Together, the mobile device will be in a heightened power consumption state for around 2.5 minutes per hour. Over the course of a day, the total time amounts to a full hour and as a consequence uses a significant part of a battery charge. Longer timeout values generate an even less favorable result.

To reduce power requirements of always-on applications on mobile devices, a number of solutions exist. In the past, some device manufacturers used proprietary mechanisms to deactivate the air interface connection without waiting for the network to take action when applications communicated only periodically with the network and exchanged only little data. As such proprietary functionality was not strictly standard conforming, it was decided to properly standardize such 'Fast Dormancy' functionality in 3GPP Release 8 [27]. The standardized feature enables the mobile devices to send a Signaling Connection Release Indication (SCRI) request message containing a new Fast Dormancy value as release reason to the network to take down the physical air interface connection to the network without waiting for the Cell-DCH and FACH timers to expire. The network can then set the mobile device into Cell-PCH or URA-PCH state. As the feature is optional on the network side, the T323 parameter was added in the System Information Broadcast 1 (SIB1) message. If T323 is present, the mobile knows that it can use the new mechanism. In practice, it can be observed today that the new Release 8 Fast Dormancy feature was quickly adopted by many network operators and significantly contributes to reducing power consumption of smartphones with many connected applications.

In addition to Fast Dormancy, the CPC feature pack described above will also help to reduce power consumption. In practice, however, only a few network operators have so far chosen to activate the CPC features. One reason for this is that the individual functionalities are complex and hence long interoperability testing and adaptation of software in the mobile devices and networks were required to ensure that CPC does not degrade network performance. As the CPC features do not only reduce power consumption but also increase the number of concurrent users a cell can serve it is likely, however, that in the near future many network operators will also activate these features because of the rising number of smartphones in their networks.

3.13.4 Web-Browsing Experience

As described in Chapter 2 for EDGE, the delay caused by the network has a considerable impact on the user experience in a web-browsing session. Independent of the underlying network technology, the following delays are experienced before a web page can be downloaded and displayed:

- The URL has to be converted into the IP address of the web server that hosts the requested page. This is done via a DNS query that causes a delay in the order of the time it takes to send one IP frame to a host in the Internet and wait for the reply. This delay is also called the RTD time.
- Once the IP address of the server has been determined, the web browser needs to establish a TCP connection. This is done via a three-way handshake. During the handshake, the client sends a synchronization TCP frame to the server, which is answered by a synchronization-ack frame. This is in turn acknowledged by the client, by sending an acknowledgement frame. As three frames are sent before the connection is established, the operation causes a delay of 1.5 times the RTD of the connection. As the first frame

containing user data is sent right after the acknowledgement frame, however, the time is reduced to approximately a single RTD time.

- Only after the TCP connection has been established can the first frame be sent to the web server, which usually contains the actual web page request. The server then analyzes the request and sends back a frame that contains the beginning of the requested web page. As the request (e.g., 300–500 B) and the first response frame (1400 B) are quite large, the network requires somewhat more time to transfer these packets than the single RTD time.

As described above, the RTD times of a UMTS Release 6 radio access network with a Category 24 device are in the order of 300–350 milliseconds for the Cell-FACH, 60–80 milliseconds for the Cell-DCH state and around 3 seconds if the mobile device is in Idle state and a radio bearer has to be established first. Owing to the keep-alive messaging of PCs, the connection is rarely in Idle state, though. Therefore, assuming an average RTD time of about 300 milliseconds for Cell-FACH state and 70 milliseconds for the Cell-DCH state, the time required between requesting a page and the browser showing the first parts of the page can be roughly calculated as follows:

Total Delay (UMTS Release 6)

$$= \text{Delay DNS query (Cell-FACH)} + \text{Delay TCP Establish (Cell-DCH/FACH)}$$

$$+ \text{Delay Request/Response}$$

$$= 300\,\text{ms} + 70\,\text{ms} + 100\,\text{ms} = 470\,\text{ms}$$

The request/response delay time is slightly larger than the normal RTD time due to the bigger packet size of the first web server response packet, which already contains a part of the web page and usually has a size of about 1400 bytes. Figure 3.52 shows the timing of a sample request for a web page in the same way as that presented in Chapter 2 for EDGE, again without the initial DNS query, which is not part of the TCP connection establishment for the actual web page. In the flow chart, the request for the web page is sent to the web server in the fourth packet, that is, the packet with a packet length of 403 bytes in the graph.

When compared to loading the same page over a high-speed DSL line, the difference is almost unnoticeable. In a side-by-side comparison, the same web page downloads in almost the same time over both connections if the radio link is already in Cell-DCH state when the web page is requested. If the radio link is in Idle state, however, the download over UMTS only starts after a noticeable delay of several seconds due to the required state change from Idle to Cell-DCH. As a result, today many network operators use the Cell-PCH or URA-PCH state instead of the Idle state.

3.14 UMTS and CDMA2000

While UMTS is the dominant 3G technology in Europe, it shares the market with a similar system called CDMA2000 in other parts of the world, for example, in North America. This section compares CDMA2000 and its evolution path to the GSM, GPRS and UMTS evolution path that has been discussed in Chapters 1–3.

IS-95A, which is also referred to as CDMAOne, was designed similar to GSM to be mostly a voice-centric mobile network. Like GSM, it offers voice and circuit-switched data services

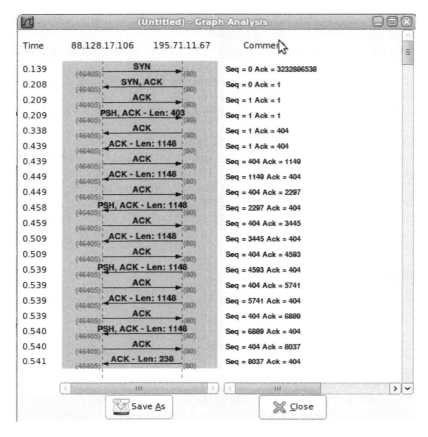

Figure 3.52 IP packet flow and delay times during the download of a web page.
Source: www.wireshark.org. Reproduced by permission of WireShark© 2010

of speeds up to 14.4 kbit/s. However, IS-95A and all evolutions of that standard are not based on GSM and as a consequence both radio and core network infrastructure and protocols are fundamentally different. In particular, the radio network is fundamentally different than GSM as it is not based on FTDMA. IS-95A was the first system to use the CDMA approach for the air interface that was later also used in the UMTS standards where it is referred to as Wideband Code Division Multiple Access or WCDMA for short.

IS-95B is a backward compatible evolution of the system which offers increased user datarates and packet data transmission of up to 64 kbit/s. Thus, it can be roughly compared to a GSM network that offers GPRS services. Similar to the earlier version of CDMAOne, it uses carriers with a bandwidth of 1.25 MHz, which multiple subscribers share by code multiplexing.

The next step in the evolution path was CDMA2000 1xRTT (Radio Transmission Technology), which can roughly be compared to UMTS Release 99. While offering theoretical datarates of 307 kbit/s in the downlink direction, most deployments limited the maximum speed to about 150 kbit/s. From the overall system point of view, there are many similarities between CDMA2000 and UMTS. These include

- the use of CDMA on the air interface;
- the use of QPSK for modulation;
- variable length codes for different datarates;
- soft handover;
- continuous uplink data transmission.

As both UMTS and CDMA2000 need to be backward compatible with their respective evolution paths, there are also many differences between them, which include the following:

- UMTS uses a WCDMA carrier with a bandwidth of 5 MHz while CDMA2000 uses a multi-carrier approach with bandwidths of multiples of 1.25 MHz. This was done to be able to use CDMA2000 in the already available spectrum for IS-95, while UMTS had no such restriction due to the completely new implementation of the air interface and availability of a dedicated frequency band for the new technology.
- UMTS uses a chip rate of 3.84 MChip/s while CDMA2000 uses a chip rate of 1.2288 MChip/s. To increase capacity, a base station can use several 1.25-MHz carriers. Up to the latest revision of the standard described in this book (1 × EV-DO see below), a subscriber is limited to a single carrier.
- UMTS uses a power control frequency of 1500 Hz compared to CDMA2000 that uses an 800 Hz cycle.
- UMTS uses unsynchronized base stations, while in CDMA2000 all base stations are synchronized using the Global Positioning System (GPS) clock.
- As UMTS uses unsynchronized base stations, a three-step synchronization process is used between the mobile device and the network, as described in Section 3.4.4. CDMA2000 achieves synchronization on the basis of a time shift process that adapts the clock of the mobile device to the network.
- While UMTS has a minimal frame length of 10 milliseconds, CDMA2000 uses 20-millisecond frames for user data and signaling and 5-millisecond frames if only signaling has to be sent.

As discussed earlier, UMTS has evolved to higher datarates with HSDPA. The corresponding upgrade path of CDMA2000 is referred to as 1 × EV-DO (Evolution-Data Only) Revision 0 and uses one or more 1.25-MHz carriers exclusively for high-speed packet data transmission. Datarates in practice range between several hundred kilobits and a few megabits. As only 1.25-MHz carriers are used, top speeds are not as high as for HSDPA or HSPA+.

In a further evolution of the standards, referred to as Revision A, uplink performance is improved to a level similar to UMTS HSUPA. Additional QoS features enabling the use of VoIP and other real-time applications over the packet-switched networks further extends the functionality.

Further enhancements of the CDMA standard were initially foreseen but at some point it was decided by network vendors and network operators that the 3GPP LTE system should become the common successor of UMTS and EV-DO. For this purpose, extensions in the LTE specifications were defined to have a standardized interface to the EV-DO core network. These enable dual mode devices to access both systems and to be able to handover ongoing sessions between the two radio network technologies in a similar way as between UMTS and LTE.

Questions

1. What are the main differences between the GSM and UMTS radio network?
2. What advantages does the UMTS radio network have compared to previous technologies for users and network operators?
3. What were the datarates for a packet-switched connection that were offered by early Release 99 UMTS networks?
4. What does OVSF mean?
5. Why is a scrambling code used in addition to the spreading code?
6. What does 'cell breathing' mean?
7. What are the differences between the Cell-DCH and the Cell-FACH RRC states?
8. In which RRC states can a mobile device be in PMM connected mode?
9. How is a UMTS soft handover performed and what are the advantages and disadvantages?
10. What is an SRNS relocation?
11. How is the mobility of a user managed in Cell-FACH state?
12. What is the compressed mode used for?
13. What are the basic HSDPA concepts to increase the user datarate?
14. How is a circuit-switched voice connection handled during an ongoing HSDPA session?
15. What are the advantages of the E-DCH concept?
16. Which options does the Node-B have to schedule the uplink traffic of different E-DCH mobile devices in a cell?

Answers to these questions can be found on the website to this book at http://www.wirelessmoves.com.

References

[1] 3GPP Release Descriptions, 3GPP, Release Descriptions, http://www.3gpp.org/ftp/Information/WORK_PLAN /Description_Releases/, Accessed in 2014.
[2] 3GPP, User Equipment (UE) Radio Transmission and Reception (FDD), TS 25.101.
[3] 3GPP, Radio Resource Control (RRC) Protocol Specification, TS 25.331.
[4] ETSI TS 102 900, Emergency Communications (EMTEL); European Public Warning System (EU-ALERT) using the Cell Broadcast Service, http://www.etsi.org/deliver/etsi_ts/102900_102999/102900/01.01.01_60 /ts_102900v010101p.pdf.
[5] 3GPP, Physical Channels and Mapping of Transport Channels onto Physical Channels (FDD), TS 25.211.
[6] M. Degermar et al., IP Header Compression, Internet Engineering Task Force, RFC 2507, February 1999.
[7] 3GPP, UTRAN Iur and Iub Interface user Plan Protocols for DCH Data Streams, TS 25.427.
[8] 3GPP, UTRAN Iu Interface Radio Access Network Application Part (RANAP) Signaling, TS 25.413.
[9] 3GPP, AMR Speech Codec; General Description, TS 26.071.
[10] 3GPP, 3G Security; Security Architecture, TS 33.102.
[11] 3GPP, UTRA High Speed Downlink Packet Access (HSDPA); Overall Description; Stage 2, TS 25.308.
[12] 3GPP, Physical Layer Aspects of UTRA High Speed Downlink Packet Access, TR 25.858, Accessed in 2010.
[13] 3GPP, Physical Layer Procedures, TS 25.214.
[14] 3GPP, High Speed Downlink Packet Access (HSDPA) Iub/Iur Protocol Aspects, TR 25.877.
[15] 3GPP, UTRAN Functions, Examples on Signaling Procedures, TS 25.931.

[16] Ferrús, R. *et al.* (June 2005) Cross Layer Scheduling Strategy for UMTS Downlink Enhancement. *IEEE Radio Communications*, 43 (6), S24–S26.

[17] L. Caponi, F. Chiti, and R. Fantacci, A Dynamic Rate Allocation Technique for Wireless Communication Systems, *IEEE International Conference on Communications*, Paris, vol. 7, pp. 20–24, June 2004.

[18] 3GPP, UE Radio Access Capabilities Definition, TS 25.306.

[19] 3GPP, Feasibility Study for Enhanced Uplink for UTRA FDD, TR 25.896.

[20] 3GPP TS 25.309, FDD Enhanced Uplink; Overall Description; Stage 2.

[21] 3GPP, Spreading and Modulation (FDD), TS 25.213.

[22] 3GPP, Continuous Connectivity for Packet Users, 3GPP TR 25.903 Version 7.0.0, 2007.

[23] 3GPP, One Tunnel Solution for Optimisation of Packet Data Traffic, TR 23.809, 2006.

[24] 3GPP, General Packet Radio Service (GPRS); Service Description; Stage 2, TS 23.060 v7.6.0, 2007.

[25] J. Bergman *et al.*, Continued HSPA Evolution of Mobile Broadband, Ericsson Review 1/2009, http://www.ericsson.com/ericsson/corpinfo/publications/review/2009_01/files/HSPA.pdf, Accessed in 2010, 2009.

[26] Sauter, M. (2013) Beyond 3G – Bringing Network Terminals and the Web Together, John Wiley & Sons Ltd, ISBN 978-1118341483..

[27] 3GPP, Enhanced SCRI Approach for Fast Dormancy, 3GPP TS 25.331 Change Request 3483, 2008.

4

Long Term Evolution (LTE) and LTE-Advanced

4.1 Introduction and Overview

Despite constant evolution, Universal Mobile Telecommunications System (UMTS), as described in Chapter 3, is approaching a number of inherent design limitations in a manner similar to what GSM and GPRS did a decade ago. The Third Generation Partnership Project (3GPP) hence decided to once again redesign both the radio network and the core network. The result is commonly referred to as 'Long-Term Evolution' or LTE for short. The main improvements over UMTS are in the following areas:

When UMTS was designed, it was a bold approach to specify an air interface with a carrier bandwidth of 5 MHz. Wideband Code Division Multiple Access (WCDMA), the air interface chosen at that time, performed very well within this limit. Unfortunately, it does not scale very well. If the bandwidth of the carrier is increased to attain higher transmission speeds, the time between two transmission steps has to decrease. The shorter a transmission step, the greater the impact of multipath fading on the received signal. Multipath fading can be observed when radio waves bounce off objects on the way from transmitter to receiver, and hence the receiver does not see one signal but several copies arriving at different times. As a result, parts of the signal of a previous transmission step that has bounced off objects and thus took longer to travel to the receiver overlap with the radio signal of the current transmission step that was received via a more direct path. The shorter a transmission step, the more the overlap that can be observed and the more difficult it gets for the receiver to correctly interpret the received signal. With LTE, a completely different air interface has been specified to overcome the effects of multipath fading. Instead of spreading one signal over the complete carrier bandwidth (e.g., 5 MHz), LTE uses Orthogonal Frequency Division Multiplexing (OFDM) that transmits the data over many narrowband carriers of 180 kHz each. Instead of a single fast transmission, a data stream is split into many slower data streams that are transmitted simultaneously. As a consequence, the attainable datarate compared to UMTS is similar in the same bandwidth but the multipath effect is greatly reduced because of the longer transmission steps.

From GSM to LTE-Advanced: An Introduction to Mobile Networks and Mobile Broadband,
Revised Second Edition. Martin Sauter.
© 2014 John Wiley & Sons, Ltd. Published 2014 by John Wiley & Sons, Ltd.

To increase the overall transmission speed, the transmission channel is enlarged by increasing the number of narrowband carriers without changing the parameters for the narrowband channels themselves. If a less than 5 MHz bandwidth is available, LTE can easily adapt and the number of narrowband carriers is simply reduced. Several bandwidths have been specified for LTE: from 1.25 MHz up to 20 MHz. All LTE devices must support all bandwidths and which one is used in practice depends on the frequency band and the amount of spectrum available to a network operator. With a 20-MHz carrier, datarates beyond 100 Mbit/s can be achieved under very good signal conditions.

Unlike in HSPA, the baseline for LTE device has been set very high. In addition to the flexible bandwidth support, all LTE devices have to support Multiple Input Multiple Output (MIMO) transmissions, a situation a situation which allows the base station to transmit several data streams over the same carrier simultaneously. Under very good signal conditions, the datarates that can be achieved this way are beyond those that can be achieved with a single-stream transmission.

In most parts of the world including Europe and the Americas, LTE uses frequency division duplex (FDD) to separate uplink and downlink transmissions. In some parts of the world, spectrum for Time Division Duplex (TDD) has been assigned to network operators. Here, the uplink and downlink transmissions use the same carrier and are separated in time. While a TDD mode already exists for UMTS, it has come to market many years after the FDD version and there are significant differences between the two air interface architectures. Hence, it has not become very popular. With LTE, both FDD and TDD have been specified in a single standard. While FDD is the dominating air interface mode today, it is likely that TDD will also gain traction in the coming years, especially in China and the United States. The differences between the two modes are mostly limited to layers 1 and 2 on the air interface. All higher layers are not affected and a higher reuse of both hardware and software on the network side is possible. On the mobile device side, early LTE devices were either FDD or TDD capable. Today, some hardware variants of LTE devices are both FDD and TDD capable of addressing market demands of countries or regions in which both air interface standards are used.

The second major change of LTE compared to previous systems has been the adoption of an all-Internet Protocol (IP) approach. While UMTS used a traditional circuit-switched packet core for voice services, for Short Messaging Service (SMS) and other services inherited from GSM, LTE solely relies on an IP-based core network. The single exception is SMS, which is transported over signaling messages. An all-IP network architecture greatly simplifies the design and implementation of the LTE air interface, the radio network and the core. With LTE, the wireless industry takes the same path as fixed-line networks with DSL, fiber and broadband IP over TV cable, where voice telephony is also transitioned to the IP side. Quality of Service (QoS) mechanisms have been standardized on all interfaces to ensure that the requirements of voice calls for a constant delay and bandwidth can still be met when capacity limits are reached. While from an architectural point of view, this is a significant advance, implementation, in practice, has proven to be difficult. As a consequence, current LTE networks use a mechanism referred to as Circuit-Switched Fallback (CSFB) to UMTS or GSM for voice calls. Over time, it is expected that networks and devices become Voice over LTE (VoLTE) capable and thus no longer require a fallback to other radio networks. More details on CSFB and VoLTE can be found at the end of this chapter.

Also, all interfaces between network nodes in LTE are now based on IP, including the backhaul connection to the radio base stations. Again, this is a great simplification compared to earlier technologies that were initially based on E-1, ATM and frame relay links, with most of them being narrowband and expensive. The standard leaves the choice of protocols to be used below the IP layer open, which means that the physical infrastructure becomes completely transparent and interchangeable. To further simplify the network architecture and to reduce user data delay, fewer logical and physical network components have been defined in LTE. In practice, this has resulted in round-trip delay times of less than 25–30 milliseconds. Optimized signaling for connection establishment and other air interface and mobility management procedures have further improved the user experience. The time required to connect to the network is in the range of only a few hundred milliseconds and power-saving states can now be entered and exited very quickly.

To be universal, LTE-capable devices must also support GSM, GPRS, EDGE and UMTS. On the network side, interfaces and protocols have been put in place so that data sessions can be moved seamlessly between GSM, UMTS and LTE when the user roams in and out of areas covered by different air interface technologies. While in the early years of deployment, LTE core network and access network nodes have often been deployed independent of the already existing GSM and UMTS network infrastructure, integrated GSM, UMTS and LTE nodes are now used in practice.

LTE is the successor technology not only of UMTS but also of CDMA2000, mostly used in the Americas. To enable seamless roaming between Code Division Multiple Access (CDMA) and LTE, interfaces between the two core networks have been specified. In practice, the user can thus also roam between these two types of access networks while maintaining his IP address and hence all established communication sessions.

Many of the ideas that have gone into the LTE standardization process have also been introduced into the evolution of UMTS. Good examples are the use of MIMO, specified for HSPA + in 3GPP Release 7, the One-Tunnel solution to reduce the number of network nodes in the user data path and the use of IP on all interfaces. These and many more features are optional in UMTS and will help satisfy the demand for higher speeds in the spectrum used for UMTS until and beyond the time LTE is available in more places. Also, it helps to combine GSM, UMTS and LTE functionalities in single nodes over time. Combined base stations with a single broadband backhaul based on IP are just one example of how the cost of operating all three radio networks simultaneously is reduced in practice today.

LTE, as specified in 3GPP Release 8, is a new beginning and also a foundation for further enhancements. With 3GPP Release 10, new ideas to further push the limits are specified as part of the LTE-Advanced project to comply with the International Telecommunication Union's (ITU's) IMT-Advanced requirements for 4G wireless networks [1].

This chapter is structured as follows: First, the general network architecture and interfaces of LTE are described. Next, the new air interface is described for both FDD and TDD systems. This is followed by a description of how user data is scheduled on the air interface as it is a major task of the LTE base station. Afterward, basic procedures are discussed to establish and maintain a data connection between a mobile device and the network, followed by an overview of mobility management and power management considerations. Network planning aspects, interconnection to GSM, UMTS and CDMA are discussed afterward, before concluding the chapter with operational topics such as VoLTE and suitability and use of different backhaul technologies for LTE.

4.2 Network Architecture and Interfaces

The general LTE network architecture is similar to that of GSM and UMTS. In principle, the network is separated into a radio network part and a core network part. The number of logical

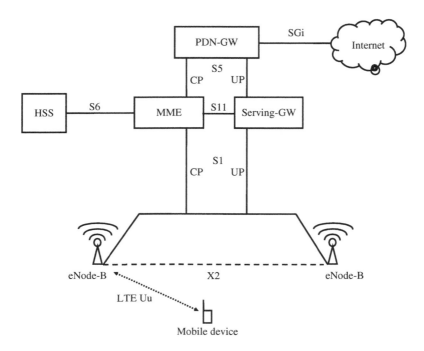

Figure 4.1 LTE Network Overview

network nodes, however, has been reduced to streamline the overall architecture and reduce cost and latency in the network. Figure 4.1 gives an overview of the LTE network and its components and the following sections give a detailed overview of the tasks of the different nodes and how they interact with each other. The subsequent sections then go on to describe the most important functionality in more detail.

4.2.1 LTE Mobile Devices and the LTE Uu Interface

In the LTE specifications, as in UMTS, the mobile device is referred to as the User Equipment (UE). In 3GPP Release 8, five different UE classes have been defined as shown in Table 4.1 and as defined in 3GPP TS 36.306 [2]. Unlike in HSPA where devices support a wide range of different modulation and coding schemes because the standard evolved over time, all LTE UEs support the very fast 64-QAM (Quadrature Amplitude Modulation) in the downlink direction and antenna diversity. Further device categories have been defined for LTE-Advanced and are described in Section 4.15.2.

In the uplink direction, only the support of the slower but more reliable 16-QAM is required for terminal classes 1–4. Class 5 devices are an exception as they have to support 64-QAM.

Except for UE category 1, which was never introduced in practice, all mobile devices have to support MIMO transmission in the downlink direction. With this advanced transmission scheme, several data streams are transmitted on the same carrier frequency from multiple antennas from the base station to multiple antennas in the mobile device. If the signals reach

Table 4.1 LTE UE categories

Category	1	2	3	4	5
Maximum downlink datarate (20 MHz carrier)	10	50	100	150	300
Maximum uplink datarate	5	25	50	50	75
Number of receive antennas	2	2	2	2	4
Number of MIMO downlink streams	1	2	2	2	4
Support for 64 QAM in the uplink_direction	No	No	No	No	Yes

the receiver via different paths, for example, because of reflections at different angles from objects owing to the spatial separation of the transmitter and receiver antennas, the receiver can distinguish between the different transmissions and recreate the original data streams. The number of transmit and receive antennas dictates the number of data streams that can be sent in parallel. Most LTE networks and devices use 2×2 MIMO, that is, two transmit and two receive antennas. In the future, 4×4 MIMO might be used with category 5 UEs in case this is also supported on the network side. It will be a challenge, however, to fit four independent antennas in a small mobile device. Further, there are many frequency bands that LTE can be used in, making antenna design in mobile devices even more challenging, especially when taking the fact that devices have to support GSM and UMTS in the same or different frequency bands into account.

In practice, most devices are of categories 3 and 4, and peak datarates observed are between 100 and 150 Mbit/s under ideal conditions when using a 20-MHz carrier. Average speeds are lower because of factors such as the presence of multiple users in the cell, interference from neighboring cells and less than ideal reception conditions. Details can be found in Sauter [3].

LTE networks are deployed in different frequency bands depending on the geographical location. Table 4.2 [4] shows the selection of frequency bands that have been defined in the LTE specifications and that are in use today. The list of used bands is by no means complete and new bands are frequently added. The band numbers shown in the table are defined in 3GPP TS 36.101 [5].

In Europe, the spectrum returned from TV broadcasters is used by LTE between 791 and 862 MHz. This band is also referred to as the digital dividend band, as TV broadcasters returned this spectrum because of reduced spectrum requirements of digital vs. analog TV signal broadcasting.

The Chinese market is an exception in the table. Unlike most other regions in the world, where FDD is used to separate uplink and downlink, China uses TDD. Hence, the frequency ranges noted in the uplink and downlink are the same.

Most LTE-capable devices also support other radio technologies such as GSM and UMTS. As a consequence, a typical LTE device today does not only support seven or more LTE frequency bands but also supports those for the other radio technologies. A device sold in Europe usually also supports 900 and 1800 MHz for GSM, 900 and 2100 MHz for UMTS and, in addition, the 850-MHz and 1900-MHz bands for international GSM and UMTS roaming. This is a challenge for the antenna design as the sensitivity of a device's antennas must be equally good in all supported non-roaming bands. Furthermore, supporting an increasing number of bands is a challenge for receiver chips as adding more input ports decreases their overall sensitivity, which needs to be compensated for by advances in receiver technology.

Table 4.2 Extract of LTE frequency bands sorted by region

Band	Downlink (DL) (MHz)	Uplink (UL) (MHz)	UL/DL separation (duplex gap in MHz)	Duplex mode	Carrier bandwidth (MHz) typically used
Europe					
3	1805–1880	1710–1785	20	FDD	20
7	2620–2690	2500–2570	50	FDD	20
20	791–821	832–862	10	FDD	10
Japan					
1	2110–2170	1920–1980	130	FDD	20
United States					
4	2110–2155	1710–1755	355	FDD	10
13	746–756	777–787	21	FDD	10
17	734–746	704–716	20	FDD	10
China					
38	2570–2620	2570–2620	–	TDD	20
39	1880–1920	1880–1920	–	TDD	20
40 according to [8]	2300–2400	2300–2400	–	TDD	20

4.2.2   The eNode-B and the S1 and X2 Interfaces

The most complex device in the LTE network is the base station, referred to as eNode-B in the specification documents. The name is derived from the name originally given to the UMTS base station (Node-B) with an 'e' referring to 'evolved'. The leading 'e' has also been added to numerous other abbreviations already used in UMTS. For example, while the UMTS radio network is referred to as the UTRAN (Universal Mobile Telecommunications System Terrestrial Radio Access Network), the LTE radio network is referred to as the eUTRAN.

eNode-Bs consist of three major elements:

- the antennas, which are the most visible parts of a mobile network;
- radio modules that modulate and demodulate all signals transmitted or received on the air interface;
- digital modules that process all signals transmitted and received on the air interface and that act as an interface to the core network over a high-speed backhaul connection.

Many vendors use an optical connection between the radio module and the digital module. This way, the radio module can be installed close to the antennas, which reduces the length of costly coaxial copper cables to the antennas. This concept is also referred to as Remote Radio Head (RRH), and significant savings can be achieved, especially if the antennas and the base station cabinet cannot be installed close to each other.

Unlike in UMTS where the base station at the beginning was little more than an intelligent modem, LTE base stations are autonomous units. Here, it was decided to integrate most of the functionality that was previously part of the radio network controller (RNC) into the base station itself. Hence, the eNode-B is not only responsible for the air interface but also for

- user management in general and scheduling air interface resources;
- for ensuring QoS such as ensuring latency and minimum bandwidth requirements for real-time bearers and maximum throughput for background applications depending on the user profile;
- for load balancing between the different simultaneous radio bearers to different users;
- mobility management;
- for interference management, that is, to reduce the impact of its downlink transmissions on neighboring base stations in cell edge scenarios. Further details are described below.

For example, the eNode-B decides on its own to hand over ongoing data transfers to a neighboring eNode-B, a novelty in 3GPP systems. It also executes the handover autonomously from higher layer nodes of the network, which are only informed of the procedure once it has taken place.

The air interface is referred to as the LTE Uu interface and is the only interface in wireless networks that is always wireless. The theoretical peak datarates that can be achieved over the air depends on the amount of spectrum used by the cell. LTE is very flexible in this regard and allows bandwidth allocations between 1.25 and 20 MHz. In 20 MHz and 2×2 MIMO configuration that is typical for current LTE networks and mobile devices, peak speeds of up to 150 Mbit/s can be reached. Speeds that can be achieved in practice depend on many factors such as the distance of a mobile device from the base station, transmission power used by the base station, interference from neighboring base stations, etc. Achievable speeds in practice are hence much lower. A full discussion can be found in Sauter [3].

The interface between the base station and the core network is referred to as the S1 interface. It is usually carried either over a high-speed copper or fiber cable, or alternatively over a high-speed microwave link. Microwave links are, for example, based on Ethernet. Transmission speeds of several hundred megabits per second or even gigabits per second are required for most eNode-Bs as they usually consist of three or more sectors. In addition, a single backhaul link might also carry traffic from colocated GSM and UMTS installations. Transmission capacity requirements for backhaul links can thus far exceed the capacity of a single sector.

The S1 interface is split into two logical parts, which are both transported over the same physical connection.

User data is transported over the S1 User Plane (S1-UP) part of the interface. IP packets of a user are tunneled through an IP link in a manner similar to that already described for GPRS to enable seamless handovers between different LTE base stations and UMTS or GPRS/EDGE. In fact, the General Packet Radio Service Tunneling Protocol (GTP) is reused for this purpose [6] as shown in Figure 4.2(b). By tunneling the user's IP data packets, they can easily be redirected to a different base station during a handover as tunneling makes this completely transparent to the end-user data flow. Only the destination IP address on layer 3 (the tunneling IP layer) is changed, while the user's IP address remains the same. For further details, the readers may refer to Chapter 2 on the Gn interface, which uses the same mechanism. The protocols on layers 1 and 2 of the S1 interface are not described in further detail in the specification and are just referred to as layer 1 (L1) and layer 2 (L2). As a consequence, any suitable protocol for transporting IP packets can be used.

The S1 Control Plane (S1-CP) protocol, as defined in 3GPP TS 36.413, [7] is required for two purposes: First, the eNode-B uses it for interaction with the core network for its own purposes, that is, to make itself known to the network, to send status and connection keep-alive

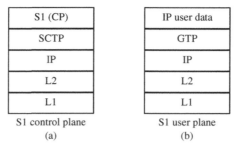

S1 control plane   S1 user plane
(a)   (b)

Figure 4.2   S1 control (a) and user (b) plane protocol stack

information and for receiving configuration information from the core network. Second, the S1-CP interface is used for transferring signaling messages that concern the users of the system. For example, when a device wants to communicate using the LTE network, an individual logical connection has to be established and the core network is responsible for authentication, for supplying keys for encrypting data on the air interface and for the establishment of a tunnel for the user's data between the eNode-B and the core network. Once the user's data tunnel is in place, the S1-CP protocol is used to maintain the connection, to organize a handover of the connection to another LTE, UMTS or GSM base station as required. Further details are discussed in Section 4.6.

Figure 4.2(a) shows the S1-CP protocol stack. IP is used as a basis. Instead of the commonly known Transmission Control Protocol (TCP) and User Datagram Protocol (UDP) protocols on layer 4, the telecom-specific Stream Control Transmission Protocol (SCTP) is used as defined in RFC 4960 [8]. It ensures that a large number of independent signaling connections can be established simultaneously with in-sequence transport, congestion management and flow control.

In previous 3GPP radio access networks, base stations were controlled by a central device. In GSM, this is the base station controller (BSC), and in UTMS it is the RNC. In these systems, the central controllers are responsible for setting up the radio links to wireless devices via the base stations, for controlling the connections while they are used, for ensuring QoS and for handing over a connection to another base station when required. In LTE, this concept was abandoned to remove latency from the user path and to distribute these management tasks as they require significant resources if concentrated in few higher layer network nodes. Especially, packet-switched connections generate a lot of signaling load because of the frequent switching of the air interface state when applications on the device only transmit and receive information in bursts with long timeouts in between. During these times of inactivity, the air interface connection to the mobile device has to be changed to use the available bandwidth efficiently and to reduce the power consumption of mobile devices. Details on this can be found in Chapter 3 for UMTS and Section 4.7 for LTE.

As a consequence of this autonomy, LTE base stations communicate directly with each other over the X2 interface for two purposes: First, handovers are now controlled by the base stations themselves. If the target cell is known and reachable over the X2 interface, the cells communicate directly with each other. Otherwise, the S1 interface and a core network are employed to perform the handover. Base station neighbor relations are either configured by the network operator in advance or can be detected by base stations themselves with the help of neighbor cell information being sent to the base station by mobile devices. This feature is referred to as Automatic Neighbor Relation (ANR) and requires the active support of mobile devices as the base stations themselves cannot directly detect each other over the air interface.

The second use of the X2 interface is for interference coordination. As in UMTS, neighboring LTE base stations use the same carrier frequency so that there are areas in the network where mobile devices can receive the signals of several base stations. If the signals of two or more base stations have a similar strength, the signals of the base stations that the mobile device does not communicate with at that moment are perceived as noise and the resulting throughput suffers significantly. As mobile devices can report the noise level at their current location and the perceived source to their serving base station, the X2 interface can then be used by that base station to contact the neighboring base station and agree on methods to mitigate or reduce the problem. Details are discussed in Section 4.8 on network planning aspects.

Like the S1 interface, the X2 interface is independent of the underlying transport network technology and IP is used on layer 3. SCTP is used for connection management, and the X2 application protocol defined in 3GPP TS 36.423 [9] encapsulates the signaling messages between the base stations. During a handover, user data packets can be forwarded between the two base stations involved in the process. For this, the GTP is used. While the X2 interface directly connects base stations with each other from a logical point of view as shown in Figure 4.1, the practical implementation is different. Here, the X2 interface is transported over the same backhaul link as the S1 interface up to the first IP aggregation router. From there, the S1 data packets are routed to the core network while X2 data packets are routed back to the radio network as shown in Figure 4.3. The main purpose of the aggregation router is to combine the traffic of many base stations into a single traffic flow. This reduces the number of links required in the field. In addition, the combined traffic flow is lower than the combined peak capacity of the backhaul links to the base stations, as in practice, the utilization of different base stations at any one time is different and variable over time.

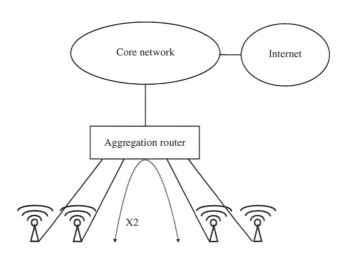

Figure 4.3 Physical routing of the S1 and the X2 interface

4.2.3 The Mobility Management Entity (MME)

While the eNode-Bs autonomously handle users and their radio bearers once they are established, overall user control is centralized in the core network. This is necessary as there needs to be a single point over which data flows between the user and the Internet. Further, a centralized user database is required, which can be accessed from anywhere in the home network and also from networks abroad in case the user is roaming.

The network node responsible for all signaling exchanges between the base stations and the core network and between the users and the core network is the Mobility Management Entity (MME). Figure 4.1 shows its location in the overall network architecture. In large networks, there are usually many MMEs to cope with the amount of signaling and due to station redundancy. As the MMEs are not involved in air interface matters, the signaling it exchanges with the radio network is referred to as Non-access Stratum (NAS) signaling. In particular, the MME is responsible for the following tasks:

- **Authentication**. When a subscriber first attaches to the LTE network, the eNode-B communicates with the MME over the S1 interface and helps to exchange authentication information between the mobile device and the MME. The MME then requests authentication information from Home Subscriber Server (HSS), which is discussed in more detail below, and authenticates the subscriber. Once done, it forwards encryption keys to the eNode-B so that further signaling and data exchanges over the air interface can be ciphered. Further details can be found in Section 4.6.2 on the attach procedure and default bearer activation.
- **Establishment of bearers**. The MME itself is not directly involved in the exchange of user data packets between the mobile device and the Internet. Instead, it communicates with other core network components to establish an IP tunnel between the eNode-B and the gateway to the Internet. However, it is responsible for selecting a gateway router to the Internet, if there is more than one gateway available.
- **NAS mobility management**. In case a mobile device is dormant for a prolonged duration of time (typical values found in practice are 10–30 seconds), the air interface connection and resources in the radio network are released. The mobile device is then free to roam between different base stations in the same Tracking Area (TA) without notifying the network to save battery capacity and signaling overhead in the network. Should new data packets from the Internet arrive for this device while it is in this state, the MME has to send paging messages to all eNode-Bs that are part of the current TA of the mobile device. Once the device responds to the paging, the bearer(s) is(are) reestablished.
- **Handover support**. In case no X2 interface is available, the MME helps to forward the handover messages between the two eNode-Bs involved. The MME is also responsible for the modification of the user data IP tunnel after a handover in case different core network routers become responsible.
- **Interworking with other radio networks**. When a mobile device reaches the limit of the LTE coverage area, the eNode-B can decide to hand over the mobile device to a GSM or UMTS network or instruct it to perform a cell change to suitable cell. In both cases, described in more detail in Section 4.9, the MME is the overall managing instance and communicates with the GSM or UMTS network components during this operation.
- **SMS and voice Support**. Despite LTE being a pure IP network, some functionality is required to support traditional services such as voice calls and SMS, which were so far part

of the GSM and UMTS circuit-switched core networks and cannot thus simply be mapped to LTE. This is discussed in more detail in Section 4.11.

For these tasks, a number of different interfaces such as the S5, S6a, S11 and SGs are used. These are shown in Figure 4.1 at the beginning of this chapter and described in the following sections.

When compared to GPRS and UMTS, the tasks of MMEs are the same as those of the SGSN. The big difference between the two entities is that while the SGSN is also responsible for forwarding the user data between the core network and the radio network, the MME deals only with the signaling tasks described above and leaves the user data to the Serving Gateway (S-GW), which is described in the next section.

Owing to this similarity, nodes that combine the functionality of a 2G SGSN, a 3G SGSN and an MME are used in networks today. As an interesting option, which is, however, not yet widely used in practice, the one-tunnel enhancement described in Chapter 3 also removes the user plane functionality from the SGSN, making a combined node a pure integrated signaling platform that lies between the access networks and a single core network for all radio technologies in the future.

4.2.4   The Serving Gateway (S-GW)

The S-GW is responsible for managing user data tunnels between the eNode-Bs in the radio network and the Packet Data Network Gateway (PDN-GW), which is the gateway router to the Internet, and is discussed in the next section. On the radio network side, it terminates the S1-UP GTP tunnels, and on the core network side, it terminates the S5-UP GTP tunnels to the gateway to the Internet. S1 and S5 tunnels for a single user are independent of each other and can be changed as required. If, for example, a handover is performed to an eNode-B under the control of the same MME and S-GW, only the S1 tunnel needs to be modified to redirect the user's data stream to and from the new base station. If the connection is handed over to an eNode-B that is under the control of a new MME and S-GW, the S5 tunnel has to be modified as well.

Tunnel creation and modification are controlled by the MME, and commands to the S-GW are sent over the S11 interface as shown in Figure 4.1. The S11 interface reuses the GTP-C (control) protocol of GPRS and UMTS by introducing new messages. The simpler UDP protocol is used as the transport protocol below instead of SCTP, and the IP protocol is used on the network layer.

In the standards, the S-GW and the MME are defined independently. Hence, the two functions can, in practice, be run on the same or different network nodes. This allows an independent evolution of signaling capacity and user data traffic. This was done because additional signaling mainly increases the processor load, while rising data consumption of users requires a continuous evolution of routing capacity and an evolution of the number and types of network interfaces that are used.

4.2.5   The PDN-Gateway

The third LTE core network node is the PDN-GW. In practice, this node is the gateway to the Internet and some network operators also use it to interconnect to intranets of large companies over an encrypted tunnel to offer employees of those companies direct access to their private internal networks. As mentioned in the previous section, the PDN-GW terminates the S5 interface.

On the user plane, this means that data packets for a user are encapsulated into an S5 GTP tunnel and forwarded to the S-GW, which is currently responsible for this user. The S-GW then forwards the data packets over the S1 interface to the eNode-B that currently serves the user, from which it is then sent over the air interface to the user's mobile device.

The PDN-GW is also responsible for assigning IP addresses to mobile devices. When a mobile device connects to the network after being switched on, the eNode-B contacts the MME as described above. The MME then authenticates the subscriber and requests an IP address from the PDN-GW for the device. For this purpose, the S5 control plane protocol is used. The procedure is similar to the procedure in GPRS and UMTS, where the SGSN requests an IP address from the GGSN as described in Chapters 2 and 3. If the PDN-GW grants access to the network, it returns the IP address to the MME, which in turn forwards it to the subscriber. Part of the process is also the establishment of corresponding S1 and S5 user data tunnels. A full message flow is presented in Section 4.6.2.

In practice, a mobile device can be assigned several IP addresses simultaneously. Several IP addresses might be necessary in cases where the device uses services that are part of the network operator's internal network such as the IP Multimedia Subsystem (IMS). A mobile device can also request the assignment of simultaneous IPv4 and IPv6 addresses. Further, it is possible to add or remove IP addresses as needed at any point in time. In most cases, however, most mobile devices only use a single IPv4 address to access the Internet.

Owing to a shortage of available IPv4 addresses, most network operators assign local IP addresses and use Network Address Translation (NAT) to map many internal IP addresses to a few public IP addresses on the Internet. This is similar to home network Asynchronous Digital Subscriber Line (ADSL) routers that are also assigned only a single public IP address from the fixed-line Internet service provider, which then assign local IP addresses to all PCs, notebooks and other devices connected to it. A downside of this approach is that services running on mobile devices cannot be directly reached from the outside world as the NAT scheme requires that the connection is always established from the local IP address. Only then can a mapping be created between the internal IP address and TCP or UDP port and the external IP address and TCP or UDP port.

An advantage of NAT is that malicious connection attempts, for example, by viruses probing the network for vulnerable hosts or data intended for the previous user of the IP address are automatically discarded at the PDN-GW. This not only protects mobile devices to a certain degree but also helps to conserve power on the mobile device's side as malicious packets cannot keep the air interface connection in a power-consuming state when no other data is transferred. Details on this topic can be found in Sauter [10].

NAT is not required for IPv6 addresses because of the large address space available. Details of how IPv6 addresses are assigned in 3GPP networks such as UMTS and LTE are discussed in Sauter [11]. Also, IPv6 might be helpful to further reduce power consumption of mobile devices as the need for keep-alive messages for many applications is reduced or removed as discussed in Sauter [12].

The PDN-GW also plays an important part in international roaming scenarios. For a seamless access to the Internet for a user while traveling abroad, roaming interfaces connect LTE, UMTS and GPRS core networks of different network operators in different countries with

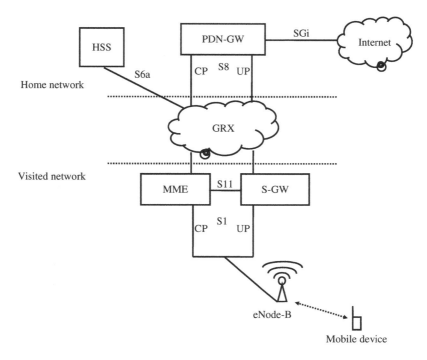

Figure 4.4   LTE international roaming with home routing

each other so that a foreign network can query the user database in the home network of a user for authentication purposes. When a bearer, for example, for Internet access, is established, a GTP tunnel is created between the S-GW in the visited network and a PDN-GW in the user's home network. The process is nearly identical to that for the establishment of a user data tunnel on the S5 interface as described before. To distinguish the scenario, however, the interface is referred to as S8. Figure 4.4 shows this setup, which is also referred to as home routing. The disadvantage of home routing is that the user's data is first transported back to the home network before it is sent to the Internet. An alternative, referred to as 'local breakout', also exists in the standards. Here, the connection to the Internet is established via a PDN-GW in the visited network. However, this is not widely used today.

Whether a standalone network node is used for the PDN-GW or a combination of several functions is embedded in a single node depends on the network operator and the size of the network. In theory, MME, S-GW and PDN-GW could all be implemented in a single device. In such a case, the S11 and S5 interfaces would be internal. In practice, the functionality is usually decoupled because of the different evolution of traffic and signaling load. In a roaming scenario, the S-GW and PDN-GW are always separate entities when default home routing is used.

4.2.6   The Home Subscriber Server (HSS)

LTE shares its subscriber database with GSM and UMTS. In these systems, the database is referred to as the Home Location Register (HLR) and the Mobile Application Part (MAP) is

used as the protocol between the Mobile Switching Center (MSC) and SGSN on the one side
and the HLR on the other side. In LTE, an IP-based protocol referred to as DIAMETER is used
to exchange information with the database. It is standardized in RFC 3588 [13] and referred to
as S6a. Further, the name of the database has been changed to HSS. In practice, however, the
HLR and the HSS are physically combined to enable seamless roaming between the different
radio access networks. Each subscriber has a record in the HLR/HSS and most properties
are applicable for communicating over all radio access networks. The most important user
parameters in the HSS are

- the user's International Mobile Subscriber Identity (IMSI), which uniquely identifies a subscriber. The IMSI implicitly includes the Mobile Country Code (MCC) and Mobile Network
  Code (MNC) and is thus used when the user is roaming abroad to find the home network of
  the user to contact the HSS. A copy of the IMSI is stored on the subscriber identity module
  (SIM) card of the subscriber;
- authentication information that is used to authenticate the subscriber and to generate encryption keys on a session basis;
- circuit-switched service properties such as the user's telephone number, referred to as the
  Mobile Subscriber Integrated Services Digital Network (MSISDN) number, and the services
  the user is allowed to use, such as SMS, call forwarding, and so on. While the MSISDN is
  used for some purposes in LTE, the other values are mainly of interest while the user is
  connected to GSM or UMTS;
- packet-switched service properties such as the Access Point Names (APNs) the subscriber
  is allowed to use, which in turn references the properties of a connection to the Internet or
  other external packet data network such as the maximum throughput;
- IMS-specific information (see Chapter 3);
- the ID of the current serving MSC so that incoming circuit-switched calls and SMS
  messages can be routed correctly;
- the ID of the SGSN or MME, which is used in case the user's HSS profile is updated to push
  the changes to those network elements.

4.2.7   Billing, Prepaid and Quality of Service

The network nodes and interfaces described in the previous sections are the main components
required to offer wireless connectivity to the user. In addition, several other supporting network
components and interfaces are usually deployed in practice to complement the network with
additional services.

To charge mobile subscribers for their use of the system, billing records are created, for
example, on the MME. These are collected and sent to a charging system, which once a month
generates an invoice that is then sent to the customer. This is also referred to as offline billing
or postpaid billing.

Another billing method is online charging, which lets subscribers buy vouchers for certain
services or a certain amount of data to be transferred within a certain time. This is also referred
to as prepaid billing and originally became popular for circuit-switched voice calls. In UMTS
and LTE, network operators can offer prepaid billing and usage tracking in real time or near
real time, which requires interaction with core network components such as the MME, S-GW
or PDN-GW. As such services are not standardized, user interaction depends on the particular

implementation. A popular implementation is to offer 'landing pages' that are automatically displayed to the user once he connects to the network for session-based billing or once the amount of data has been used up. Through the landing page, the user can subsequently buy additional credit or enter an ID from a voucher that was previously bought in a shop.

Usage tracking for billing purposes is often also used for postpaid subscribers which are charged on a monthly basis and who have subscribed to an Internet option, which is throttled to a very low speed once a subscribed data bucket has been used up. This requires a function in the network that can monitor the data usage of the subscriber and a device that can limit his data rate once he has used up his monthly data volume. Some network operators also offer the option to postpaid subscribers to buy additional data volume which is then invoiced via his monthly bill. Typically, such systems are located behind the SGi interface.

For real-time applications such as Voice over Internet Protocol (VoIP), ensuring constant delay and a minimal bandwidth on all interfaces within the LTE network are crucial during times of high traffic load. This can be done via a standardized QoS node, the Policy Control Resource Function (PCRF). Applications can use the standardized Rx interface to request a certain QoS profile for a data flow. The PCRF then translates this request and sends commands to the PDN-GW and the S-GW, which in turn enforce the QoS request in the core and access network. The PCRF is part of the 3GPP IMS specifications and was originally intended for use by IMS services. In practice, it can also be used by other network operator-deployed services, but is rarely used currently. It is important to note that only network operator services can access the PCRF. Internet-based services have no means to request any kind of QoS from the LTE network.

4.3 FDD Air Interface and Radio Network

The major evolution in LTE compared to previous 3GPP wireless systems is the completely revised air interface. To understand why a new approach was taken, a quick look back at how data was transmitted in previous generation systems is necessary:

GSM is based on narrow 200 -kHz carriers that are split into eight repeating timeslots for voice calls. One timeslot carries the data of one voice call, thus limiting the number of simultaneous voice calls on one carrier to a maximum of eight. Base stations use several carriers to increase the number of simultaneous calls. Later on, the system was enhanced with GPRS for packet-switched data transmission. The decision to use 200 -kHz carriers, however, remained the limiting factor.

With UMTS, this restriction was lifted by the introduction of carriers with a bandwidth of 5 MHz. Instead of using dedicated timeslots, CDMA, where data streams are continuous and separated with different codes, was used. At the receiving end, the transmission codes are known and the different streams can hence be separated again. With HSPA, the CDMA approach was continued but a timeslot structure was introduced again to improve user data scheduling. The timeslots, however, were not optimized for voice calls but for quickly transporting packet-switched data traffic.

With today's hardware and processing capabilities, higher datarates can be achieved by using an increased carrier bandwidth. UMTS, however, is very inflexible in this regard as the CDMA transmission scheme is not ideal for wider channels. When the carrier bandwidth is increased, the transmission steps need to become shorter to take advantage of the additional bandwidth. While this can be done from a signal processing point of view, this is very disadvantageous in a practical environment where the radio signal is reflected by objects, and the signal reaches the receiver via several paths. As a result, the receiver sees not just one signal per transmission step but several, each arriving at a slightly different time. When increasing the transmission speed, which results in a decrease in the time of each transmission step, the negative effect of the delayed signal paths increases. As a consequence, CDMA is not suitable for carrier bandwidths beyond 5 MHz. Multicarrier operation has been defined for UMTS to mitigate the problem to some degree at the expense of rising complexity.

The following sections now describe how LTE enables the use of much larger bandwidths in the downlink and the uplink directions.

4.3.1   OFDMA for Downlink Transmission

In the downlink direction, LTE uses Orthogonal Frequency Division Multiple Access (OFDMA). Instead of sending a data stream at a very high speed over a single carrier as in UMTS, OFDMA splits the data stream into many slower data streams that are transported over many carriers simultaneously. The advantage of many slow but parallel data streams is that transmission steps can be sufficiently long to avoid the issues of multipath transmission on fast data streams as discussed above. Table 4.3 shows the number of subcarriers (i.e., data streams) used, depending on the bandwidth used for LTE. Not included in the count is the center carrier, which is always left empty. The more bandwidth is available for the overall LTE carrier, the more the number of subcarriers used. As an example, a total of 600 subcarriers are used with an overall signal bandwidth of 10 MHz. In other words, the overall datarate can be up to 600 times the datarate of each subcarrier.

To save bandwidth, the subcarriers are spaced in such a way that the side lobes of each subcarrier wave are exactly zero at the center of the neighboring subcarrier. This property is referred to as 'orthogonality'. To decode data transmitted in this way, a mathematical function referred to as Inverse Fast Fourier Transformation (IFFT) is used. In essence, the input to an IFFT is a frequency domain signal that is converted into a time domain signal. As each subcarrier uses a different frequency, the receiver uses an FFT that shows which signal was sent in each of the subcarriers at a specific instant in time.

Figure 4.5 shows how the concept works in practice. At the top left, the digital data stream is delivered to the transmitter. The data stream is then put on parallel streams, each of which is then mapped to subcarriers in the frequency domain. An IFFT function is then used to convert the result into a time domain signal, which can then be modulated and sent over the air to

Table 4.3   Defined bandwidths for LTE

Bandwidth (MHz)	Number of subcarriers	FFT size
1.25	76	128
2.5	150	256
5	300	512
10	600	1024
15	900	1536
20	1200	2048

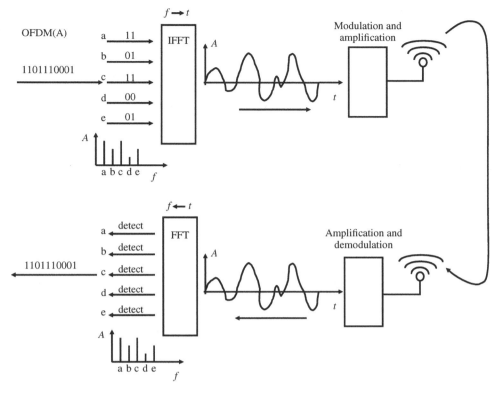

Figure 4.5 Principles of OFDMA for downlink transmission. Source: Wiley 2009. Reproduced with permission from John Wiley and Sons Ltd

the receiver. The receiving end is shown in the lower part of the figure. After demodulation of the signal, it is fed into the FFT function that converts the time domain signal back into a frequency domain representation in which the individual subcarrier frequencies can then be detected. Finally, the slow data streams from each subcarrier are assembled again into the single fast data stream, which is then forwarded to higher layers of the protocol stack.

LTE uses the following physical parameters for the subcarriers:

- **subcarrier spacing**: 15 kHz;
- **length of each transmission step** (**OFDM symbol duration**): 66.667 microseconds;
- **standard cyclic prefix**: 4.7 microseconds. The cyclic prefix is transmitted before each OFDM symbol to prevent intersymbol interference due to different lengths of several transmission paths. For difficult environments with highly diverse transmission paths, a longer cyclic prefix of 16.67 microseconds has been specified as well. The downside of using a longer cyclic prefix is a reduced user data speed since the symbol duration remains the same and, hence, fewer symbols can be sent per time interval.

It is interesting to compare the very narrow subcarrier spacing of 15 kHz to the 200-kHz channels used in GSM to see just how narrow the individual subcarriers are. Further the subcarrier spacing remains the same regardless of the overall channel bandwidth. For a wider channel, the number of subcarriers is increased while the individual subcarrier bandwidth remains the same. This is an important concept as this enables and preserves the channel bandwidth flexibility even beyond the maximum of 20 MHz specified for LTE in 3GPP Release 8.

In UMTS, users are scheduled in the code domain and, with HSPA, additionally in the time domain. In LTE, users are scheduled in the frequency domain, that is, at a certain point in time several users can receive data on a different set of subcarriers. In addition, users are scheduled in the time domain.

4.3.2   SC-FDMA for Uplink Transmission

For uplink data transmissions, the use of OFDMA is not ideal because of its high Peak to Average Power Ratio (PAPR) when the signals from multiple subcarriers are combined. In practice, the amplifier in a radio transmitter circuit has to support the peak power output required to transmit the data and this value defines the power consumption of the PA device regardless of the current transmission power level required.

With OFDM, the maximum power is seldom used and the average output power required for the signal to reach the base station is much lower. Hence, it can be said that the PAPR is very high. The overall throughput of the device, however, does not correspond to the peak power but instead corresponds to the average power, as this reflects the average throughput. Therefore, a low PAPR would be beneficial to balance power requirements of the transmitter with the achievable datarates.

For a base station, a high PAPR can be tolerated as power is abundant. For a mobile device that is battery driven, however, the transmitter should be as efficient as possible. 3GPP has hence decided to use a different transmission scheme, referred to as Single-Carrier Frequency Division Multiple Access (SC-FDMA). SC-FDMA is similar to OFDMA but contains additional processing steps as shown in Figure 4.6. In the first step, shown in the top left of the figure, the input signal is delivered. Instead of dividing the data stream and putting the resulting substreams directly on the individual subcarriers, the time-based signal is converted to a frequency-based signal with an FFT function. This distributes the information of each bit onto all subcarriers that will be used for the transmission and thus reduces the power differences between the subcarriers. The number of subcarriers used depends on the signal conditions, the transmission power of the device and the number of simultaneous users in the uplink. Subchannels used for uplink transmissions are encoded with 0. This frequency vector is then fed to the IFFT as in OFDMA, which converts the information back into a timebase signal. For such a signal, it can be mathematically shown that the PAPR is much lower than that obtained without the additional FFT.

At the receiving end shown in the lower part of Figure 4.6, the signal is first demodulated and then sent to the FFT function as in OFDMA. To get the original signal back, the resulting frequency signal is given to an IFFT function that reverses the initial processing step of the transmitter side.

Apart from the additional processing step on the transmitting and receiving sides, the same physical parameters are used as for the downlink direction.

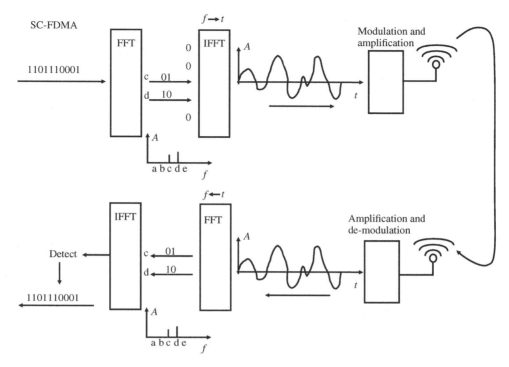

Figure 4.6   Principles of SC-FDMA for uplink transmission. Source: Wiley 2009. Reproduced with permission from John Wiley and Sons Ltd

4.3.3   Symbols, Slots, Radio Blocks and Frames

Data transmission in LTE is organized as follows: The smallest transmission unit on each subcarrier is a single transmission step with a length of 66.667 microseconds. A transmission step is also referred to as a symbol and several bits can be transmitted per symbol depending on the modulation scheme. If radio conditions are excellent, 64-QAM is used to transfer 6 bits ($2^6 = 64$) per symbol. Under less ideal signal conditions, 16-QAM or QPSK (Quadrature Phase Shift Keying) modulation is used to transfer 4 or 2 bits per symbol. A symbol is also referred to as a Resource Element (RE).

As the overhead involved in assigning each individual symbol to a certain user or to a certain purpose would be too great, the symbols are grouped together in a number of different steps as shown in Figure 4.7. First, seven consecutive symbols on 12 subcarriers are grouped into a Resource Block (RB). An RB occupies exactly one slot with a duration of 0.5 milliseconds.

Two slots form a subframe with a duration of 1 millisecond. A subframe represents the LTE scheduling time, which means that at each millisecond the eNode-B decides as to which users are to be scheduled and which RBs are assigned to which user. The number of parallel RBs in each subframe depends on the system bandwidth. If a 10 -MHz carrier is used, 600

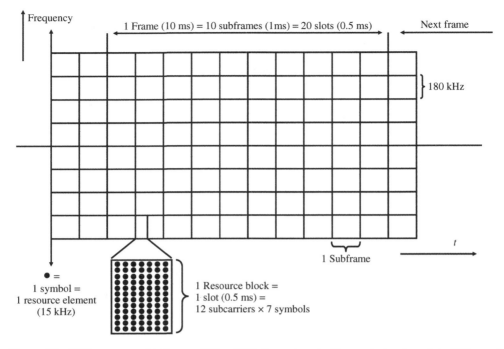

Figure 4.7   LTE resource grid. Source: Wiley 2009. Reproduced with permission from John Wiley and Sons Ltd

subcarriers are available (see Table 4.3). As an RB bundles 12 subcarriers, a total of 50 RBs can be scheduled for one or more users per subframe.

The network has two options to transmit a subframe. The first option is Localized Virtual Resource Blocks (LVRBs), which are transmitted in a coherent group as shown in Figure 4.7. In this transmit mode, the eNode-B requires a narrowband channel feedback from the mobile device to schedule the RBs on subcarriers that do not suffer from narrowband fading. The second option is to transfer data in Distributed Virtual Resource Blocks (DVRBs), where the symbols that form a block are scattered over the whole carrier bandwidth. In this case, the mobile device returns either no channel feedback or a wideband channel feedback over the whole bandwidth.

And finally, 10 subframes are combined into an LTE radio frame, which has a length of 10 milliseconds. Frames are important, for example, for the scheduling of periodic system information (SI) as is discussed further. At this point, it should be noted that Figure 4.7 is a simplification as only eight RBs are shown in the y-axis. On a 10 -MHz carrier, for example, 50 RBs are used.

4.3.4   Reference and Synchronization Signals

As described above, the network assigns a number of RBs to a user for each new subframe, that is, once a millisecond. However, not all symbols of an RB can be used to transmit user

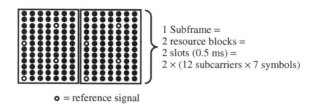

o = reference signal

Figure 4.8   Symbols in a resource block used for the reference signal

data. Which of the symbols are used for other purposes depends on the location of the RB in the overall resource grid.

To enable mobile devices to detect LTE carriers during a network search and to estimate the channel quality later on, reference symbols, also referred to as reference signals, are embedded in a predefined pattern over the entire channel bandwidth. Reference signals are inserted on every seventh symbol on the time axis and on every 6th subcarrier on the frequency axis as shown in Figure 4.8. Details are given in 3GPP TS 36.211 [14]. A total of 504 different reference signal sequences exist, which help a mobile device to distinguish transmissions of different base stations. These patterns are also referred to as the Physical Cell Identity (PCI). Neighboring base stations need to use different symbols for the reference signals for the mobile device to properly distinguish them. Hence, six PCI groups have been defined, each shifted by one subcarrier.

For initial synchronization, two additional signal types are used. These are referred to as the primary and secondary synchronization signals (PSSs and SSSs) and they are transmitted in every first and sixth subframe on the inner 72 subcarriers of the channel. On each of those subcarriers, one symbol is used for each synchronization signal. Hence, synchronization signals are transmitted every 5 milliseconds. Further details can be found in Section 4.6.1 where the initial cell search procedure is described.

4.3.5   The LTE Channel Model in Downlink Direction

All higher layer signaling and user data traffic are organized in channels. As in UMTS, logical channels, transport channels and physical channels have been defined as shown in Figure 4.9. Their aim is to offer different pipes for different kinds of data on the logical layer and to separate the logical data flows from the properties of the physical channel below.

On the logical layer, data for each user is transmitted in a logical Dedicated Traffic Channel (DTCH). Each user has an individual DTCH. On the air interface, however, all dedicated channels are mapped to a single shared channel that occupies all RBs. As described above, some symbols in each RB are assigned for other purposes and hence cannot be used for user data. Which RBs are assigned to which user is decided by the scheduler in the eNode-B for each subframe, that is, once per millisecond.

Mapping DTCHs to a single shared channel is done in two steps. First, the logical DTCHs of all users are mapped to a transport layer Downlink Shared Channel (DL-SCH). In the second step, this data stream is then mapped to the Physical Downlink Shared Channel (PDSCH).

Transport channels are able to not only multiplex data streams from several users but also multiplex several logical channels of a single user before they are finally mapped to a physical

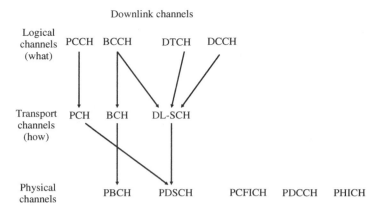

Figure 4.9   LTE downlink channel structure

channel. An example is as follows: A UE that has been assigned a DTCH also requires a control channel for the management of the connection. Here, the messages that are required, for example, for handover control, neighbor cell measurements and channel reconfigurations are sent. The DTCH and the DCCH are multiplexed on the DL-SCH before they are mapped to the PDSCH, that is, to individual RBs. In addition, even most of the cell-specific information that is sent on the logical broadcast control channel (BCCH) is also multiplexed on the transport downlink shared channel as shown in Figure 4.9.

In LTE, all higher layer data flows are eventually mapped to the physical shared channel, including the Paging Control Channel (PCCH), which is used for contacting mobile devices that are in a dormant state to inform them of new IP packets arriving from the network side. The PCCH is first mapped to the transport layer Paging Channel (PCH), which is then mapped to the PDSCH.

The only exception to the general mapping of all higher layer information to the shared channel is the transmission of a small number of system parameters that are required by mobile devices to synchronize to the cell. They are transmitted on the Physical Broadcast Channel (PBCH), which occupies three symbols on 72 subcarriers (= 6 RBs) in the middle of a channel every fourth frame. Hence, it is broadcast every 40 milliseconds and follows the PSSs and SSSs.

4.3.6   Downlink Management Channels

As discussed in the previous section, most channels are mapped to a single PDSCH, which occupies all RBs in the downlink direction except for those symbols in each RB which are statically assigned for other purposes. Consequently, a mechanism is required to indicate to each mobile device as to when, where and what kind of data is scheduled for them on the shared channel and which RBs they are allowed to use in the uplink direction. This is done via Physical Downlink Control Channel (PDCCH) messages.

The downlink control information occupies the first one to four symbols over the whole channel bandwidth in each subframe. The number of symbols that are used for this purpose is broadcast via the Physical Control Format Indicator Channel (PCFICH), which occupies 16 symbols. This flexibility was introduced so that the system can react on changing signaling bandwidth requirements, that is, the number of users that are to be scheduled in a subframe. This topic is discussed in more detail in Section 4.5.

Downlink control data is organized in Control Channel Elements (CCEs). One or more CCEs contain one signaling message that is either addressed to one device, or, in the case of broadcast information, to all mobile devices in the cell. To reduce the processing requirements and power consumption of mobile devices for decoding the control information, the control region is split into search spaces. A mobile device therefore does not have to decode all CCEs to find a message addressed to itself but only those in the search spaces assigned to it.

And finally, some symbols are reserved to acknowledge the proper reception of uplink data blocks or to signal to the mobile device that a block was not received correctly. This functionality is referred to as Hybrid Automatic Retransmission Request (HARQ) and the corresponding channel is the Physical Hybrid Automatic Retransmission Request Indicator Channel (PHICH).

In summary, it can be said that the PDSCH is transmitted in all RBs over the complete system bandwidth. In each RB, however, some symbols are reserved for other purposes such as the reference signals, the synchronization signals, the broadcast channel, the control channel, the PCFICH and the HARQ indicator channel. The number of symbols that are not available to the shared channel depends on the RB and its location in the resource grid. For each signal and channel, a mathematical formula is given in 3GPP TS 36.211 [14] so that the mobile device can calculate where it can find a particular kind of information.

4.3.7 System Information Messages

As in GSM and UMTS, LTE uses SI messages to convey information that is required by all mobile devices that are currently in the cell. Unlike in previous systems, however, only the Master Information Block (MIB) is transported over the broadcast channel. All other SI is scheduled in the PDSCH and their presence is announced on the PDCCH in a search space that has to be observed by all mobile devices.

Table 4.4 gives an overview of the different system information blocks (SIBs), their content and example repetition periods as described in 3GPP TS 36.331 [15]. The most important SI is contained in the MIB and is repeated every 40 milliseconds. Cell-related parameters are contained in the SIB 1 that is repeated every 80 milliseconds. All other SIBs are grouped into SI messages whose periodicities are variable. Which SIBs are contained in which SI message and their periodicities are announced in SIB 1.

Not all of the SIBs shown in Table 4.4 are usually broadcast as some of them are functionality dependent. In practice, the MIB and, in addition, SIB-1 and SIB-2 are always broadcast because they are mandatory. They are followed by SIB-3 and optionally SIB-4. SIB-5 is required only if the LTE network uses more than one carrier frequency. The broadcast of SIB-5 to −7 then depends on the other radio technologies used by the network operator.

4.3.8 The LTE Channel Model in Uplink Direction

In the uplink direction, a similar channel model is used as in the downlink direction. There are again logical, transport and physical channels to separate logical data streams from the physical transmission over the air interface and to multiplex different data streams onto a single channel. As shown in Figure 4.10, the most important channel is the Physical Uplink Shared

Table 4.4   System information blocks and content overview

Message	Content
MIB	Most essential parameters required for initial access
SIB 1	Cell identity and access-related parameters and scheduling information of system information messages containing the other SIBs
SIB 2	Common and shared channel configuration parameters
SIB 3	General parameters for intrafrequency cell reselection
SIB 4	Intrafrequency neighbor cell reselection information with information about individual cells
SIB 5	Interfrequency neighbor cell reselection parameters
SIB 6	UMTS inter-RAT cell reselection information to UMTS
SIB 7	GSM inter-RAT cell reselection information to GSM
SIB 8	CDMA2000 inter-RAT cell reselection information
SIB 9	If the cell is a femto cell, i.e., a small home eNode-B, this SIB announces its name
SIB 10	Earthquake and tsunami warning system (ETWS) information
SIB 11	Secondary ETWS information
SIB 12	Commercial mobile alert system (CMAS) information

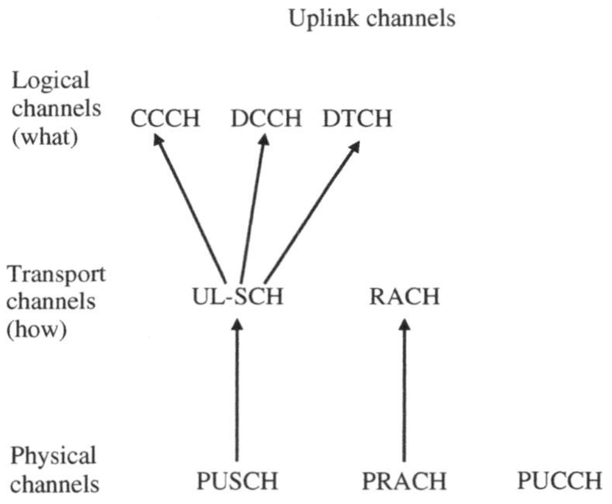

Uplink channels

Logical channels (what)   CCCH   DCCH   DTCH

Transport channels (how)   UL-SCH   RACH

Physical channels   PUSCH   PRACH   PUCCH

Figure 4.10   LTE Channel Uplink Structure

Channel (PUSCH). Its main task is to carry user data in addition to signaling information and signal quality feedback.

Data from the PUSCH are split into three different logical channels. The channel that transports the user data is referred to as the DTCH. In addition, the DCCH is used for higher layer signaling information. During connection establishment, signaling messages are transported over the Common Control Channel (CCCH).

Before a mobile device can send data in the uplink direction, it needs to synchronize with the network and has to request the assignment of resources on the PUSCH. This is required in the following scenarios:

- The mobile has been dormant for some time and wants to reestablish the connection.
- A radio link failure has occurred and the mobile has found a suitable cell again.

- During a handover process, the mobile needs to synchronize with a new cell before user data traffic can be resumed.
- Optionally for requesting uplink resources.

Synchronizing and requesting initial uplink resources is performed with a random access procedure on the Physical Random Access Channel (PRACH). In most cases, the network does not know in advance that a mobile device wants to establish communication. In these cases, a contention-based procedure is performed as it is possible that several devices try to access the network with the same Random Access Channel (RACH) parameters at the same time. This will result in either only one signal being received or, in the network, no transmissions being received at all. In both cases, a contention resolution procedure ensures that the connection establishment attempt is repeated.

Figure 4.11 shows the message exchange during a random access procedure. In the first step, the mobile sends one of the 64 possible random access preambles on the RACH. If correctly received, the network replies with a random access response on the PDSCH which includes

- a timing advance value so that the mobile can synchronize its transmissions;
- a scheduling grant to send data via the PUSCH;
- a temporary identifier for the UE that is only valid in the cell to identify further messages between a particular mobile device and the eNode-B.

The mobile device then returns the received temporary UE identifier to the network. If properly received, that is, only a single mobile tries to reply with the temporary UE identifier, the network finalizes the contention resolution procedure with a random access response message.

During handovers, the new eNode-B is aware that the mobile is attempting a random access procedure and can thus reserve dedicated resources for the process. In this case, there is no risk of several devices using the same resources for the random access procedure and no contention

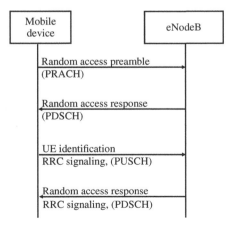

Figure 4.11   Random access procedure

resolution is required. As this random access procedure is contention free, only the first two messages shown in Figure 4.11 are required.

On the network side, the eNode-B measures the time of the incoming transmission relative to its own timing. The farther the mobile device is from the base station, the later the signal arrives. As in GSM, the network then informs the mobile device to adjust its transmission time. The farther the mobile device is from the base station, the earlier it has to start its transmissions for the signal to reach the base station at the right time. This is also referred to as the timing advance and the system can compensate transmission distances of up to 100 km.

When a mobile device has been granted resources, that is, it has been assigned RBs on the PUSCH, the shared channel is used for transmitting user data and also for transmitting lower layer signaling data, which is required to keep the uplink connection in place and to optimize the data transmission over it. The following lower layer information is sent alongside user data packets:

- The Channel Quality Indicator (CQI) that the eNode-B uses to adapt the modulation and coding scheme for the downlink direction.
- MIMO-related parameters (see Section 4.3.9).
- HARQ acknowledgments so that the network can quickly retransmit faulty packets (see Section 4.3.10).

In some cases, lower layer signaling information has to be transferred in the uplink direction, while no resources are assigned to a mobile device on the uplink shared channel. In this case, the mobile can send its signaling data such as HARQ acknowledgments and scheduling requests on the Physical Uplink Control Channel (PUCCH), which uses the first and the last RBs of a carrier. In other words, the PUCCH is located at the edges of the transmission channel.

As in the downlink direction, some information is required for the receiver, in this case the eNode-B, to estimate the uplink channel characteristics per mobile device. This is done in two ways: During uplink transmissions, Demodulation Reference Signals (DRS) are embedded in all RBs for which a mobile has received uplink scheduling grants. The symbols use the fourth symbol row of RB for the purpose. As the eNode-B knows the content of the DRS symbols, it can estimate the way the data transmission on each subcarrier is altered by the transmission path. As the same frequency is used by all mobile devices, independent of the eNode-B with which they communicate, different phase shifts are used for the DRS symbols depending on the eNode-B for which the transmission is intended. This way, the eNode-B can filter out transmissions intended for other eNode-Bs, which are, from its point of view, noise.

A second optional reference is the Sounding Reference Signal (SRS) that allows the network to estimate the uplink channel quality in different parts of the overall channel for each mobile device. As the uplink quality is not necessarily homogeneous for a mobile device over the complete channel, this can help the scheduler to select the best RBs for each mobile device. When activated by the network, mobile devices transmit the SRS in every last symbol of a configured subframe. The bandwidth and number of SRS stripes for a UE can be chosen by the network. Further, the SRS interval can be configured between 2 and 160 milliseconds.

4.3.9 MIMO Transmission

In addition to higher order modulation schemes such as 64-QAM, which encodes 6 bits in a single transmission step, 3GPP Release 8 specifies and requires the use of multiantenna techniques, also referred to as MIMO in the downlink direction. While this functionality has also been specified for HSPA in the meantime, it is not yet widely used for this technology because of backward compatibility issues and the necessity to upgrade the hardware of already installed UMTS base stations. With LTE, however, MIMO has been rolled out with the first network installations.

The basic idea behind MIMO techniques is to send several independent data streams over the same air interface channel simultaneously. In 3GPP Release 8, the use of two or four simultaneous streams is specified. In practice, up to two data streams are used today. MIMO is only used for the shared channel and only to transmit those RBs assigned to users that experience very good signal conditions. For other channels, only a single-stream operation with a robust modulation and coding is used as the eNode-B has to ensure that the data transmitted over those channels can reach all mobile devices independent of their location and current signal conditions.

Transmitting simultaneous data streams over the same channel is possible only if the streams remain largely independent of each other on the way from the transmitter to the receiver. This can be achieved if two basic requirements are met.

On the transmitter side, two or four independent hardware transmit chains are required to create the simultaneous data streams. In addition, each data stream requires its own antenna. For two streams, two antennas are required. In practice, this is done within a single antenna casing by having one internal antenna that transmits a vertically polarized signal while the other antenna is positioned in such a way as to transmit its signal with a horizontal polarization. It should be noted at this point that polarized signals are already used today in other radio technologies such as UMTS to create diversity, that is, to improve the reception of a single signal stream.

A MIMO receiver also requires two or four antennas and two or four independent reception chains. For small mobile devices such as smartphones, this is challenging because of their limited size. For other mobile devices, such as notebooks or netbooks, antennas for MIMO operation with good performance are much easier to design and integrate. Here, antennas do not have to be printed on the circuit board but can, for example, be placed around the screen or through the casing of the device. The matter is further complicated because each radio interface has to support more than one frequency band and possibly other radio technologies such as GSM, UMTS and CDMA, which have their own frequencies and bandwidths.

The second requirement that has to be fulfilled for MIMO transmission is that the signals have to remain as independent as possible on the transmission path between the transmitter and the receiver. This can be achieved, for example, as shown in Figure 4.12, if the simultaneous transmissions reach the mobile device via several independent paths. This is possible even in environments where no direct line of sight exists between the transmitter and the receiver. Figure 4.12 is a simplification; however, as in most environments, the simultaneous transmissions interfere with each other to some degree, which reduces the achievable speeds. In theory, using two independent transmission paths can double the achievable throughput and four independent transmission paths can quadruple the throughput. In practice, however, throughput gains will be lower because of the signals interfering with each other. Once the interference gets too strong, the modulation scheme has to be lowered, that is, instead of using 64-QAM and MIMO together, the modulation is reduced to 16-QAM. Whether it is more favorable to use only one stream with 64-QAM or two streams with 16-QAM and MIMO depends on the characteristics of the channel, and it is the eNode-B's task to make a proper decision on how to transmit the data. Only in very ideal conditions, with very short distances

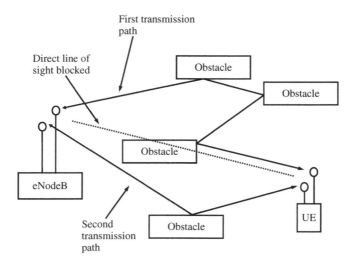

First transmission
path

Direct line of
sight blocked

Obstacle

Obstacle

Obstacle

eNodeB

Obstacle

UE

Second
transmission
path

Obstacle

Figure 4.12 Simplified illustration of MIMO operation

between the transmitter and the receiver, can 64-QAM and MIMO be used simultaneously. As modulation, coding and the use of MIMO can be changed every millisecond on a per device basis, the system can react very quickly to changing radio conditions.

In some documents, the terms 2×2 MIMO or 4×4 MIMO are used to describe the two or four transmitter chains and two or four receiver chains. In the LTE specifications, the term 'rank' is often used to describe the use of MIMO. Rank 1 transmission mode signifies a single-stream transmission while rank 2 signifies a two-stream MIMO transmission for an RB. The term is derived from the rank of the channel matrix that describes the effects that the independent data streams have on each other when they interfere with each other on the way from the transmitter to the receiver.

On a somewhat more abstract level, the 3GPP specifications use the term 'layers' to describe the number of simultaneous data streams. MIMO is not used when only a single layer is trans-mitted. A 2×2 MIMO requires two layers. Each transmission layer uses its own reference symbols as described in Section 4.3.4. The corresponding symbols on the other layer(s) have to be left empty. This slightly reduces the overall throughput of MIMO as the number of symbols that can be used in a RB for the shared channel is lower compared to that in single-stream transmission.

Two different MIMO operation schemes can be used by the eNode-B for downlink trans-missions. In the closed-loop MIMO operation mode, a precoding matrix is applied on the data streams before they are sent over the air to change the modulation of the two signal paths in a favorable way to increase the overall throughput at the receiver side. The mobile device can inform the eNode-B as to which parameters in the precoding matrix would give the best results. This is done by sending a Precoding Matrix Indicator (PMI), a Rank Indicator (RI) and a CQI over the control channel. The RI informs the eNode-B about the number of data streams that can be sent over the channel from the receiver's point of view. The CQI information is used by the eNode-B to decide as to which modulation (QPSK, 16-QAM, 64-QAM) and which coding rate, that is, the ratio between user data bits and error detection bits in the data stream, should be used for the transmission.

For fast moving users, it is difficult to adapt the precoding matrix quickly enough. Such scenarios are thus better handled with open-loop MIMO, for which only the RI and the CQI are reported by the mobile device to the network.

And finally, multiple antennas can also be used for transmit and receive diversity. Here, the same data stream is transmitted over several antennas with a different coding scheme on each. This does not increase the transmission speed beyond what is possible with a single stream but it helps the receiver to better decode the signal and, as a result, enhances datarates beyond what would be possible with a single transmit antenna. In total, 12 different possibilities exist for using MIMO or transmit diversity subfeatures so that the eNode-B has a wide range of options to adapt to changing signal conditions.

In the uplink direction, only single-stream transmission from the point of view of the mobile device has been defined for LTE in 3GPP Release 8. In addition, methods have been specified to enable the eNode-B to instruct several mobile devices to transmit within the same RB. The receiver then uses MIMO techniques to separate the individually transmitted signals. This is not visible to the mobile devices and is referred to as multiuser MIMO. The eNode-B is able to tell the data streams coming from different mobile devices apart by instructing them to apply a different cyclic shift to their reference signals. In practice, however, multiuser MIMO is not used so far. In subsequent versions of the specification, single-user MIMO for the uplink has been specified as well. Like multiuser MIMO, it is also not used in practice so far.

Table 4.5 shows the different transmission modes specified in 3GPP Release 8, TS 36.213 [16] for the downlink shared channel. The decision of which of these are implemented and under which signal conditions these are used is left to the discretion of the network vendor. On the mobile device side, however, all modes have to be supported.

4.3.10   HARQ and Other Retransmission Mechanisms

Despite adaptive modulation and coding schemes, it is always possible that some of the transmitted data packets are not received correctly. In fact, it is even desired that not all packets are received correctly as this would indicate that the modulation and coding scheme is too conservative and hence capacity on the air interface is wasted. In practice, the air interface

Table 4.5   LTE transmission modes

Transmission mode	Transmission scheme of the PDSCH
Mode 1	Single-antenna port
Mode 2	Transmit diversity
Mode 3	Transmit diversity or large delay CDD
Mode 4	Transmit diversity or closed-loop MIMO
Mode 5	Transmit diversity or multiuser MIMO
Mode 6	Transmit diversity or closed-loop MIMO
Mode 7	Single antenna port 0 or port 5

is utilized best if about 10% of the packets have to be retransmitted because they have not been received correctly. The challenge of this approach is to report transmission errors quickly and to ensure that packets are retransmitted as quickly as possible to minimize the resulting delay and jitter. Further, the scheduler must adapt the modulation and coding scheme quickly to keep the error rate within reasonable limits. As in HSPA, the HARQ scheme is used in the Medium Access Control (MAC) layer for fast reporting and retransmission. In LTE, the mechanism works as described in the following sections.

HARQ Operation in the MAC Layer

In the downlink direction, asynchronous HARQ is used, which means that faulty data does not have to be retransmitted straight away. The eNode-B expects the mobile device to send an acknowledgment (ACK) if the data within each 1-millisecond subframe has been received correctly. A negative acknowledgment (NACK) is sent if the data could not be decoded correctly. HARQ feedback is sent either via the PUSCH or via the PUCCH if the mobile device has not been assigned any uplink resources at the time the feedback is required. This can be the case, for example, if more data is transmitted to a mobile device in the downlink direction than the mobile device itself has to send in the uplink direction.

If an ACK is received, the eNode-B removes the subframe data from its transmission buffer and sends the next chunk of data if there is more data waiting in the buffer. In case a NACK is received, the eNode-B attempts to retransmit the previous data block. The retransmission can occur immediately or can be deferred, for example, owing to the channel currently being in a deep fading situation for a particular user.

Before a data block is sent, redundancy bits are added to the data stream that can be used to detect and correct transmission errors to a certain degree. How many of those redundancy bits are actually sent depends on the radio conditions and the scheduler. For good radio conditions, most redundancy is removed from the data stream again before transmission. This is also referred to as puncturing the data stream. If a transmission error has occurred and the added redundancy is not sufficient to correct the data, the eNode-B has several options for the retransmission:

- It can simply repeat the data block.
- It sends a different redundancy version (RV) that contains a different set of redundancy bits, that is, some of those bits that were previously punctured from the data stream. On the receiver side, this data stream can then be combined with the previous one, thus increasing the number of available error detection and correction information.
- The network can also decide to change the modulation and coding scheme for the transmission to increase the chances for proper reception.

Repeating a data block requires time for both the indication of the faulty reception and the repetition of the data itself. In LTE, the ACK/NACK for a downlink transmission is sent after four subframes to give the receiver enough time to decode the data. The earliest repetition of a faulty block can thus take place five subframes or 5 milliseconds after the initial transmission. The eNode-B can also defer the transmission to a later subframe if necessary. Depending on the radio conditions and the modulation and coding selected by the scheduler, some data blocks might have to be retransmitted several times. This, however, is rather undesirable as it reduces the overall effectiveness. The eNode-B can set an upper limit for the number of retransmissions and can discard the data block if the transmission is not successful even after several attempts. It is then left to higher layers to detect the missing data and initiate a retransmission if necessary or desired. This is not the case for all kinds of data streams. For VoIP transmissions, it can be better to discard some data if it does not arrive in time as it is not needed anymore anyway. As described in Chapter 1, voice codecs can mask missing or faulty data to some degree.

As faulty data can only be repeated after 5 milliseconds at the earliest, several HARQ processes must operate in parallel to fill the gap between the transmission of a data block and the ACK. Up to eight HARQ processes can thus run in parallel to also cover cases in which the eNode-B does not immediately repeat the faulty data. Figure 4.13 shows five HARQ processes transmitting data. When downlink data is scheduled via the PDCCH, the scheduling grant message has to describe the modulation and coding, the HARQ process number to which the data belong, whether it is a new transmission or a repetition of faulty data and which RV of the data stream is used.

At this point, it is interesting to note that the shortest HARQ delay in HSPA is 10 milliseconds because of a block size of 2 milliseconds and thus twice as long. In practice, LTE has hence even shorter jitter and round-trip delay times compared to the already good values in HSPA. Together with the shorter LTE HARQ delay in the uplink direction as described below, overall round-trip delay times of the complete LTE system of less than 20 milliseconds can be achieved.

For the data stream in the uplink direction, synchronous HARQ is used where the repetition of a faulty data block follows the initial transmission after a fixed amount of time. If uplink data of a 1-millisecond subframe has been received by the eNode-B correctly, it acknowledges the proper receipt to the mobile device four subframes later. The ACK is given via the PHICH, which is transmitted over a number of symbols in the first symbol row of each subframe. As several mobile devices can get a scheduling grant for different RBs during a subframe interval, a mathematical function is used to describe which symbols of the PHICH contain the feedback for which mobile device. Once the positive ACK has been received by the mobile device, the next data block of a HARQ process can be sent in the uplink direction.

If the network did not receive the data of a subframe correctly, it has to request a retransmission. This can be done in two ways. The first possibility for the network is to send a NACK and order a retransmission in a new format and possibly different location of the resource grid via a scheduling grant on the PDCCH. The second option is to send only a NACK without

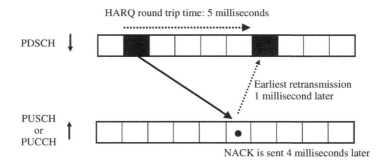

Figure 4.13   Synchronous HARQ in the downlink direction

any further information given via the PDCCH. In this case, the mobile device repeats the transmission on the same resources that were assigned for the original transmission.

ARQ on the RLC layer

Packets lost despite the HARQ mechanism can be recovered via the Automatic Retransmission Request (ARQ) feature on the next higher protocol layer, the radio link control (RLC) layer which is specified in 3GPP TS 36.322 [17]. While used for most bearers, its use is optional and might not be activated, for example, for radio bearers used by VoIP applications such as the IMS. ARQ is split into two main functions:

- As soon as the receiver detects a transmission failure, it sends a report to the transmitting side which then repeats the missing RLC frame. A sliding window approach is used so that the transfer of frames is not interrupted if a frame is not received. Only if the missing frame has not been received when the window size is met, the overall transmission is stopped until the missing RLC frame has been received. This ensures that the RLC buffer on the receiver side, that is, either in the eNode-B or in the UE, does not overrun.
- During normal operation, the sender periodically requests an ARQ status report by setting the polling indicator bit in the RLC header of a data frame. This way, unnecessary status reports do not have to be sent while it ensures that no RLC error message is missed.

While HARQ and ARQ ensure the correct delivery of data over the air interface, higher layer protocols also have functionality built in to ensure the correct delivery from an end-to-end perspective. The IP protocol, for example, contains a CRC field so that each router in the path between sender and receiver can check the integrity of the packet. And finally, the TCP uses a sliding window mechanism similar to that of ARQ to detect missing IP packets and to request a retransmission.

4.3.11 PDCP Compression and Ciphering

Above the MAC and RLC layers, discussed in the previous sections, is the Packet Data Convergence Protocol (PDCP) layer. Its main task is to cipher user and signaling traffic over the air interface. Further, it ensures the integrity of signaling messages by protecting against various man-in-the-middle attack scenarios. Several ciphering algorithms have been defined for LTE and the network decides the one to use during the bearer establishment procedure.

Another important but optional task of the PDCP layer is IP header compression. Depending on the size of the user data in a packet, a more or less significant percentage of the overall air interface capacity is taken up by the various headers in the IP stack. This is especially the case for VoIP packets, which are usually sent every 20 milliseconds to minimize the speech delay and are hence very short. With efficient codecs such as AMR (see Chapter 1), each IP packet has a size of around 90 Bytes and two-thirds of the packet is taken up by the various headers (IP, UDP and RTP). Other applications such as web browsing, for which large chunks of data are transferred, IP packets typically have a size of over 1400 Bytes. Here, header compression is less important but still beneficial. As a consequence, header compression is not widely used in practice today and will only gain more traction when mobile operators introduce VoLTE.

For the LTE air interface, Robust Header Compression (RoHC) can be used if supported by both the network and the mobile device. It is specified in RFC 4995 and 5795 [18]. The basic idea of RoHC is not to apply header compression to all IP packets in the same manner but to group them into streams. IP packets in a stream have common properties such as the same target IP address and TCP or UDP port number and the same destination IP address and port number. For a detected stream of similar IP packets, one of several RoHC profiles is then used for compression. For VoIP packets, one of the profiles described in RFC 5225 [19] can be used to compress the IP, UDP and Real-time Transport Protocol (RTP for voice and other multimedia codecs) headers. For other streams, more general profiles that only compress the IP and TCP headers can be used. Once a stream is selected, an RoHC session is established and several sessions can be established over the same bearer simultaneously. After the RoHC session setup phase, static header parameters such as the IP addresses, UDP port numbers, and so on. are completely omitted and variable parameters such as counters are compressed. A checksum protects against transmission errors.

In addition, the PDCP layer ensures that during a handover, the user and control plane contexts are properly transferred to the new eNode-B. For signaling and RLC unacknowledged data streams (e.g., for IMS VoIP), a seamless handover procedure is performed. Here, PDCP helps to transfer the contexts to the new eNode-B, but PDCP packets that are transferred just before the handover and not received correctly might be lost. For VoIP packets such a loss is even preferred to later recovery due to the delay sensitivity of the application. For RLC acknowledged mode (AM) user data streams, a lossless handover is performed. Here, PDCP sequence numbers are used to detect packets that were lost in transit during the handover; these are then repeated in the new cell.

4.3.12   Protocol Layer Overview

In the previous sections, some of the most important functions of the different air interface protocol layers such as HARQ, ARQ and ciphering have been discussed. These functions are spread over different layers of the stack. Figure 4.14 shows how these layers are built on each other and where the different functions are located.

On the vertical axis, the protocol stack is split into two parts. On the left side of Figure 4.14, the control protocols are shown. The top layer is the NAS protocol that is used for mobility management and other purposes between the mobile device and the MME. NAS messages are tunneled through the radio network, and the eNode-B just forwards them transparently. NAS messages are always encapsulated in Radio Resource Control (RRC) messages over the air interface. The other purpose of RRC messages is to manage the air interface connection and they are used, for example, for handover or bearer modification signaling. As a consequence, an RRC message does not necessarily have to include a NAS message. This is different on the user data plane shown on the right of Figure 4.14. Here, IP packets are always transporting user data and are sent only if an application wants to transfer data.

The first unifying protocol layer to transport IP, RRC, and NAS signaling messages, is the PDCP layer. As discussed in the previous section, it is responsible for encapsulating IP packets and signaling messages, for ciphering, header compression and lossless handover support. One layer below is the RLC. It is responsible for segmentation and reassembly of higher layer packets to adapt them to a packet size that can be sent over the air interface. Further, it is responsible for detecting and retransmitting lost packets (ARQ). Just above the physical layer

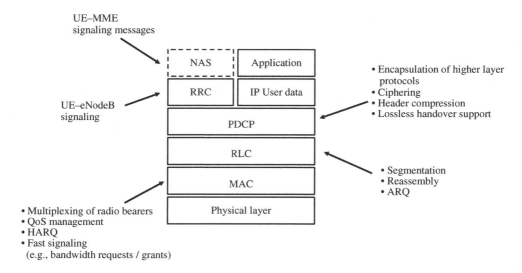

Figure 4.14   Air interface protocol stack and main functions

is the MAC. It multiplexes data from different radio bearers and ensures QoS by instructing the RLC layer about the number and the size of packets to be provided. In addition, the MAC layer is responsible for the HARQ packet retransmission functionality. And finally, the MAC header provides fields for addressing individual mobile devices and for functionalities such as bandwidth requests and grants, power management and time advance control.

4.4   TD-LTE Air Interface

Most initial LTE deployments around the world use FDD, which means that data is transmitted and received simultaneously on two separate channels. In addition, a TDD variant of the LTE air interface has been specified. It is used, for example, in China. Apart from using a single channel for both uplink and downlink transmissions, most other parameters of the air interface are identical to the FDD implementation described earlier. Some differences exist, however, which are described in this section.

As Time Division Long-Term Evolution (TD-LTE) only uses a single channel, a transmission gap is required when switching from transmission mode to reception mode. This is also referred to as the guard period and its length is determined by the time it takes to switch the operation mode of the transceiver unit and the maximum timing advance that can be encountered in a cell. The timing advance influences the guard period because the more distant a mobile device is from the center of the cell, the earlier it has to start its transmissions so that they are synchronized with the transmissions of devices that are closer to the center of the cell. The guard period has to be long enough to accommodate such earlier transmissions in relation to mobile devices closer to the cell. The guard period can be up to 10 OFDM symbol durations. In urban environments with a cell radius of 2 km or less, a single OFDM symbol duration is usually sufficient and only little capacity is wasted.

Two switching intervals between transmission and reception have been defined in the specifications: 5 milliseconds and 10 milliseconds. These correspond to 5 and 10 subframes, respectively. After each interval, a guard period is inserted. Four TDD configurations exist for the 5-milliseconds interval and three for the 10-milliseconds interval to flexibly assign subframes to the uplink and the downlink direction. TDD configuration 3, for example, assigns three subframes for the uplink direction and six subframes for the downlink direction. The 10th subframe is special as the first part is used for downlink transmissions of the previous frame, the middle part is used for the guard period and the remainder is used for uplink transmissions of the next frame. This way, the 10-milliseconds timing of a radio frame, which comprises 10 subframes, is kept the same as in the FDD version, independent of the number of OFDM symbols used for the guard period.

Owing to the different number of uplink and downlink subframes, it was necessary to introduce a flexible resource assignment concept. In FDD, where the number of uplink and downlink subframes are identical owing to the simultaneous transmission and reception on two channels, an uplink transmission opportunity assigned in a certain downlink subframe implicitly grants access to the uplink four subframes later. As the number of uplink and downlink subframes is not necessarily the same in TDD, it is necessary to specify individual allocation schemes for each TDD configuration.

A similar complication exists for the HARQ mechanism. Owing to the non-symmetric configuration of uplink and downlink subframes, some subframes have to contain the ACK/NACK for transmissions in several subframes of the other direction. How this is done again depends on the TDD configuration used.

4.5 Scheduling

Data transmissions in LTE in both the uplink and the downlink directions are controlled by the network. This is similar to other technologies such as GSM and UMTS. In these systems, some or all of the scheduling control is located in centralized network components such as the PCU (see Chapter 2) or the RNC (see Chapter 3). In LTE, the scheduling is fully controlled by the eNode-Bs as higher layer radio network control instances were removed from the overall network design. Some of the advantages of network-based scheduling are as follows:

- The network can react to changing radio conditions of each user and optimize the overall throughput.
- The network can ensure the QoS for each user.
- Overload situations can be dealt with.

Other technologies such as Wi-Fi do not use centralized control and leave it to the devices communicating in the network to cooperatively use the air interface. Here, central control is not necessary as the number of devices simultaneously communicating over a Wi-Fi access point is much lower and the coverage area is much smaller. Details are discussed in Chapter 5.

4.5.1 Downlink Scheduling

In the downlink direction, the eNode-B's scheduler is responsible for forwarding the data that it receives from the network, for all users it serves, over the air interface. In practice, a single logical default bearer is usually assigned to a mobile device, over which the data is transported. To ensure QoS for applications such as VoIP via the IMS (see Chapter 3), it is also possible to assign more than one logical bearer to a mobile device. The VoIP data stream then uses a dedicated bearer for which the requested bandwidth and a low time variation between two packets (jitter) is ensured by the network.

Dynamic Scheduling

Scheduling is a simple task if there is only one user and if there is less data waiting in the transmission buffer than can be sent over the air interface. When the eNode-B serves several users, or several bearers to be precise, and the amount of data in the downlink buffer exceeds that which can be sent in a subframe, then the scheduler has to decide which users and bearers are given an assignment grant for the next subframe and how much capacity is allocated to each. This decision is influenced by several factors.

If a certain bandwidth, delay and jitter have been granted for a bearer to a particular user then the scheduler has to ensure that this is met. The data of this bearer is then given preference over the data arriving from the network for other bearers for the same or a different user. In practice, however, such QoS attributes are not widely used and hence most bearers have the same priority on the radio interface.

For bearers with the same priority, other factors may influence the scheduler's decision when to schedule a user and how many RBs are allocated to him in each subframe. If each bearer of the same priority was treated equally, some capacity on the air interface would be wasted. With this approach, mobile devices that currently or permanently experience bad radio conditions, for example, at the cell edge, would have to be assigned a disproportional number of RBs because of the low modulation and coding scheme required. The other extreme is to always prefer users that experience very good radio conditions as this would lead to very low datarates for users experiencing bad radio conditions. As a consequence, proportional fair schedulers take the overall radio conditions into account, observe changes for each user over time and try to find a balance between the best use of the cell's overall capacity and the throughput for each user.

Scheduling downlink data for a user works as follows: For each subframe the eNode-B decides the number of users it wants to schedule and the number of RBs that are assigned to each user. This then determines the required number of symbols on the time axis in each subframe for the control region. As shown in Figure 4.7, there are a total of $2 \times 7 = 14$ symbols available on the time axis if a short cyclic prefix is used. Depending on the system configuration and the number of users to schedule, one to four symbols are used across the complete carrier bandwidth for the control region. The number of symbols can either be fixed or changed as per the demand.

The eNode-B informs mobile devices about the size of the control region via the PCFICH, which is broadcast with a very robust modulation and coding scheme. The 2 bits describing the length of the control region are secured with a code rate of 1/16, which results in 32 output bits. QPSK modulation is then used to map these bits to 16 symbols in the first symbol column of each subframe.

With the mobile device aware of the length of the control region, it can then calculate where its search spaces are located. As described above, search spaces have been introduced to reduce the mobile device's processing load to save battery capacity. In mobile device (UE) specific search spaces that are shared by a subset of mobile devices or in common search spaces that have to be observed by all mobile devices for broadcast messages, the mobile decodes all PDCCH messages. Each message has a checksum in which the mobile's identity is implicitly embedded. If the mobile can correctly calculate the checksum, it knows that it is the intended recipient of the message. Otherwise the message is discarded.

The length of a PDCCH message is variable and depends on the content. For easier decoding, a number of fixed-length PDCCH messages have been defined. A message is assembled as follows: On the lowest layer, four symbols on the frequency axis are grouped into a Resource Element Group (REG). Nine REGs form a CCE. These are further aggregated into PDCCH messages, which can consist of 1, 2, 4 or 8 CCEs. A PDCCH message consisting of two CCEs, for example, contains 18 REGs and $18 \times 4 = 72$ symbols. In other words, it occupies 72 subcarriers. With QPSK modulation, the message has a length of 144 bits. The largest PDCCH message with eight CCEs has a length of 576 bits.

A PDCCH message can be used for several purposes and as their lengths differ, several message types exist. In the standards, the message type is referred to as Downlink Control Information (DCI) format. Table 4.6 shows the 10 different message types or DCI formats that have been defined.

If the message describes a downlink assignment for a mobile device on the downlink shared channel, the message contains the following information:

- the type of resource allocation (see below);
- a power control command for the PUCCH;
- HARQ information (new data bit, RV);
- modulation and coding scheme;
- number of spatial layers for MIMO operation;
- precoding information (how to prepare the data for transmission).

The eNode-B has several ways to indicate a resource allocation:

- Type 0 resource allocations give a bitmap of assigned RB groups. For 10-MHz channels, the group size is three RBs. For the 50 RBs of a 10-MHz channel, the bitmap has a length of

Table 4.6   Downlink control channel message types (DCI formats)

Message type (DCI format)	Content
0	Uplink scheduling grants for the PUSCH (see Section 4.5.2)
1	PDSCH assignments with a single codeword
1a	PDSCH compact downlink assignment
1b	PDSCH compact downlink assignment with precoding vector
1c	PDSCH assignments using a very compact format
1d	PDSCH assignments for multiuser MIMO
2	PDSCH assignments for closed-loop MIMO
2a	PDSCH assignments for open-loop MIMO
3	Transmit power control for the UL with 2-bit power adjustments (see Section 4.5.2)
3a	Transmit power control for the UL with 1-bit power adjustments (see Section 4.5.2)

17 bits. For 20-MHz carriers, a group size of four RBs is used and the 100 RBs are addressed with a bitmap of 25 bits.

- Type 1 resource allocations also use a bitmap, but instead of assigning full groups to a mobile device with a '1' in the bitmap, only one of the RBs of a group is assigned. This way, resource assignments can be spread over the complete band and frequency diversity can thus be exploited.
- Type 2 resource allocations give a starting point in the frequency domain and the number of allocated resources. The resources can either be continuous or spread over the complete channel.

In summary, the allocation of downlink resources from the mobile device's point of view works as shown in Figure 4.15. At the beginning of each subframe, the mobile device first reads the PCFICH to detect the number of columns of the subframe that are reserved for the control region in which the PDCCH messages are sent. On the basis of its ID, it then computes the search spaces in which it has to look for assignment messages. It then decodes these areas to see if it can find messages for which it can successfully calculate the CRC checksum, which implicitly contains the device's ID. If a downlink assignment message is found, the resource allocation type then determines how the information is given (bitmap or a starting point with length). With this information, the mobile device can now go ahead and attempt to decode all areas of the subframe in which its data is transmitted.

Depending on the current activity and the amount of data that arrives for a user from the network, it might not be necessary to schedule data for a user in every subframe. In such cases, it is also not necessary for the mobile device to search for potential scheduling grants in every subframe. To reduce the power consumption of a mobile device, the network can thus configure discontinuous reception (DRX) periods during which the control channel does not have to be observed. Further details are discussed in Section 4.7.

Semipersistent Scheduling

While dynamic scheduling is ideal for bursty, infrequent and bandwidth-consuming data transmissions such as web surfing, video streaming and e-mails, it is less suited for real-time streaming applications such as voice calls. Here, data is sent in short bursts at regular intervals. If the datarate of the stream is very low, as is the case for voice calls, the overhead of the dynamic scheduling messages is very high as only little data is sent for each scheduling message. The solution for this is semipersistent scheduling. Instead of scheduling each uplink or downlink transmission, a transmission pattern is defined instead of single transmission opportunities. This significantly reduces the scheduling assignment overhead.

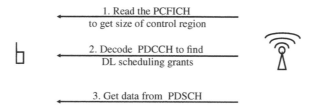

Figure 4.15 Downlink data reception overview

During silence periods, wireless voice codecs stop transmitting voice data and only send silence description information with much longer time intervals in between. During these silence times persistent scheduling can be switched off. Hence, grants are semipersistent. In the uplink direction, the semipersistent grant scheme is implicitly canceled if no data is sent for a network-configured number of empty uplink transmission opportunities. In the downlink direction, semipersistent scheduling is canceled with an RRC message. Details can be found in Section 5.10 of the LTE MAC protocol specification [20].

In practice, it is left to the network vendor to decide when to use which scheduling type. The challenge with semipersistent scheduling for the network is to detect when it can be used. For network operator-based VoIP services such as Voice over LTE (see Section 4.13), the voice service can request the LTE network to establish a logical dedicated bearer for the data flow. All IP packets that meet certain criteria such as source and destination IP addresses or certain UDP port numbers are then always sent over the air interface in this logical dedicated bearer for which a certain bandwidth, delay and priority are ensured by the network. On the air interface, packets flowing through this dedicated bearer can then be scheduled using semipersistent grants. All other data to and from the device would continue to use the default bearer for which the eNode-B uses dynamic scheduling. Internet-based voice services such as Skype, however, have no means to interact with the LTE network to request a certain QoS. For such services, the IP transport channel is transparent and VoIP packets use the same default bearer together with IP packets to and from all other connected services on the mobile device. A potential solution for the future to optimize the use of scheduling grants might thus be that the network analyzes the traffic flow and then adapts its use of the different types of scheduling methods accordingly.

4.5.2   Uplink Scheduling

To get resources assigned on the PUSCH, the mobile device has to send an assignment request to the eNode-B. If no physical connection currently exists with the eNode-B, the mobile first needs to reestablish the link. This is done as already described above by sending an RRC connection request message on the RACH. The network then establishes a channel and assigns resources on the PUSCH so that the mobile device can transmit signaling and user data in the uplink direction. The assignment of uplink resources is performed via PDCCH messages in the control section of each subframe.

While a mobile device is in active communication with the network and has resources assigned on the uplink shared channel, it includes buffer status reports in the header of each packet. This information is then used by the eNode-B to assign uplink resources in the following subframes. If the connection is active but the mobile device currently has no resources on the uplink shared channel, it has to send its bandwidth requests via the PUCCH as described earlier.

Uplink scheduling grants are sent as PDCCH messages in the same way as described earlier for downlink scheduling grants. For this purpose, DCI messages of type 0 are used as shown above in Table 4.6. In fact, while scanning the search spaces in the control section of each subframe, both uplink and downlink scheduling grants that can be given as full-duplex FDD devices can transmit and receive data at the same time.

To make the best use of the resources of the uplink physical shared channel, the eNode-B must be aware of how much transmission power a mobile device has left compared to its current power output. It can then select an appropriate modulation and coding scheme and the number of RBs on the frequency axis. This is done via power headroom reports, which are periodically sent by the mobile device in the uplink direction.

4.6 Basic Procedures

After the introduction of various reference signals and channels of the air interface, the following sections now give an overview of the different procedures required for communication with the network with references to the previous section for details on the channels.

4.6.1 Cell Search

When a mobile device is powered on, its first task from a radio point of view is to search for a suitable network and then attempt to register. To speed up the task, it is guided by information on the SIM card stored in the home network with access technology field (EF-HPLMNwAcT). With this field the network operator can instruct the mobile device for which radio access technology (GSM, UMTS, LTE) to search first and then use for registration. Older SIM cards that have not been updated still contain UMTS in this field, which means that the mobile device will first search for and use the UMTS network even if an LTE network is available and only switch to LTE once registration has been performed. Newer SIM cards or cards that have been updated over the air instruct the mobile device to first search for an LTE network of the network operator and use this radio access technology for registration.

To shorten the search process, the mobile device stores the parameters of the last cell it used before it was switched off. After the device is powered on, it can go straight to the last known band and use the last known cell parameters to see if the cell can still be found. This significantly speeds up the cell search procedure if the device has not been carried to another place while it was switched off and the last used radio access technology is the same as the network operator preference stored on the SIM card.

In case the previous cell was not found with the stored information, it performs a full search. UMTS and GSM cell search has been described in the previous chapters; so, this section focuses only on the LTE cell search mechanism.

For the first step, the mobile device searches on all channels in all supported frequency bands for an initial signal and tries to pick up a primary synchronization signal (PSS) that is broadcast every 5 milliseconds, that is, twice per air interface frame. Once found, the device remains on the channel and locates the SSS, which is also broadcast once every 5 milliseconds. While the content of the PSS is always the same, the content of the SSS is alternated in every frame so that the mobile device can detect from the pattern as to where to find the beginning of the frame. Figure 4.16 shows where the synchronization signals can be found in a frame on the time axis.

To make cell detection easier, the PSS and SSS are broadcast only on the inner 1.25 MHz of the channel, irrespective of the total channel bandwidth. This way, a simpler FFT analysis can be performed to detect the signals. Also, the initial cell search is not dependent on the channel bandwidth. Hence, this speeds up the cell search process.

The PSSs and SSSs implicitly contain the PCI. The PCI is not equal to the cell-ID as previously introduced in GSM and UMTS but is simply a lower layer physical identity of the cell. It can thus be best compared to the Primary Scrambling Code (PSC) in UMTS. Like GSM

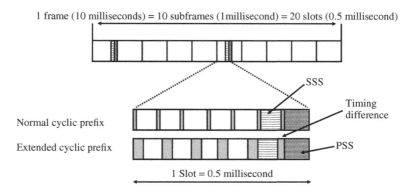

Figure 4.16   PSS and SSS in an LTE FDD frame

and UMTS, LTE also knows a cell identity on higher layers, which is discussed later on. The PCI is important to distinguish neighboring cells transmitting on the same frequency. In practice, mobile devices, especially in cell edge scenarios, receive several PSS and SSS and hence detect several PCIs on the same frequency.

After detection of the PSS and SSS, the mobile device is also aware if the cell uses a normal or an extended cyclic prefix. As shown in Figure 4.16, there is a different timing of the two signals depending on the length of the prefix as only six symbols form a slot when the extended cyclic prefix is used compared to seven symbols with a normal cyclic prefix.

The signals transmitted from the different cells on the same channel interfere with each other. As a channel is used only by one operator except at national borders, the mobile device would attempt to start communication only with the cell with the strongest synchronization signals and ignore the other cells on the same frequency. In case the mobile device has found the cell it used before it was switched off, it may go directly to this cell and stop searching for other cells on different channels in the current frequency band, even if the cell is not the strongest on the current channel. After a successful attach procedure as described below, the cell reselection mechanism or a handover will ensure that the mobile device is served by the strongest cell it receives.

The next step in the cell search procedure is to read the MIB from the PBCH, which is broadcast every 40 milliseconds in the inner 1.25 MHz of the channel. The MIB contains the most important information about the configuration of the channel, which is essential for the mobile before it can proceed. Very conservative modulation and strong error detection and correction information is added to allow successful decoding of this information even under very unfavorable reception conditions. The first information that the mobile device gets from the MIB is the total bandwidth used for the channel since all decoding attempts so far were only performed in the inner 1.25 MHz of the channel. Further, the MIB contains the structure of the HARQ indicator channel (PHICH, see Section 'HARQ Operation in the MAC Layer') and the System Frame Number (SFN), which is required, for example, for ciphering and calculation of paging opportunities as described later on.

With the information from the MIB, the mobile device can then begin to search for the SIB-1. As it is broadcast on the downlink shared channel every 80 milliseconds, the mobile device needs to decode the 'common' search space in the control region of a subframe to find a downlink control channel (PDCCH) message that announces the presence and location of the SIB-1 in the subframe. Once found, the SIB-1 message contains the following information:

- The MCC and MNC of the cell. These parameters tell the mobile device if the cell belongs to the home network or not.
- The NAS cell identifier which is similar to the cell-ID in GSM and UMTS.
- The Tracking Area Code (TAC), which corresponds to the location and routing areas in GSM and UMTS.
- Cell barring status, that is, whether the cell can be used or not.
- Minimum reception level (q_RxLevMin) that the mobile device must receive the cell with. If the level is lower, the mobile device must not try to establish communication with the cell.
- A scheduling list of other SIBs that are sent and their intervals.

With the information provided in SIB-1, the mobile device can decide if it wants to start communicating with this cell. If so, for example, since the cell belongs to the home network, the mobile device then continues to search and decode further SI messages.

SIB-2 contains further parameters that are required to communicate with a cell, such as

- the configuration of the RACH;
- the paging channel configuration;
- the downlink shared channel configuration;
- the PUCCH configuration;
- the SRS configuration in the uplink;
- uplink power control information;
- timers and constants (e.g., how long to wait for an answer to certain messages, etc.);
- uplink channel bandwidth.

Further SIBs contain information that is mainly relevant for cell reselection once the mobile device has established a connection with the network. Hence, they are discussed in more detail in Section 4.7.2 on cell reselection and Idle state procedures.

If the cell is not part of the home network or does not belong to the last used network stored on the mobile device (e.g., during international roaming), the device then goes on and searches other channels on the current frequency band and also on other frequency bands. If the frequency band can be used by more than one radio technology, such as the 1800-MHz band which can be used by GSM and LTE, the mobile device would try to detect transmissions from different radio systems in the same band.

4.6.2 Attach and Default Bearer Activation

Once the mobile device has all the required information to access the network for the first time after it has been powered on, it performs an attach procedure. From a higher layer point of view, the attach procedure delivers an IP address and the mobile device is then able to send and receive data from the network. In GSM and UMTS, a device can be connected to the network without an IP address. For LTE, however, this has been changed and a mobile device hence always has an IP address when it is connected to the network. Further, the attach process including the assignment of an IP address has been streamlined compared to GSM and UMTS to shorten the time from power on to providing service as much as possible.

Initial Connection Establishment

Figure 4.17 gives an overview of the first part of the attach procedure as per 3GPP TS 23.401 [21]. The first step is to find a suitable cell and detect all necessary parameters for accessing the network as described in the previous section. The attach procedure then begins with requesting resources on the uplink shared channel via a request on the RACH as shown in Figure 4.11. Once this procedure has been performed, the mobile device is known to the eNode-B and has been assigned a Cell Radio Network Temporary Identity (C-RNTI). This MAC layer ID is used, for example, in scheduling grants that are sent in downlink control channel (PDCCH) messages.

Next, the mobile device has to establish an RRC channel so that it can exchange signaling messages with the eNode-B and the core network. This is done by sending an RRC connection request message to the network. The message contains the reason for the connection establishment request, for example, mobile-originated signaling, and the mobile's temporary core network (NAS) identity, the SAE (Service Architecture Evolution) Temporary Mobile Subscriber Identity (S-TMSI). As the name implies, this parameter identifies the mobile device in the core network, specifically in the MME as described at the beginning of this chapter.

If access is granted, which is usually the case except during overload situations, the network responds with an RRC connection setup message, which contains the assignment parameters

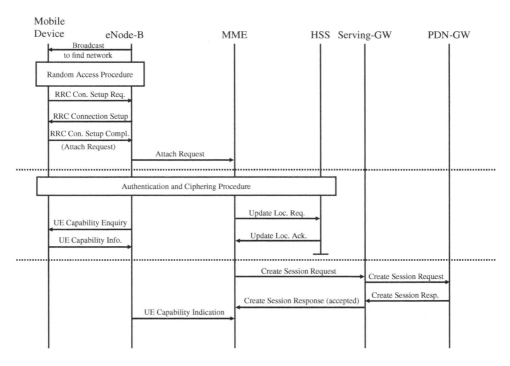

Figure 4.17   Attach and default bearer activation message flow – part 1

for a dedicated radio signaling bearer (SRB-1) that is from that moment onward used to transfer RRC messages to the eNode-B. In addition, the SRB-1 is also used to transfer NAS signaling to and from the MME. These messages are encapsulated in RRC messages. Further, the message contains MAC and physical layer parameters such as the uplink shared channel configuration, uplink power control, use of SRSs in uplink and how scheduling requests should be sent in the uplink direction.

In the next step, the mobile device returns an RRC connection setup complete message to the eNode-B. In the RRC part of the message, the mobile device informs the eNode-B to which MME it was last connected. In LTE, an eNode-B can communicate with more than a single MME for load balancing and redundancy reasons. If no information about the previous MME is given, the eNode-B selects one on its own.

The RRC connection setup complete message also contains an embedded NAS message, the actual Attach Request message, which the eNode-B transparently forwards to the MME it has selected. Part of the message is the Globally Unique Temporary Identity, or GUTI for short. It is the same as the Packet TMSI in UMTS and is a temporary identifier that the device was assigned when it was previously connected to the network. This enables the MME to locate the subscriber's record in its cache or to find the MME to which the device was previously connected to so that it can inform the old MME that the device has changed its location and to retrieve the user's subscription profile.

The signaling connection is then used for mutual authentication between the network and the mobile device. As in UMTS, mutual authentication ensures that the network can be sure about the identity of the device and that the device can validate that it is communicating to a network that has properly obtained the authentication information from the HSS. This effectively prevents a man-in-the-middle attack. After the authentication procedure, the MME then sends a Security Mode Command message to activate integrity checking and optionally encryption of all messages between the MME and the mobile device. Integrity checking ensures that signaling messages between a mobile device and the MME cannot be modified by an attacker. A Security Command Complete message completes the transaction, and all further signaling messages are sent with an integrity checksum and are optionally encrypted.

Once the subscriber is authenticated, the MME confirms the successful authentication to the HSS by sending an update location request message to the HSS, which responds with an update location acknowledge.

To also protect user data packets and signaling messages that are exchanged between the mobile device and the eNodeB requires an additional Security Mode Command/Complete procedure. This procedure is not performed with the MME but directly between the mobile device and the eNodeB.

As further shown in Figure 4.17, the eNode-B then asks the mobile device to provide a list of its supported air interface functionalities with a UE capability inquiry. The mobile device responds to the message with a UE capability information message which contains information such as the supported radio technologies (GSM, UMTS, CDMA, etc.,), frequency band support of each technology, RoHC header compression support (e.g., for VoIP) and information on optional feature support. This information helps the eNode-B later on to select the best air interface parameters for the device and also helps to select the interband and interradio technology measurements that it should configure so that the device can detect other networks for a handover when it leaves the LTE coverage area. This information is also forwarded to the MME.

Session Creation

Once the MME has received the update location acknowledge message from the HSS, it starts the session establishment process in the core network that results in the creation of a tunnel over which the user's IP packets can be sent. This is done by sending a create session request message to the serving-GW of its choice. For load balancing, capacity and redundancy reasons, an MME can communicate with more than one serving-GW. The serving-GW in turn forwards the request to a PDN-gateway which is located between the LTE core network and the Internet. The PDN-GW then selects an IP address from a pool and responds to the serving-GW with a create session response message. The serving-GW then returns the message to the MME and the tunnel for the IP packets of the user between the serving-GW and the PDN-GW is ready to be used. This tunnel is necessary as the user's location and hence its serving-GW can change during the lifetime of the connection. By tunneling the user's IP data traffic, the routing of the data packets in the LTE network can be changed at any time without assigning a new IP address to the user. For further details on user data tunneling, see Chapter 2. Establishing a tunnel is also referred to in the specification as establishing a context.

Establishing a Context in the Radio Network

After the context for the user has been established in the core network, the MME responds to the initial Attach Request with an Initial Context Setup Request message, which includes the Attach Accept message as shown in Figure 4.18. On the S1 interface between the MME and eNode-B, this message starts the establishment procedure for a user data tunnel between the eNode-B and the serving-GW. It includes the Tunnel Endpoint Identity (TEID) used on the serving-GW for this connection.

The final link that has to be set up now is the bearer for the user's IP packets on the air interface. This is done by the eNode-B by sending an RRC Connection Reconfiguration message to the mobile device. Earlier during the attach process, a signaling radio bearer (SRB-1) was established for the signaling messages. With this connection reconfiguration, a second signaling radio bearer is established for lower priority signaling messages and a Data Radio Bearer (DRB) for the user's IP packets. The message also includes two further NAS messages, the Attach Accept message and the Activate Default Bearer Context Request message. These messages configure the device's higher protocol layers on the radio protocol stack. In addition, this step is also used to assign the IP address to the mobile device for communication with the Internet and other parameters such as the IP address of the DNS server, which is required to translate URLs (e.g., www.wirelessmoves.com) into IP addresses. Also, the message includes the QoS profile for the default bearer context.

Once the RRC part of the protocol stack has been configured, the mobile device returns an RRC Connection Reconfiguration Complete message to the eNode-B. This triggers the confirmation of the session establishment on the S1 interface with an Initial Context Setup Response message.

After the mobile has also configured the user plane part of the protocol stack, it returns an Attach Complete message to the eNode-B, which includes the Activate Default Bearer Complete message. Both messages are destined for the MME.

The final step of the attach procedure is to finalize the user data tunnel establishment on the S1 interface between the serving-GW and the eNode-B. So far, only the eNode-B is

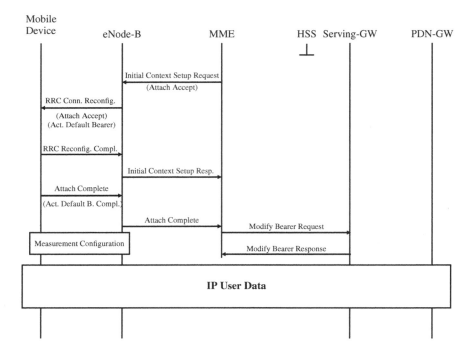

Figure 4.18   Attach and default bearer activation message flow – part 2

aware of the TEID on the serving-GW, because it has received this information in the Initial
Context Setup Request message. At this point in the procedure, the MME can now inform the
serving-GW of the TEID that the eNode-B has assigned on its side for the user data tunnel
with a Modify Bearer Request message. The serving-GW stores the TEID of the eNode-B for
this tunnel and can now forward any incoming IP packets for the user over the correct tunnel
to the eNode-B.

At this point, the connection is fully established and the mobile device can now send and
receive IP packets to and from the Internet via the eNode-B, the serving-GW and the PDN-GW
as shown in Figure 4.18. As the MME is only responsible for the overall session management,
it is not part of the user data path. Despite the overall complexity of the procedure it is usually executed in only a fraction of a second and is thus performed more quickly than similar
procedures in GSM and UMTS. This is due to simplification on all network interfaces and
bundling of messages of several protocol layers into a single message that is then sent over the
air interface.

Once the connection is established, the eNode-B exchanges a number of additional RRC
Reconfiguration messages to configure measurements and reporting of neighbor cells so that
the connection can be handed over to a different cell later on if required. This is not part of
the attach procedure as such but is mentioned here anyway to give a complete picture of the
overall process.

4.6.3   Handover Scenarios

On the basis of the measurement and reporting configuration that the mobile device has received from the eNode-B, it starts measuring the signal strength of neighboring cells. Once a configured reporting criterion has been met, it reports the current values for the signal strength of the active cells and neighboring cells to the eNode-B. Details on measurements and reporting are discussed in Section 4.7. On the basis of this input, the eNode-B can take a decision if a handover of the connection to a neighboring cell with a better signal is necessary. Apart from ensuring that the connection does not fall, a handover usually also improves the data throughput of the mobile device in the uplink direction as well as in the downlink direction. At the same time, it also reduces the amount of power required for uplink transmissions and hence decreases the overall interference.

In LTE, there are two types of handovers. The most efficient one is a handover where the source eNode-B and the target eNode-B directly communicate with each other over the X2 interface. This handover is referred to as an X2 handover. If for some reason the two eNode-Bs cannot communicate with each other, for example, because they have not been configured for direct communication, the handover signaling will take place over the S1 interface and the MME assists in the process. Such a handover is referred to as an S1 handover.

X2 Handover

On the basis of the measurement reports from the mobile device on the reception level of the current cell and the neighboring cells, the eNode-B can take the decision to hand over the ongoing connection to another eNode-B. As shown in Figure 4.19, the first step in this process is a Handover Request message from the source eNode-B to the target eNode-B, which contains all relevant information about the subscriber and all relevant information about the connection to the mobile device as described in 3GPP TS 36.423 [22]. The target eNode-B then checks if it still has the resources required to handle the additional subscriber. Particularly, if the connection of the subscriber requires a certain QoS, the target eNode-B might not have enough capacity on the air interface left during a congestion situation and might thus reject the request. In a well-dimensioned network, however, this should rarely, if at all, be the case. Also, it should be noted at this point that in practice, no specific QoS requirements are used except for network operator-based voice calls.

If the target eNode-B grants access, it prepares itself by selecting a new C-RNTI for the mobile device and reserves resources on the uplink so that the mobile device can perform a noncontention-based random access procedure once it tries to access the new cell. This is necessary as the mobile device is not synchronized, that is, it is not yet aware of the timing advance necessary to communicate with the new cell. Afterward, the target eNode-B confirms the request to the source eNode-B with a Handover Request Acknowledge message. The message contains all the information that the mobile device requires to access the new cell. As the handover needs to be executed as fast as possible, the mobile device should not be required to read the SI messages in the target cell. Hence, the confirmation message contains all the system parameters that the mobile device needs to configure itself to communicate with the target cell. As was described in more detail earlier in this chapter, the information required includes

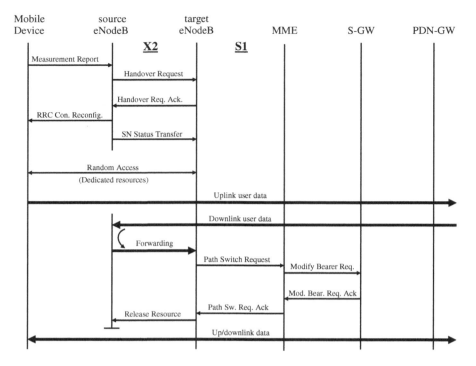

Figure 4.19   X2-based handover message flow

the PCI, the carrier bandwidth, RACH parameters, the uplink shared channel configuration, reference signal configuration, PHICH configuration, SRS parameters, and so on.

Once the source eNode-B receives the confirmation, it immediately issues a handover command to the mobile device and ceases to transmit user data in the downlink direction. Data arriving from the network over the S1 interface after the handover command has been issued is forwarded over the X2 interface to the target eNode-B.

In LTE, there is no dedicated handover command. Instead, an RRC Connection Reconfiguration message is used that contains all the parameters necessary to connect to the new cell. Upon receiving the reconfiguration message, the mobile device stops sending data in the uplink direction and reconfigures itself for communication with the new eNode-B. At the same time, the source eNode-B stops accepting uplink data traffic and sends an SN Status Transfer message to the target eNode-B with the sequence number of the last valid uplink data block. This helps the target eNode-B to request an uplink retransmission if it detects that there are some data blocks missing and allows for data transmission continuity.

As the mobile device has already performed measurements, there is no need to search for the new cell. Hence, the device can immediately transmit a random access preamble on the PRACH as shown in Figure 4.11. As dedicated resources are used for the RACH sequence, the device does not have to identify itself and only the first two messages shown in the figure are required.

The Random Access Response message from the new eNode-B ends the handover procedure from the mobile point of view, and it can immediately resume transmitting data in the uplink direction.

As the eNode-B was given the serving-GW's IP address and the TEID for the connection in the initial handover request, it can forward all uplink data directly to the serving-GW without any detour over the source eNode-B. Downlink data that continues to be forwarded from the source eNode-B to the target eNode-B can now also be delivered to the mobile device. In the radio and core network, however, additional steps are required to redirect the S1 tunnel from the source eNode-B to the target eNode-B. Figure 4.19 shows the simplest variant in which the MME and serving-GW do not change in the process.

The S1 user data tunnel redirection and an MME context update are invoked with a Path Switch Request message that the target eNode-B sends to the MME. The MME then updates the subscriber's mobility management record and checks if the target eNode-B should continue to be served by the current serving-GW or if this should be changed as well, for example, for a better load balancing or to optimize the path between the core network and the radio network. In this example, the serving-GW remains the same; so only a Modify Bearer Request message has to be sent to the current serving-GW to inform it of the new tunnel endpoint of the target eNode-B. The serving-GW makes the necessary changes and returns a Modify Bearer Response message to the MME. The MME in turn confirms the operation to the target eNode-B with a Path Switch Request Acknowledge message. Finally, the target eNode-B informs the source eNode-B that the handover has been performed successfully and that the user data tunnel on the S1 interface has been redirected with a Release Resource message. The source eNode-B can then delete the user's context from its database and release all resources for the connection.

S1 Handover

It is also possible that the source eNode-B is not directly connected to the target eNode-B; so a direct X2 handover is not possible. In such cases, the source eNode-B requests the help of the MME. All signaling exchanges and user data forwarding are then performed over the S1 interface as shown in Figure 4.20. Consequently, such a handover is referred to as an S1 handover.

From a mobile device point of view, there is no difference between an X2 and an S1 handover. When radio conditions trigger a measurement report, the source eNode-B can decide to initiate the handover. As the X2 link is missing, it will send a Handover Request message to the MME. On the basis of the TA ID of the new eNode-B, the MME can decide if it is responsible by itself for the new cell or if another MME should take over the connection. In the scenario shown in Figure 4.20, the same MME remains responsible for the connection so that no further messaging is required at this stage to contact another MME. In this example, the MME also decides that the current serving-GW remains in the user data path after the handover so that no further signaling is required.

In the next step, the MME contacts the target eNode-B with a Handover Request message. If the eNode-B has enough capacity to handle the additional connection, it returns a Handover Request Acknowledge message to the MME which, as in the previous examples, contains all

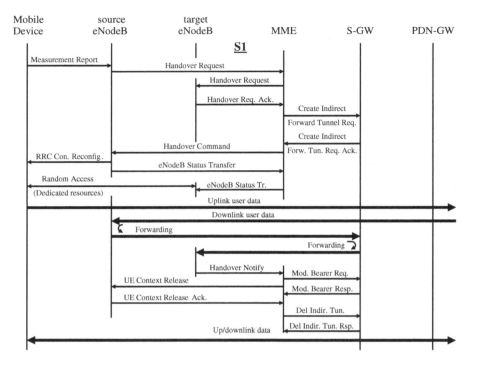

Figure 4.20   Basic S1-based handover

the parameters required for the mobile device to make the handover. This includes information on dedicated resources in the uplink direction to perform a noncontention-based random
access procedure to speed up the handover. For details, see the description of the X2 handover above.

Before the handover can be executed, a temporary tunnel for downlink user data is established to ensure that no packets are lost during the handover. There are two options to forward
data during an S1 handover. Even if no signaling message exchange is possible over the X2
interface between the source and the target eNode-B, it might still be possible that the user
data can be forwarded directly. This is called direct forwarding. This is not the case in the
scenario shown in Figure 4.20 and hence the MME requests a serving-GW to create a temporary indirect tunnel between the source and the target eNode-B with a Create Indirect Forward
Tunnel Request message. The serving-GW that supports the indirect tunnel does not have to
be the same gateway that is responsible for the default tunnel of the user. In this example,
however, this is the case to reduce complexity, and the serving-GW responds with a Create
Indirect Forward Tunnel Request Acknowledge message.

Once the indirect tunnel is created, the MME confirms the handover with a Handover Command to the source eNode-B. The source eNode-B then executes the handover by issuing an
RRC Reconfiguration Command to the mobile device, which includes the parameters of the
target eNode-B. It then stops forwarding downlink data to the mobile device and sends its
current user plane state such as the packet counter to the target eNode-B via the MME in an
eNode-B Status Transfer message.

User data packets still received in the uplink direction are forwarded directly to the serving-GW. User data in the downlink direction, arriving from the current serving-GW, is sent back to the core network to the serving-GW that supports the indirect forward tunnel. This serving-GW then forwards the user data packets to the target eNode-B, where they are buffered until the radio connection to the mobile device has been restored. Redirecting the downlink user data in this way creates additional traffic on the S1 interface, but as only the eNode-B is exactly aware of the time the handover takes place it ensures the smoothest handover possible.

Once the mobile device has contacted the target eNode-B with a contention-free Random Access procedure, that is, the resources allocated for the Random Access are dedicated to the mobile device and hence no identification of the mobile is necessary, the target eNode-B contacts the MME and confirms the handover with a Handover Notify message. The MME then redirects the downlink user data tunnel to the target eNode-B by modifying eNode-B tunnel ID of the bearer context on the serving-GW with a Modify Bearer Request message. Once the operation is confirmed, the MME then releases the user's context in the source eNode-B with a UE Context Release message, which is answered with a UE Context Release Acknowledge message.

In the final step, the indirect forwarding tunnel on the serving-GW is also removed. At this point, all resources that are no longer needed are removed and the user data directly flows to and from the target eNode-B.

While from a mobile device's point of view this handover takes slightly longer than a pure X2 handover, which is a bit less complicated and requires fewer messages between the different entities in the network, an S1 handover is also executed within just a few hundred milliseconds. Owing to the optional user data rerouting during the handover procedure itself the outage experienced by the mobile device is further minimized. Also, as previously stated, the complexity of a handover in the network is completely hidden from the mobile device as it only sees the RRC Reconfiguration message and then uses the Random Access procedure to get access to the new cell, independently of whether the handover is executed over the X2 or S1 interface.

Not shown in the X2 and S1 handover examples above is a potential TA update procedure that has to be executed by the mobile device after the handover in case the target eNode-B is part of a new TA. This is necessary so that the network can locate the mobile device when the eNode-B later decides to remove the physical connection to the device because of prolonged inactivity. This concept is described in more detail in Section 4.7.2 on mobility management and cell reselection in the RRC Idle state.

MME and S-GW Changes

In the X2 and S1 handover examples above, no core network node changes were shown. Under some circumstances, however, these have to be changed during or after a handover as well:

- for load balancing, processing and user plane capacity reasons;
- to optimize the user data path between the radio network and the core network;
- when the target eNode-B is in a TA that is not served by the source MME.

In such cases, the handover processes described above are extended with additional procedures to include the network elements becoming newly responsible for the connection in the overall process. From a mobile device's point of view, however, this is transparent and just increases the time between the initial measurement report and the execution of the handover with the RRC Reconfiguration message. For further details, refer to 3GPP TS 23.401 [21].

4.6.4   Default and Dedicated Bearers

Despite the similarities in the processes among GSM, UMTS and LTE, there is nevertheless one major difference between LTE and earlier technologies. As mentioned above, the attach process already includes the assignment of an IP address. This is unlike in GSM and UMTS where the device could attach to the packet-switched network and only later request for the assignment of an IP address with a separate procedure. This second procedure was often also referred to as a 'packet call' to compare the establishment of an Internet session with the establishment of a voice call. In LTE, this comparison no longer fits, as LTE devices get an IP address straight away, similar to a computer that immediately attaches to a Wi-Fi network or a cabled Local Area Network (LAN) once it is detected and configured.

The IP connection that is automatically established during the Attach procedure uses a default bearer. For network operator-based applications such as VoIP, special QoS requirements such as a constant delay and minimum bandwidth can be ensured for a particular traffic flow by establishing a dedicated bearer. IP packets flowing over a dedicated bearer use the same source IP address as packets flowing through the default bearer.

In practice, a mobile device can also be assigned several IP addresses. This can be useful, for example, to separate services offered by the mobile network operator from general Internet access. For each IP address assigned to the mobile device, a new default bearer is established. A device can hence have more than one default bearer. For each default bearer, it is possible to establish one or more dedicated bearers in addition if they are required for some time for data streams with stringent QoS requirements. The establishment of dedicated bearers is controlled by the application itself, which is why only network operator-deployed applications can make use of them at this time.

It is also theoretically possible to allow Internet-based applications to request the establishment of a dedicated bearer or for the network to detect multimedia streams automatically by analyzing IP addresses, port numbers, etc., and to act accordingly. This is not standardized, however, and also not widely used at the time of publication.

4.7   Mobility Management and Power Optimization

After the major LTE procedures have been introduced in the previous sections, the following sections now take a look at the general mobility management and power consumption optimization functionality. LTE knows two general activity states for mobile devices. These are the RRC Connected state and the RRC Idle state. This state model is much simpler than the one used in UMTS, which has many more states such as Cell-DCH, Cell-FACH, Cell-PCH, URA-PCH and Idle.

4.7.1   Mobility Management in Connected State

While the mobile device is in RRC Connected state, it is usually fully synchronized with the network in the uplink and the downlink directions and can hence transmit and receive data at any time. While the mobile device is in this state, a user data tunnel is established on the S1 interface between the eNode-B and the serving-GW and another tunnel is established between the serving-GW and the PDN-GW. Data arriving for the mobile device can be immediately forwarded to the device. Data waiting to be transmitted in the uplink direction can also be sent immediately, either over continuously allocated RBs on the uplink shared channel or, during times of lower activities, after a quick scheduling request via the uplink control channel. Furthermore, the mobile device actively monitors the signal quality of the serving cell and the signal quality of neighboring cells and reports the measurements to the network. The network can then perform a handover procedure when another cell is better suited to serve the mobile device.

Measurements for Handover

A handover is controlled autonomously by each eNode-B and it also decides if and when mobile devices should send measurement reports either periodically or event triggered. The standard is flexible in this regard so that different eNode-B vendors can use different strategies for measurement reporting. Measurement report parameters are sent to the mobile device after an RRC connection has been established as shown in Figure 4.18 and after a handover to a new cell with an RRC Connection Reconfiguration message. The measurement parameter set includes information on how to measure, and also it contains report triggers as well as basic parameters of neighboring cells so that they can be found quickly.

While mobile devices can easily measure the signal quality of neighboring cells on the same channel, transmission gaps are required to measure the signal quality of LTE, UMTS and GSM neighboring cells on other channels. Such measurements are thus only configured if the eNode-B detects that the signal quality of the current cell decreases and no other intrafrequency cell is available to take over the connection.

Unlike in GSM, where only the Received Signal Strength Indication (RSSI) is used for the decision, LTE uses two criteria. This is necessary as neighboring base stations transmit on the same channel. A mobile device thus receives not only the signal of the current serving cell but also the signals of neighboring cells, which, from its point of view, are noise for the ongoing data transfer. In LTE, the following criteria are used to describe the current reception conditions:

- **RSRP**: The Reference Signal Received Power, expressed in dBm (the power relative to 1 mW on a logarithmic scale). With this parameter, different cells using the same carrier frequency can be compared and handover or cell reselection decisions can be taken. For example, a strong and hence very good RSRP value equals -50 dBm on a logarithmic scale or 0.000001 mW (10^{-9} W) on a linear scale. A weak RSRP value, which still allows reception in practice but at lower speeds, is -90 dBm, which equals 0.000000001 mW (10^{-12} W) on a linear scale.
- **RSSI**: The Received Signal Strength Indication. This value includes the total power received, including the interference from neighboring cells and other sources.
- **RSRQ**: The Reference Signal Received Quality . It equals the RSRP divided by the RSSI. The better this value the better can the signal of the cell be received compared to the interference generated by other cells. The RSRQ is usually expressed on a logarithmic scale in decibel (dB) and is negative as the reference signal power is smaller than the overall

power received. The closer the negative value is to 0, the better the RSRQ. In practice, an RSRQ of −10 results in very low transmission speeds. An RSRQ of −3 or higher results in very good transmission speeds if the overall signal strength (RSRP) of the cell is also high.

Network optimizations try to improve both the RSRP and RSRQ values. This means that in as many places as possible there should always be only one dominant cell with a strong signal. This means that the RSRP is high (e.g., −50 dBm) and the RSRQ is also high (e.g., −3). If two cells are received with an equal signal power, the RSRPs of the two cells might be strong while the resulting RSRQ for each cell is very low (e.g., −8), as the signals interfere with each other.

In practice, both the RSRP and the RSRQ are used for handover decisions. On the one hand, the observed neighbor cell should have a strong signal, that is, the received reference signals should be strong. Hence, the RSRP should be high. On the other hand, the interference should be as low as possible, which means that the quality expressed with the RSRQ should be as high as possible. This is not always the case. At the edge of the LTE coverage area, for example, the signal quality (RSRQ) might be high as there are no or only weak neighboring LTE cells that interfere with the signal of an observed cell. The RSRP of the cell, however, is very low owing to the high attenuation caused by the distance between the mobile device and the cell. In such a scenario, it does not make sense to hand over the connection to a cell that only has a better RSRQ if there are other alternatives. Instead, the eNode-B could decide to redirect the connection to a UMTS access network if the reception conditions of such a cell are better. This is discussed in more detail in Section 4.9.

Discontinuous Reception (DRX) in the Connected State to Save Power

Continuously scanning for scheduling grants in each subframe once a millisecond is power consuming and should be avoided if the overall throughput required by a device at one time is far below that which could be transferred if the device was scheduled in every subframe. In LTE, it is possible to configure a device to only periodically check for scheduling assignments. This functionality is referred to as DRX and works as follows.

When the network configures DRX for a device, it defines the value for a timer that starts running after each data block has been sent. If new data is sent, the timer is restarted. If no data was sent by the time the timer expires, the device enters DRX mode with a (optional) short DRX cycle. This means that it will go to sleep and wake up after a short time. If new data arrives from the network, it can be delivered quite quickly and with relatively little latency as the device only sleeps for short periods. The short DRX cycle mode also has a configurable timer and once it expires, that is, no data was received during the short cycle mode, the device implicitly enters the long DRX cycle. This is even more power efficient, but it increases the latency time. If a scheduling grant is received during the times when the mobile device scans the control region, all timers are reset and the device enters the full activity state again until the short DRX cycle timer expires again. Figure 4.21 shows how a connection is switched between the different DRX states.

4.7.2 Mobility Management in Idle State

During long times of inactivity, it is advantageous for both the network and the mobile device to put the air interface connection into the RRC Idle state. This reduces the amount of signaling

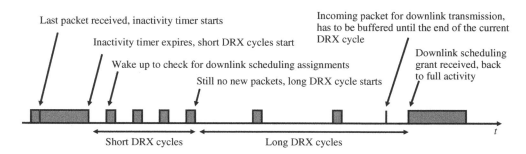

Figure 4.21   Short and long DRX cycles

and the amount of power required for the mobile device to maintain the connection. In this state, the mobile device autonomously performs cell reselections, that is, it changes on its own from cell to cell whenever required by signal conditions. The network is contacted only when a cell is in a new TA. As a consequence, the MME in the core network is only aware of the TA that usually comprises many cells. The LTE TA concept is hence similar to the concept of location and routing areas in GPRS and UMTS and reduces the location signaling, which helps to reduce the mobile device's power consumption.

In RRC Idle state, no user data tunnel is present on the S1 interface between the eNode-B and the serving-GW. The user data tunnel between the serving-GW and the PDN-GW, however, remains in place. From a logical point of view, the connection is still established and all logical bearers remain in place. This means that the IP address or the IP addresses assigned to the mobile device remain in place. Whenever there is renewed activity, the physical radio bearer and the S1 user data tunnel have to be reestablished.

When the mobile device wants to send a new IP packet to a server on the Internet, the tunnel reestablishment is straightforward. After the mobile device has connected to the eNode-B, the S1 tunnel is recreated and the device enters RRC Connected mode again.

In case an IP packet arrives from the Internet while the mobile device is in RRC Idle state, it can be routed through the core network up to the serving-GW. As the serving-GW has no S1 tunnel for the user, it contacts the MME and requests it to reestablish the tunnel. As the MME is only aware of the TA and not the individual cell in which the mobile is currently located, it sends a paging message to all eNode-Bs that the belong to TA. The eNode-Bs in turn forward the paging message over the air interface to inform the mobile device that data is waiting in the network.

When in RRC Idle state, the mobile device deactivates the radio module for most of the time. Only at the end of the paging interval, usually in the range of $1-2$ seconds, does it temporarily activate the radio receiver to check if the eNode-B is transmitting a paging message with its identity. This behavior is also referred to as DRX in RRC Idle state, which is different from the DRX mode in the RRC Connected state that was described earlier.

If the mobile device finds a paging message addressed to itself, it reestablishes a connection with the eNode-B with a random access procedure and requests the reestablishment of the connection. The eNode-B that receives the mobile's request then answers the paging message from the MME and both the air interface connection and the S1 tunnel are reestablished. Once both are in place, the MME contacts the serving-GW, which then forwards the waiting

IP packets to the mobile device. At this point, the process of moving the device through the different activity states starts from the beginning. The following list summarizes the different activity states:

- RRC Connected state with an observation of the control region for assignment grants in every subframe.
- RRC Connected state with an observation of the control region for assignment grants in a short DRX cycle pattern. The receiver is switched off for short periods of time.
- RRC Connected state with an observation of the control region for assignment grants in a long DRX cycle pattern. The receiver is switched off for longer periods of time.
- RRC Idle state in which the mobile scans only periodically for incoming paging messages.

While the mobile device is in RRC Idle state, it decides on its own as to when to change the serving cell. If a new cell is in the same TA as the previous cell, no interaction with the network is required. If the new cell is in a new TA, the mobile device needs to perform a TA update. For this purpose, a temporary RRC connection is established with the eNode-B, which is then used to perform the TA update with the MME. Once the update is finished, the RRC connection is released and the mobile device goes back into a full RRC Idle state, only observing incoming paging messages at the end of each paging interval.

While the mobile device decides on its own as to when to change cells without interaction with the network, the parameters used for the decision are given to the mobile device by the eNode-B via SI messages. Each eNode-B may have a different cell reselection configuration. When the device changes from one cell to another, it not only has to check if the new cell is in a new TA but also has to read the SI messages and decode all messages that contain parameters for the cell reselection mechanism. Only afterward can it go back to the Idle state and monitor the paging channel for incoming paging messages. For cell reselection, the following parameters are important:

- **Cell barring status in SIB 1**. If the cell is barred, the mobile device must not use it as its new serving cell.
- **A serving cell hysteresis in SIB 3**. The degree by which the current cell should be preferred to neighboring cells (in dB).
- **Speed state selection in SIB 3**. Depending on the speed of the mobile, that is, whether it is stationary or in a car, train, etc., different cell reselection parameter settings can be defined. When moving, the cell search mechanism could be started while the reception level is still relatively high to prevent the loss of coverage due to fast cell changes and not enough time to make appropriate measurements. When the mobile is stationary, the cell search could be started when reception levels are lower. Thresholds can be set higher to prevent unnecessary cell changes. As neighbor cell measurements consume additional power when the mobile device is stationary, there is a good possibility of being able to reduce the energy consumption when reception conditions are good.
- **Start of intrafrequency search in SIB 3**. Defines the signal quality level of the serving cell at which the mobile device should start looking for neighboring cells.
- **Start of interfrequency and inter-RAT (Radio Access Technology) search in SIB 3**. Defines the signal quality level of the serving cell at which the mobile device should, in addition, start looking for neighboring cells on other LTE frequencies and cells of other

RATs such as GSM, UMTS and CDMA. Usually, this is set at a somewhat lower value than the intrafrequency search value since finding an LTE cell is preferred.

- **Neighbor cell information in SIB 4–8**. These SI messages contain further details about neighboring cells on the same frequency, on another frequency and other RAT cells. Table 4.4 at the beginning of the chapter contains additional details. SIB 4 with intracell neighbor information is optional. If not present, the mobile device performs a blind search.

4.7.3   Mobility Management And State Changes In Practice

In practice, many factors influence how network operators configure the air interface connection to a mobile device and when reconfigurations take place. On the one hand being in a fully connected state without DRX results in the fastest response times and generates no signaling overhead between the base stations and the core network. On the other hand, being in a connected state even when no data is transferred is inefficient on the mobile side as observing the downlink control channels and continuously transmitting control information in the uplink requires significant power on the mobile side thus draining the battery quickly. The disadvantage on the network side is the reduced capacity in the uplink direction due to many devices transmitting control information in parallel. A compromise therefore has to be found of how long a mobile device is in fully connected state before it enters connected DRX and how long it takes afterward before the network sets the mobile device into the Idle state. The following list gives typical ranges of values found in networks today:

- Timer to switch from Connected Mode to Connected with DRX Mode after inactivity: 100–500 milliseconds
- Short-DRX Cycle: 20 milliseconds (not used by many networks and not supported by all devices)
- Long-DRX Cycle: 40–320 milliseconds
- On-Duration Timer: 2–20 milliseconds
- Timer to switch from Connected to Idle state after inactivity: 5–10 seconds

4.8   LTE Security Architecture

The LTE security architecture is similar to the mechanisms already used in UMTS and discussed in Section 3.9. The architecture is based on a secret key which is stored on the SIM card of the subscriber and in the HSS in the network. The same key is used for GSM, UMTS and LTE. It is therefore possible to efficiently move the security context between network nodes when the user roams between different RATs.

During the initial contact with the LTE network, that is, during the attach procedure described earlier, security procedures are invoked between the UE, the MME and the HSS. During this process, the UE authenticates to the network and the network authenticates to the UE. This prevents man-in-the-middle attacks. The authentication algorithms required for the process are stored and executed in the SIM card and in the HSS. This way, the secret key remains in a protected environment and cannot be read by potential attackers eavesdropping on the message exchange on an interface between the SIM and the mobile device or the HSS and the MME. SIM cards must be capable of performing UMTS authentication. Consequently, old GSM-only SIM cards cannot be used for authentication in LTE and the attach procedure is rejected with such SIM cards. A disadvantage of this approach is that many network operators still use GSM-only SIM cards today for cost reasons, which have to be exchanged when the network operator launches an LTE network. If a GSM-only SIM card is used in an LTE-capable device that then tries to access an LTE network, the MME at first queries the HSS for authentication and ciphering keys. As the HSS receives the request from an LTE network node, it rejects the request as the subscriber's HSS entry contains only GSM authentication information. The MME then terminates the attach process with a reject cause 15 (no suitable cells in this tracking area) that triggers the mobile device to change to UMTS or GSM and to perform a new attach procedure there.

Once authentication has been performed, a set of session keys are generated as described in more detail in [23]. Afterward, ciphering and integrity protection can be activated for all NAS messages between the UE and the MME. While integrity checking is mandatory, the use of ciphering for signaling messages between the mobile device and the MME is optional. Once the corresponding keys are known by the eNode-B, it will also activate integrity checking and ciphering for RRC messages and ciphering for the user data bearer over the air interface. As NAS messages are carried inside RRC messages, they are ciphered twice if encryption for signaling messages was activated in the previous MME/UE security exchange. In any case, two integrity checks are performed, one between the UE and the eNodeB, and another one between the UE and the MME.

When ciphering and integrity checking are activated, the UE, MME and eNode-B can select an appropriate EPS Encryption Algorithm (eea0, eea1, eea2, etc.) and an EPS Integrity Algorithm (eia1, eia2, etc.) from a list of supported algorithms that is supported by both sides. Eea0 corresponds to no encryption being used. Therefore, in operational networks, the use of eea0 between the mobile device and the eNodeB should be the exception. Integrity checking is always used even if encryption is not activated. This is why eia0 does not exist. Eea1/eia1 corresponds to the algorithms introduced in 3GPP Release 7 for UMTS (UEA2, SNOW3G).

4.9   Interconnection with UMTS and GSM

When a mobile device is at the border of the coverage area of the LTE network, it should switch to another network layer such as UMTS and GSM to ensure connectivity. In the worst case, the mobile device loses the LTE network coverage and if it does not find a suitable LTE cell on the current channel it will search for LTE cells on other channels and also switch to other frequency bands and other RATs to regain contact with the network. These actions take a significant amount of time, typically between 10 and 30 seconds, during which the device is not reachable for services trying to contact it such as push e-mail or incoming voice calls. It is therefore better if the network supports the mobile device to find other suitable channels, bands or radio technologies. There are three basic procedures for these purposes that are described in the following section:

- cell reselection from LTE to UMTS or GSM;
- RRC connection release with redirect from LTE to UMTS or GSM;
- inter-RAT handover from LTE to UMTS or GSM.

Irrespective of whether the mobile has to find a GSM or UMTS network by itself or if it is supported by the network, the LTE network has to be connected with the GSM and UMTS networks so that the subscriber's context, that is, the assigned IP address, QoS settings, authentication and ciphering keys, and so on. can be seamlessly exchanged between all core network components involved. One option connects the LTE core network and the GSM/UMTS core networks by using LTE-based interfaces as described in 3GPP TS 23.401 [21]. The use of these interfaces requires software updates on the GSM/UMTS core network nodes. As many network operators are reluctant to upgrade service network elements until stability and quality of enhancements are proven, 3GPP offers an alternative in annex D of [21]. The alternative is based on reusing the well known and stable Gn interface from GSM and UMTS (see Chapters 2 and 3). In such a setup, the LTE core network components behave as classic SGSNs and GGSNs toward the GSM and UMTS core networks. Figure 4.22 shows this alternative setup. How the Gn interface is used with LTE is now described for different scenarios in the following sections.

4.9.1   Cell Reselection between LTE and GSM/UMTS

The simplest way from a network and signaling point of view to move from LTE to another RAT is cell reselection in RRC Idle state. For this purpose, the eNode-Bs broadcast information on neighboring GSM, UMTS and CDMA cells in their SI messages as described above. When a network-configured signal level threshold is reached, the mobile device starts searching for non-LTE cells and reselects to them based on their reception level and usage priority.

With LTE, an optional usage priority scheme for different frequencies and RATs was introduced as it might be, for example, preferable in practice to remain with an LTE cell with a weak signal for as long as possible rather than to reselect to a GSM cell with a strong signal but with a very low bandwidth compared to an LTE cell. If priorities are broadcast in SI messages or sent in dedicated signaling messages to the mobile device, it would furthermore

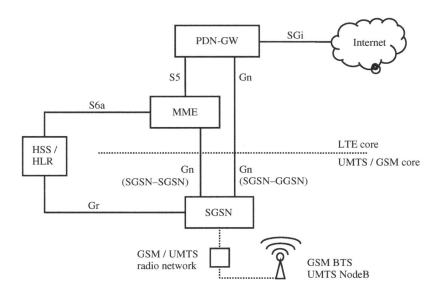

Figure 4.22   Interconnection of LTE to GSM and UMTS networks

also try to reselect to frequencies and RATs with a higher priority when a configured threshold is reached, even if the current serving cell has a stronger signal. Details of this mechanism are described in 3GPP TS 36.304 [24]. At this point, it should be noted that before the introduction of LTE, a usage priority, was not necessary for GSM and UMTS, as UMTS cells were generally preferred to GSM cells. With three radio technologies and the rising number of frequency bands in use, however, LTE cells in one band might in some situations be preferred to LTE cells in another band, which are preferred over UMTS cells, which in turn are preferred to GSM cells.

Once the mobile device decides to move from an LTE cell to a GSM or UMTS cell, it performs a location area update with the circuit-switched side of the core network if it has circuit-switched voice and SMS capabilities. For details, see Section 4.12. If no operator-based VoIP solution is used in the LTE network, the MSC will not be able to retrieve the subscription information from a previous MSC and will hence request the mobile device to identify itself so that it can retrieve the user's subscription record from the HLR to complete the location update.

On the packet-switched side of the GSM or UMTS network, the mobile device performs a routing area update. The request includes information on the previously used MME/S-GW, which, from the point of view of the GSM/UMTS SGSN, is another SGSN. On the basis of this information, the SGSN derives the IP address of the previous SGSN (MME) and requests the subscriber's current context over the Gn interface as shown in Figure 4.22. From the SGSN's point of view, this is a standard inter-SGSN routing area update procedure that is also used to retrieve a subscriber's context that has performed a cell reselection from a GSM or UMTS cell connected to a different SGSN. As a consequence, no software changes are needed in the SGSN to support an LTE inter-RAT cell reselection.

With the information provided in the MME, the SGSN then authenticates the subscriber and requests a user data tunnel modification from the PDN-GW, which acts as a GGSN over the Gn interface. The procedure is finished with an update of the subscriber's location in the HLR and the confirmation of the routing area update to the mobile device. The device then goes back to the Idle state. The details of this procedure are described in 3GPP TS 23.060 [25].

When an LTE-capable mobile device roams through a GSM, UMTS or CDMA network, it should return to an available LTE network as soon as it roams into an LTE-covered area. Again, SI messages are used to inform the mobile device of nearby LTE cells. In GSM, this information is broadcast in SI 2-quarter messages and in UMTS in SIB 19. Again, usage priorities can be used to steer the network selection, in addition to taking signal strength into account, so that LTE cells with a lower signal level can be preferred to cells of other radio technologies.

Once the mobile device performs a cell reselection to an LTE cell, it performs a TA Update procedure with the MME. In the TA update message, the mobile device includes information about the last used location and the routing area so that the MME can determine the previously used SGSN. It will then contact the SGSN over the Gn interface to request the subscriber's context to preserve the session. The 2G/3G SGSN will again see this as a standard inter-SGSN routing area update. Once the subscriber has been authenticated the MME contacts a serving-GW and the PDN-GW that has so far acted as an SGSN to redirect the core network user data tunnel. Finally, the MME also contacts the HSS to inform it of the new location of the subscriber. Afterward, the TA update is confirmed to the mobile device and it can return to RRC Idle state.

4.9.2   RRC Connection Release with Redirect between LTE and GSM/UMTS

While the mobile device is in the LTE RRC Connected state, the network is responsible for the mobility management. The network needs to coordinate times during which the mobile device is not available to receive data in the downlink direction because of measurements on other channels and on other frequency bands. Network control of the mobility management process in the connected state is also important so that no data is lost when the mobile is handed over to a different cell.

Depending on the capabilities of the mobile device and the network, the eNode-B can instruct the mobile device to start a search for neighboring LTE, UMTS and GSM cells on other channels and frequency bands once the mobile device reports a deteriorating signal quality. The eNode-B then sends the mobile device a list of frequencies and bands in which to search for other cells. Furthermore, a transmission gap pattern in which the measurements are to be performed is given. The pattern can be adapted to the number of cells, the types of radio networks and possibly also to the current throughput required by the user. The more frequent and longer the gaps, the faster can the neighboring cells be found and reported to the network.

Once the signal level reaches a point where a data transfer cannot be sustained any longer, the easiest method to send the mobile device to a UMTS or GSM cell is to release the RRC connection with a redirection order. If inter-RAT measurements have taken place the network can select the cell with the best radio conditions. If no measurements have taken place, the network simply informs the mobile device of the UMTS channel number on which to search for suitable UMTS cells. In practice, this takes somewhat longer compared to a release for which measurements have taken place before. However, it can be observed that the reselection is still performed quickly.

Once the redirection order has been received, the mobile device terminates the communication, changes to the new frequency and RAT and tries to find and synchronize to the new cell. If the new cell is not found, the mobile device performs a general cell search and selects a suitable cell on the basis of signal strength and the priority given for certain bands and radio technologies in the SI of the last cell.

Once the mobile device has synchronized to the new cell, it establishes a signaling connection, and a location area update and a routing area update procedure will be performed as was previously described for Idle state Inter-RAT cell reselections. If the intended cell is found immediately, the redirection procedure will typically take less than 4 seconds under ideal conditions.

A similar procedure is used to redirect a mobile device currently served by a UMTS Node-B while it is in Cell-DCH state. To find neighboring LTE and GSM cells, the RNC instructs the mobile device regarding what to search for and where and defines a measurement gap pattern. In GSM, the procedure is a bit different. Here, the mobile device searches autonomously for the inter-RAT cells it knows about from SI 2-quarter messages even while in GPRS ready state and reselects to LTE or UMTS without the help of the network.

In all cases, the procedure takes several seconds in which the data transfer is interrupted. In addition, data that was buffered in the network during the interruption is discarded and the TCP/IP layer or the applications have to ensure that the packets are repeated.

4.9.3 Handover between LTE and GSM/UMTS

As the outage time during a Cell Change Order procedure is in the range of several seconds, it is not suitable for a number of applications such as VoIP. For other applications such as web browsing, outage times of several seconds are acceptable but they have a negative impact on the overall user experience. As a consequence, the LTE standards also define handover procedures between LTE, GSM and UMTS in 3GPP TS 23.304 [24] with which almost seamless handovers are possible. From a network point of view, handovers are much more complicated than Cell Change Orders and hence the functionality is not used in all networks.

In general, an inter-RAT handover procedure from LTE is performed in a similar way as a handover from one LTE cell to another as described in the Section 'S1 Handover' and Figure 4.20. In addition to the steps discussed for an intrafrequency LTE handover, the following actions are required when handing over an ongoing connection to UMTS:

- For measurements on other frequencies, the eNode-B needs to reconfigure the radio connection so that reception and transmission gaps for measurements on other channels can be inserted.
- The handover command in the RRC reconfiguration message contains the new frequency, RAT technologies and other parameters required to change to another RAT.
- Once the connection has been handed over, the mobile device has to perform a routing area update procedure to update the core network nodes and the HSS with its current position.
- Depending on the support of network operator voice and SMS service, a location update with the CS core network might have to be performed so that the mobile remains reachable for these services after the handover.

The standard also contains a procedure to hand over a mobile device in RRC Connected state to GSM. In practice, however, such functionality is not widely used from UMTS to GSM presently. It is therefore likely that such functionality is also not introduced in the foreseeable future with LTE.

Additional measures have to be taken if a voice call from a network operator-based service is ongoing over an LTE connection when running out of network coverage. This is especially the case if only a GSM network is available, that is, there is no alternative but to hand over the call to the circuit-switched part of the network. More details are discussed in Section 4.13.

4.10 Interworking with CDMA2000 Networks

In some countries, for example, the United States, some CDMA2000 network operators have chosen LTE as their next generation RAT. To allow roaming between the legacy radio access network and LTE, a functionality similar to that for GSM and UMTS described in Section 4.9 is required:

- cell reselection between LTE and CDMA2000;
- RRC connection release with redirect from LTE to CDMA2000;
- inter-RAT handover between LTE and CDMA2000.

To support the seamless roaming in the different access networks, 3GPP TS 23.402 [26] describes a number of new network interfaces to interconnect the two systems with each other. These are shown in Figure 4.23 and described in the following sections.

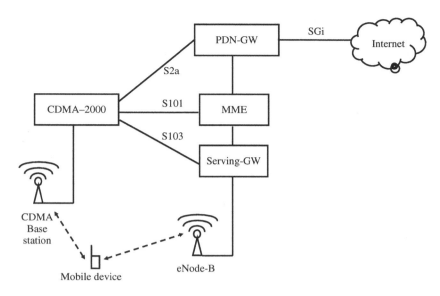

Figure 4.23   Interconnection of LTE to CDMA2000 networks

4.10.1   Cell Reselection between LTE and CDMA2000 Networks

When the mobile device is in RRC Idle state, it can decide on its own as to when to move between the two radio access network technologies. On the LTE side, it is supported by neighbor cell information in the SIB-8 message.

To keep the IP address when changing between the two radio technologies and core networks, it is necessary that the PDN-GW is selected as a mobile IP home agent (HA) by the CDMA2000 network when the connection is first established. For this purpose, the S2a interface is used as shown in Figure 4.23.

When reselecting from LTE to CDMA2000, the mobile device logically disconnects from the LTE network and establishes a signaling connection with the legacy network. The network then contacts the PDN-GW via the S2a interface and moves the subscriber's context to the CDMA2000 side.

4.10.2   RRC Connection Release with Redirect between LTE and CDMA2000

While in LTE RRC Connected state, the network decides as to when to move a connection to another LTE cell or another radio access system in case the user is at the edge of the LTE coverage area. If supported by the mobile device, the network can be aided by signal strength measurements. The simplest form of such a transfer is to issue an RRC connection release with redirect information to a CDMA2000 cell. Two variants are described in 3GPP TS 23.402 [26]:

- **Non-optimized**. The mobile device changes the RAT and performs the same signaling message exchange as described above for cell reselection.

- **Optimized**. If supported by the network, the mobile device performs a preregistration with the CDMA2000 access network at the time it initially registers with the LTE network. This way, a session is already established in the CDMA2000 access network and then put into dormant state until the time the mobile device reselects into the CDMA network. This minimizes the delay during the reselection process, as fewer actions have to be taken at the time the mobile changes from one radio network technology to the other. As shown in Figure 4.23, the S101 interface has to be in place for this purpose. This interface is used to transparently tunnel CDMA messages between the mobile device in the LTE network over the S1 interface to the MME and from there to the CDMA access network.

4.10.3   Handover between LTE and CDMA2000

To further reduce the interruption time when changing between the two RATs, it is also possible to perform a handover of the ongoing data session from LTE to CDMA2000 and vice versa. As in the previous section, the network can instruct the mobile device to perform signal strength measurements, for example, once the signal strength or quality of the current cell reaches a certain threshold. Once a handover has to be made, the eNode-B sends an Evolved-UMTS Terrestrial Radio Access (E-UTRA) handover preparation request message to the UE. The UE then communicates with the CDMA radio network over the S101 interface to prepare the other radio network for the incoming connection. As part of this preparation phase, the MME prepares a transparent indirect user data tunnel between the current eNode-B, a serving-GW and the CDMA2000 radio network via the S103 interface to forward any incoming data from the Internet to the CDMA network until the handover has been executed. When the S103 user data tunnel is in place, the handover is executed and the mobile device moves to the CDMA network. Once there, the CDMA network informs the MME that the mobile device has performed the handover successfully. The user data tunnel between the PDN-GW and the S-GW is then redirected to the CDMA access network. In addition, the temporary data forwarding tunnel over the S103 interface and user's context in the eNode-B are removed.

4.11   Network Planning Aspects

As in GSM, UMTS and CDMA, meticulous network planning is essential to ensure a high performing network in as many places as possible and to reduce the effect of interference from neighboring cells and other mobile devices. The following sections describe some of the challenges faced and discuss potential solutions.

4.11.1   Single Frequency Network

Like UMTS and CDMA, the LTE radio access network reuses the same carrier frequencies for all cells. To extend capacity, it is possible to operate several carriers in the same frequency band, although network operators usually take advantage from the up to 20-MHz channel bandwidth that has been specified for LTE. In some bands, 20-MHz channels might not be feasible, however, for a number of reasons:

- Enough spectrum is not available because of several network operators sharing the available spectrum in a small band. An example is band 20, the European digital dividend band. As shown at the beginning of this chapter in Table 4.2, only 30 MHz is available for each direction. If used by more than two operators, the maximum channel bandwidth per network operator is 10 MHz at best.
- Certain bands are not suitable for 20-MHz channel, for example, because of a narrow duplex gap between uplink and downlink. This makes it difficult for filters in mobile devices to properly separate the uplink and the downlink data streams in the transceiver. In such bands, more than one carrier might be used if a network operator has been able to acquire more bandwidth.

4.11.2   Cell Edge Performance

Owing to neighboring cells using the same channel, mobile devices can receive the signals of several cells. While they are close to one cell, the signals of other cells are much lower and hence their interference is limited. When a mobile device is at the center of the coverage areas of several cells; however, two or even more cells might be received with similar signal strength. If all cells are also heavily loaded in the downlink direction, the resulting interference at the location of the mobile device can be significant. The resulting datarate in the downlink direction to this particular user is then very limited because a robust modulation and coding scheme with good error protection has to be used. This also impacts the overall capacity of the cell as more time has to be spent to transmit data to devices at the cell edge with a low speed that cannot be used to send data much faster to devices that experience a better signal quality.

To improve cell edge performance and the overall throughput of an eNode-B and the radio network in general, a load indication message has been defined in 3GPP TS 36.423 [9] for Intercell Interference Coordination (ICIC). As eNode-Bs autonomously decide as to how they use their air interface, the X2 interface can be used to exchange interference-related information between neighboring eNode-Bs, which can then be used to configure transmissions in such a way as to reduce the problem.

To reduce interference in the downlink direction, eNode-Bs can inform their neighbors of the power used for RBs on the frequency axis. This way, an eNode-B could, for example, use the highest power only for a limited number of RBs to serve users at the edge of the cell, while for most other RBs, less power would be used to serve users that are closer to the center of the cell. As neighboring eNode-Bs cannot directly measure the strength of downlink transmissions from neighboring cells, they can use this information as an indicator of the RBs in which there is likely to be high interference at the edge of the cell and hence schedule a different set of RBs for use at the edge of the cell. This in effect creates two-tiered cells as shown for a simplified two-cell scenario in Figure 4.24. Such a scheme is also referred to as Fractional Frequency Reuse (FFR) as only a nonoverlapping fraction of the spectrum is used to serve cell edge users.

In the uplink direction, an eNode-B can measure interference from mobile devices communicating with another eNode-B directly and take measures to avoid scheduling those RBs for its own users. In addition, the load indication message on the X2 interface can be used to inform neighboring eNode-Bs regarding the RBs in which high interference is experienced so that neighbors can change, limit or adapt the power usage of mobile devices on certain RBs.

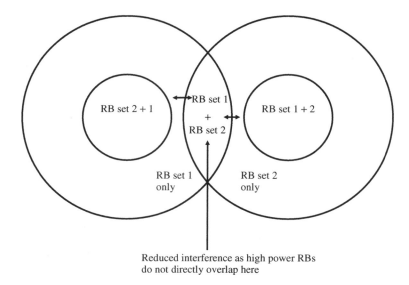

Figure 4.24   Fractional frequency reuse for reducing cell edge interference

How eNode-Bs use the information they receive from their neighbors is implementation specific and not described in the 3GPP standards.

4.11.3   Self-Organizing Network Functionality

In addition to interference management, there are a variety of other tasks that have to be performed manually during the buildup of the network and afterward while maintaining it to ensure high capacity and high availability. For LTE, 3GPP has created a work item referred to as Self-configuring and Self-Organizing Network (SON) to define ways to optimize and automate many labor-intensive tasks. 3GPP Technical Recommendation (TR) 36.902 [27] gives an overview of a number of tasks that can be fully or partially automated in the future:

- **Initial self-configuration**. Retrieval of basic operational parameters from a centralized configuration server when the cell is first activated.
- **ANR**. Mobile devices can report cells that are not in the neighbor list to the base station they are currently served by. This information can then be used by the network to automatically establish neighbor relationships for handovers.
- **Coverage and capacity optimization**. Interference between cells due to extensively overlapping coverage areas and coverage holes is one of the main issues of deployed networks. Such conditions are usually detected with drive tests. Here, SON aims to use mobile device and base station measurements to detect these issues. While interference can potentially be reduced automatically, unintended coverage holes can sometimes be fixed only with additional base stations. In such a case, the equipment could at least notify the network operator.

- **Energy saving**. Reduction of transmission power in case it is not needed – automatic shut-down and reinitialization of femtocells when the user arrives in or leaves the coverage area of a femtocell.
- **PCI configuration**. As described above, every LTE cell has a PCI that is used during the cell search procedure to distinguish the transmissions of several cells on the same carrier from each other. Only 504 IDs are available and neighboring base stations should use a certain combination for easier detection. As it is sometimes difficult to predict all cell neighbors, an auto-configuration functionality is highly desirable. Again, the mobile is required to report to the network as to which cells it looks out for the automated configuration process.
- **Handover optimization**. By analyzing the causes of failure of handovers, coverage holes or wrong handover decisions can be detected and changed.
- **Load balancing**. If a cell already experiences high load from many users, users at the cell edge can be redirected to other nearby cells.
- **RACH optimization**. The RACH is needed for the initial communication between nonsynchronized mobile devices and the network. Depending on the load, the number of resources dedicated to the RACH can be changed dynamically.

Further details and references to specifications on some of these topics can be found in Sauter [28].

4.12   CS-Fallback for Voice and SMS Services with LTE

One of the major design choices of LTE was to focus on the development of a packet-based core and access network infrastructure. The circuit-switched core network and dedicated telephony features of GSM and UMTS radio access networks have not been adapted for LTE. This significantly reduces the overall complexity of the network and follows the direction that has been taken in fixed-line networks many years earlier. Here, a clear trend toward IP and voice services over IP is well underway. At the homes of customers or in offices, multifunctional gateways that include a DSL modem, a Wi-Fi access point, fixed-line Ethernet ports and also RJ-11 ports to connect ordinary telephones are now common. Inside the device, Session Initiation Protocol (SIP)-based IP telephony data streams and signaling are converted into the classic analog or ISDN format and the user can thus continue to use his legacy devices.

With LTE, a reuse of legacy equipment is not possible, and hence, other ways have to be found to offer voice services. Another major complication that is not found in fixed-line networks is the necessity for voice and other previously circuit-switched services such as SMS to be backward compatible to the services offered in fixed-line networks. For a user, it should be invisible if the service is offered over the circuit-switched part of the GSM or UMTS network or the packet-switched IP-based LTE network. Also, an ongoing voice call over LTE should be seamlessly handed over to GSM or UMTS if the user leaves the LTE coverage area. In other words, the IP-based voice call must be converted to a circuit-switched voice call on the fly as otherwise the overall user experience will be unsatisfactory. The system designed for LTE to tackle these challenges is referred to as VoLTE and is based on the IP Multimedia Subsystem (IMS) that was first introduced with 3GPP Release 5. Many additions and enhancements were necessary over time. However, when the first LTE networks appeared in practice, stable and fully functional VoLTE systems were still not available. As a consequence, it was decided to continue using GSM and UMTS for voice and SMS services, despite being incompatible with LTE. This solution is referred to as Circuit-Switched Fallback (CSFB) and is described in this section. The following section then takes a closer look at the fully IP-based VoLTE system that will gradually take over from CSFB in the coming years.

4.12.1 SMS over SGs

One of the most popular services besides voice telephony in wireless networks is the SMS. In GSM, SMS uses the signaling channels of the circuit-switched side of the network. In addition, SMS is important to deliver information on international roaming prices and bill shock warning messages to customers. While at first it was envisaged to bring voice service and SMS as a single function to LTE, it was later decided to speed up the deployment of SMS. The result of this is the SMS over SGs functionality as specified in 3GPP TS 23.272 [29]. As shown in Figure 4.25, the SGs interface has been specified to forward SMS messages between a GSM/UMTS circuit-switched MSC and the MME of the LTE core network. It is similar to the Gs interface that connects the circuit-switched MSC to the packet-switched SGSN in a GSM/GPRS network to exchange paging notifications and SMS messages as described in Chapter 2. From the MME, the SMS message is delivered in an NAS signaling message to the mobile device. Mobile-originated messages take the reverse path. As in GSM and UMTS, the SMS service remains a non-IP-based service as it continues to be transmitted over signaling channels. On the LTE side of the network, however, the signaling channel is transported over the S1 link, which is based on IP. From an end-to-end point of view, however, SMS remains a non-IP service as the message over the air interface is not embedded in an IP packet but in an RRC signaling message. As a consequence, no IP-based higher layer application is required to send and receive SMS messages.

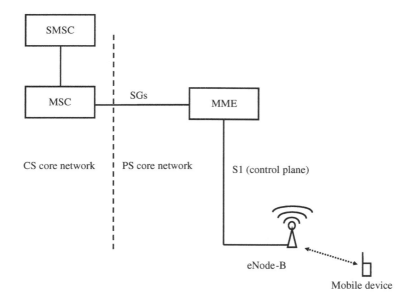

Figure 4.25   SGs interconnection for delivery of SMS messages

To send and receive SMS messages while in the LTE network, a mobile device has to inform the MME during the attach procedure of its SMS capabilities. Instead of performing a standard attach procedure, the mobile device sets a flag that the MME should also register the mobile device with the circuit-switched GSM or UMTS core network. This is also referred to as a 'combined attach SMS only' in the specification documents and is mainly used today for non-voice LTE devices such as tablets that can send and receive SMS messages.

To deliver SMS messages over the SGs interface, the MME registers itself with the HLR for the delivery of SMS messages during the attach procedure. When a subscriber sends an SMS message to another subscriber that is currently located in the LTE network, the message is first delivered to the SMS service center. The SMS center then queries the HLR to get the address of the network node to which the SMS should be forwarded for delivery over the radio network. In this case, it will receive the address of the MME and then forward the SMS message to a SGs-capable MSC. From there, it is routed over the IP-based SGs interface to the MME.

If the subscriber is in RRC Connected state at the time the SMS is delivered to the MME, it can be forwarded immediately to the mobile device in an NAS signaling message. If the mobile device is currently in RRC Idle state the MME needs to first page the device in the last reported TA. Once the mobile device responds and a signaling connection to the MME has been reestablished, the SMS is forwarded and the device returns to RRC Idle state.

4.12.2 CS Fallback

In addition to SMS messages, the SGs interface can also be used to deliver paging messages that inform the mobile device of an incoming call. The call itself, however, is not delivered over the LTE interface and the mobile device has to fall back to a GSM or UMTS network where a circuit-switched connection is then established for the call. This method of delivering voice calls is therefore referred to as CS (circuit-switched) fallback and is executed as follows. Further details can be found in 3GPP TS 23.272 [29].

The Preparation Phase

- When the GSM/UMTS/LTE-capable device first connects to the EPS (that is, to LTE), it indicates to the network that it wants to perform a 'combined attach' in the same way as for the SMS over SGs functionality described above. In practice, this means that it requests the network to also register its presence in the 2G/3G circuit-switched network.
- Registration of the mobile in the 2G/3G network is performed on behalf of the mobile device by the MME. From the MSC point of view, the MME acts as an SGSN. To the MSC, the mobile device seems to be attached to the 2G/3G network via an SGSN by performing a combined circuit-switched/packet-switched location update.
- For registration in the network, the MME has to inform the MSC of the 2G/3G Location Area Identity (LAI) in which the mobile device is currently 'theoretically' located. Since this is only a theoretical value, it has to be computed out of the Tracking Area Identity (TAI), which is the corresponding identifier in LTE. In practice, this creates a dependency between the TAI and the LAI, that is, the location areas that describe a group of base stations in 2G/3G and LTE must be configured in a geographically similar way for the fallback to work later on.

The Execution Phase: Mobile-Terminated Call

- When a circuit-switched call for a subscriber arrives at the MSC, it signals the incoming call via the SGs interface to the MME, which is, in its eyes, a 2G or 3G SGSN. From here, the notification is forwarded to the mobile device. From the MSC point of view, this is a legacy procedure that already exists.
- If the mobile is in RRC Connected state, the MME can forward the request immediately. If the mobile wants to receive the call, it signals to the MME that it would like to be handed over to the 2G or 3G network in which it can receive the call. The MME then informs the eNode-B that the mobile has to be handed over to the 2G/3G network.
- Since there might still be an ongoing IP data transfer at the time of the handover, the standard contains two options on how to proceed: Either the data transfer is suspended or the packet-switched connection is handed over to the 2G/3G network. This is possible only for UMTS as most 2G networks are not able to handle voice and data connections simultaneously.
- If the mobile is in RRC Idle state when the voice call is signaled, the MME has to page the mobile device to reestablish radio contact. Once contact has been reestablished, it forwards the information about the call. Since there is no ongoing data transfer at this time, no handover of the IP connection is required as the mobile can reestablish the packet-switched connection by itself once it is in the 2G/3G network.
- The eNode-B has the possibility to request 2G/3G measurements from the device to have a better idea as to which cell to hand over the mobile, or it can do so blindly by sending it information about a preconfigured cell.
- Once the mobile device is in the 2G or 3G cell, it answers to the initial paging via the legacy cell. There are two variants of the procedure depending on how the core network is set up. While introducing LTE, many network operators choose to introduce an additional MSC to the network with an SGs interface but without connectivity to the radio network. The advantage of this approach is that existing MSCs do not have to be updated for the launch of LTE. In this case, the location area of the target 2G or 3G cell is different from the one to which the mobile device is registered. As a consequence, the mobile device needs to perform a location area update that triggers a forwarding of the call from the core network from the SGs MSC to the MSC that controls the target cell. The disadvantage of this approach is that the procedure increases the call setup time by around 2.5 seconds. The call establishment time of a call between two mobile devices thus increases from around 5 seconds if both devices are located in a 3G network to 10 seconds if both are located in the LTE network and have to perform a fallback first. As this has a significant impact on user experience, most network operators therefore choose to upgrade all MSCs in their network over time with SGs capabilities. This means that the location area the mobile device is registered with while in LTE is the same as that of the target cell. With this setup, no location update procedure is required and the CS-fallback procedure takes only around half a second longer than a conventional 3G call setup.

The Execution Phase: Mobile-Originated Call

This procedure is similar to the mobile-terminated example above. The difference is that no paging is sent by the network, unlike in the case of an incoming call, and there is no paging response to the MSC after the device is in the legacy cell.

SMS and Call-Independent Supplementary Services (CISS)

- For receiving SMS text messages, the mobile device can remain in the LTE network as the text message is forwarded by the MSC to the MME via the SGs interface and from there via RRC signaling over the LTE radio network to the mobile device. Sending text messages works in a similar way and hence there is also no need to fall back to a legacy network.
- For call-independent supplementary services (CISS) such as changing call forwarding configuration, checking prepaid balance via USSD messaging, and so on, a fallback to the legacy network is required.

While only the support of the SGs interface has to be added to the circuit-switched core network, the solution is relatively simple to implement. However, there are a number of significant drawbacks. These include

- The fallback to a GSM or UMTS network takes several seconds, which needs to be added to an already increased call setup time compared to fixed-line networks. This has a negative impact on the overall user experience compared to fixed-line networks and mobile voice calls established in GSM or UMTS networks.
- If a GSM network is used for the voice call, no packet-switched services can be used during the conversation as most GSM networks do not support the dual transfer mode (DTM) functionality for simultaneous voice and data transmission. In addition, even if DTM is supported, datarates will be very low compared to those in the LTE network.
- If a UMTS network is used for a voice call, it is optionally possible to move an ongoing data session from LTE to UMTS during the fallback procedure. However, this takes additional time.
- After the voice call, the mobile device has to return to the LTE network that again consumes many seconds, during which time no data transfers can take place.

4.13 Voice in Combined LTE and CDMA 2000 Networks (SV-LTE)

Although GSM/UMTS network operators have deployed CS-fallback for voice services as described in the previous sections, some CDMA network operators decided to take a different route. Instead of falling back to a legacy network, most CDMA/LTE-capable devices are designed to cope up with two cellular radios transmitting and receiving simultaneously. They can thus be active in LTE for data transfers and active in the CDMA network for voice calls simultaneously. This however means that the mobile device needs to keep track of two cellular networks even while no voice call is ongoing which results in a higher battery consumption to some extent. For further details, see Klug [30] and [31]. Also, the dual-radio approach requires a close coordination between the LTE and the CDMA part of the device while a voice call is ongoing to keep the power output of the device within acceptable limits.

It should be noted at this point that the main reason why GSM/UMTS network operators have not chosen this approach is likely because of the fact that UMTS networks also offer, besides circuit-switched voice services, excellent and high-speed data services so that there is no need for a dual-radio approach.

4.14   Voice over LTE (VoLTE)

In Chapter 1, the structure of the GSM network was described in combination with voice telephony. It is difficult to separate the GSM voice service from the GSM network as the (voice-) service and the network are completely integrated. Even in UMTS, this is still the case to some degree. For LTE, however, 3GPP decided, with only few exceptions, to completely separate the network from any kind of higher layer service, including voice telephony. This is the reason why the description of LTE in this chapter could be done without mentioning voice telephony.

Voice telephony, however, is still an important service and the CS-fallback solution described in the previous chapter should only be a temporary solution on the path to an All-IP network in which all services including voice telephony are based on the Internet protocol. This may be accomplished over the next few years with the VoLTE system, which is based on the SIP.

In practice, it can be observed today that LTE networks have not yet reached the same level of geographic coverage as that of GSM and UMTS networks and it is likely to remain that way for some time to come. Therefore, a fallback of an ongoing voice call to a classic circuit-switched channel is still required. This functionality referred to as Single-Radio Voice Call Continuity (SR-VCC) is also described below.

4.14.1   The Session Initiation Protocol (SIP)

A telephony service has to fulfill two basic tasks, independently of being implemented as a circuit-switched service or being based on IP. The first task of the service when a user makes a call is to locate the destination party and to signal the incoming call. The second task is to establish a direct or an indirect connection, also referred to as a session. In the case of voice telephony, the session is then used to transport a voice data stream in both directions. In practice, IP-based voice services, such as Skype, have become popular. Also, systems that use the standardized Session Initiation Protocol (SIP) are in wide use, especially as PBX systems in companies. An open-source PBX implementation that uses SIP and has become quite popular is the Asterisk platform (https://www.asterisk.org).

SIP is a generic protocol and can therefore be used for the establishment of a connection between two or more parties for many different session types. In this chapter, SIP is mainly described as a protocol for establishing a voice session. Details can be found in IETF RFC 3261 [32] specification as well as in various 3GPP documents that are freely available on the Internet.

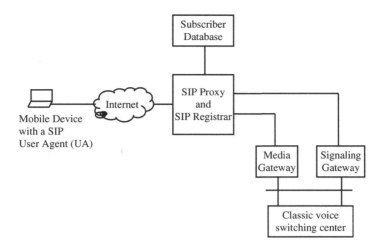

Figure 4.26   The Basic SIP Infrastructure

The core of an SIP-based telephony system is the SIP Registrar and the SIP Proxy as shown in Figure 4.26. When powered on, a device has to register with the SIP system to be reachable by others and also to establish outgoing calls. The SIP software on a user's device is referred to as an SIP User Agent (UA). On the network side, the SIP Registrar is responsible for the authentication and registration of devices, that is, UAs. Figure 4.27 shows how registration is performed in practice. At the beginning, the device sends a request to the Domain Name System (DNS) server to retrieve the IP address of the SIP Registrar server, whose domain name together with the user's identity and authentication information has been configured in the user's device. Afterward, the UA sends an SIP 'Register' message to the Registrar. The Registrar then searches its user data base for the SIP-ID of the UA and the corresponding authentication information and then requests the UA to authenticate by responding with an SIP '401 Unauthorized' message. As described before for other systems, authentication is based on a common key/password and an algorithm that uses a random number on both sides for generating challenge/response values. As only the random number and the result of the calculation are exchanged between the UA and the Registrar, proper authentication is possible over an insecure and non-encrypted connection. When the UA receives the random number, it uses its private key to calculate a result and returns it in a second 'Register' message to the Registrar. If the result of the UA is the same as the result calculated by the Registrar, it answers with an 'SIP 200 OK' message and the subscriber is registered with the system. The Registrar also saves the IP address and the UDP port used on the UAs side in the subscriber database so that it can forward incoming session requests to the subscriber.

It should be noted at this point that the result codes in the answer messages above (401, 200, etc.) are based on the result codes of the Hypertext Transfer Protocol (HTTP) that is used to request web pages from a web server.

Figure 4.27 shows an SIP 'Register' message recorded with Wireshark [33] after a '401 Unauthorized' response has been received. In the central part of the figure, the SIP-ID of the subscriber can be seen (5415468) together with the SIP domain (@sipgate.de). Together they form the Universal Resource Identifier (URI). Furthermore, the random number (Nonce) that

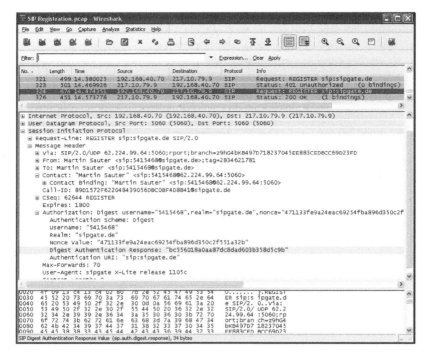

Figure 4.27 A SIP Register message. Source: www.wireshark.org. Reproduced by permission of WireShark© 2013

is sent by the network is also part of the message as well as the authentication response value (Digest Authentication Response).

Figure 4.27 also shows the '200 OK' of the Registrar server that reports the presence of '1 binding'. This means that this is the only UA that has registered against the SIP-ID at this point in time. SIP also allows to register multiple UAs to a single Public User ID. For incoming calls, both devices are then notified. Once the UA is registered, voice sessions can be established to other subscribers, and incoming sessions can be received. Figure 4.28 shows how a voice session is established: as the user knows the SIP-ID of the subscriber, he wants to contact but not the current IP address of his device; signaling messages always traverse the SIP Proxy Server in the network. The SIP Proxy is first contacted with an SIP 'Invite' message whose ultimate destination is the terminating device. Before the session is established, the SIP proxy authenticates the originator by returning a '408 Authentication Request' message. Only once authentication has been performed will the SIP Invite request be forwarded to the destination. If the destination is a customer of another network operator, the 'Invite' message cannot be delivered to the destination directly but is routed to the SIP proxy of the other network operator (SIP Proxy B). The other proxy then searches its database for the IP address of the terminating device that has registered for the SIP-ID received in the 'Invite' message and then forwards the message to the destination device. To ensure that SIP messages in the other direction also traverse both SIP proxies, both proxies add their IP address to each SIP message before forwarding it.

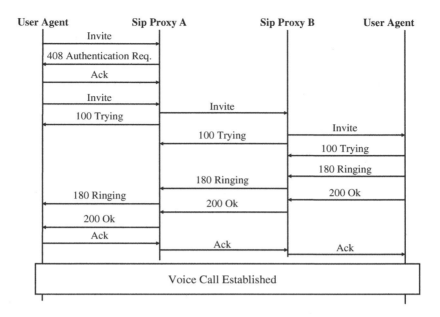

Figure 4.28   SIP call establishment

After the device has received the 'Invite' message, it responds with a '100 Trying' message and prepares itself to receive the call. Once it is ready, it sends a '180 Ringing' message and thus signals the caller that the user is alerted about the call. If the user accepts the call, a '200 OK' message is sent over the two proxy servers to the calling party, and both devices activate the speech path and enable the loudspeaker and microphone. Which codec is used depends on the codecs supported by each device. Codec negotiation is initiated by the calling party by including a codec list in the 'Invite' message. The other side then selects a compatible codec from the list and informs the calling device during the call establishment which codec it has selected.

While signaling messages always traverse the SIP proxies, the speech channel can be established directly between the two devices. In practice, however, it is often the case that both devices are behind a Network Address Translation (NAT) router and thus use local IP addresses and UDP ports that are translated in the NAT routers into a public IP address and a different UDP port. A direct exchange of speech packets is thus possible only if a device detects the address translation and is able to inform the other device of the public IP address and UDP port to which the speech packets have to be sent. As the public addresses are not visible to the UA, it sends a number of probe messages to an STUN (Session Traversal Utilities for NAT) server on the Internet. The STUN server receives the packet with the public IP address and UDP port that were generated by the NAT router and returns those values to the UA. The UA then tries to find a rule how the private UDP port is mapped to its public counterpart and then uses that knowledge to determine the likely UDP port number used during a call establishment. Unfortunately, there are quite a number of different NAT implementations found in practice and it is thus not always possible to find a rule. This is why SIP providers often use media gateways. Instead of direct interaction between the two devices, each device sends its media stream to the media gateway and receives the corresponding stream from the other direction also from there. As communication to the media gateway is initiated from the device, there is no need to detect the mapping between internal and external UDP port number.

Independent of whether there is a direct connection between devices or whether the connection uses a media gateway, the Real-Time Protocol (RTP) is used for transporting the voice data. It is specified in RFC 3550 [34]. In fixed-line SIP implementations, the G.711 codec is mostly used, which is also used in circuit-switched networks. If supported by both devices, the G.722 wideband codec is preferred as it offers a much better voice quality. Both codecs transmit at a rate of 64 kbit/s and packets are split into chunks of 20 milliseconds that are then transmitted over UDP. With the protocol overhead, this results into a data rate of around 100 kbit/s in each direction. It should be noted at this point that the G.722 wideband codec is not compatible to the G.722.2 wideband codec used in cellular networks as the data rate of this codec is only 12.2 kbit/s. Wideband calls between fixed and wireless networks are thus only be made by transcoding from one wideband codec to the other on a gateway in the network. Figure 4.29 shows how the list of speech codecs is sent to the other subscriber in an SIP 'Invite' message. The list is part of the Session Description Protocol (SDP) section specified in RFC 4566 [35].

In addition to looking up the IP address of a destination subscriber and to forward messages to them or other proxies, proxies are also allowed to modify messages. If, for example, a user is already busy in another call and thus rejects another 'Invite' request, the SIP proxy can replace the incoming 'Busy' message and generate another 'Invite' message that is then sent to a voice mail system in the network. SIP proxies can thus offer services similar to those of the MSC architecture for circuit-switched networks that were described in Chapter 1. To communicate

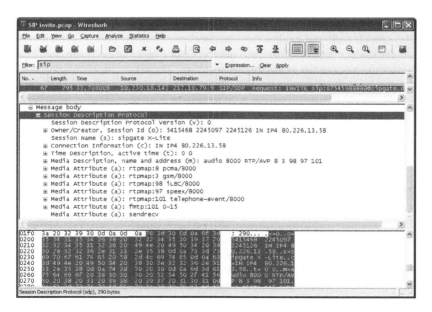

Figure 4.29   List of codecs in the SDP section of a SIP 'Invite' message. Source: www.wireshark.org. Reproduced by permission of WireShark© 2013

with subscribers on circuit-switched networks, SIP proxies can interact with signaling and media gateways. The signaling gateway translates the SIP messages into classic SS-7 messages and vice versa (cp. Chapter 1), whereas the media gateway translates an IP-based media stream into an E-1-based stream and vice versa.

4.14.2 The IP Multimedia Subsystem (IMS) and VoLTE

For mobile networks, the SIP system has been significantly extended by 3GPP and is referred to as the IMS. Figure 4.30 shows the central components of the IMS. The core of the system is the Serving Call Session Control Function (S-CSCF) that implements the role of SIP Registrar and SIP Proxy. To communicate with the central database, the HSS, the S-CSCF uses the Cx-Interface and the Diameter Protocol as described in RFC 6733 [36].

As there can be several S-CSCF in a large IMS system, a distribution function is required for incoming requests. This is performed by the Interrogating-CSCF (I-CSCF), which also has a connection to the HSS via the Cx interface to be able to retrieve subscriber information that is relevant for the routing of the signaling messages to the responsible S-CSCF. At the border of the IMS system to the mobile network, the Proxy-CSCF (P-CSCF) plays another important role. Its role is to act as an SIP Proxy but it also has a role to represent the user toward the IMS system. This is necessary as the connection between the network and a mobile device can be interrupted during a session because of loss of network coverage. In such a case, the mobile device is not able to send an SIP 'Bye' message to close the session properly. This is done by the P-CSCF once it is informed by the LTE network that the connection to the subscriber has been lost. Between each other, all CSCF elements communicate over the Mw interface and the SIP protocol. A mobile device communicates with the P-CSCF via the LTE network and the PDN-GW over the already existing SGi-Interface, which is also used for non-IMS applications for direct Internet access.

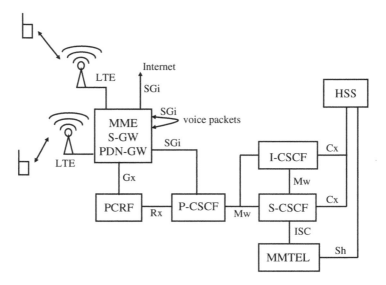

Figure 4.30   The basic IMS components

In addition, the P-CSCF is also responsible for managing the QoS settings for the voice session. For this purpose, a dedicated radio bearer will be created during session establishment that is then used exclusively for voice data packets. This bearer then gets preference in the network and on the radio interface to ensure timely packet delivery and to guarantee the required data rate for speech packets for as long as radio network conditions permit. For this purpose, the P-CSCF is connected to the Policy and Charging Rules Function (PCRF) via the Rx-interface, which translates the requirements of the IMS system for the speech bearer into commands for the LTE network over the Gx-interface as shown in Figure 4.30.

In addition, the IMS architecture defines Application Servers (AS) that extend the SIP architecture. ASs can be put in place to control the establishment and maintenance of a session by modifying SIP messages of a subscriber, which are forwarded from the S-CSCF. For further flexibility, a user's profile can contain a configuration to forward SIP messages to several ASs. For VoLTE, an AS is used that implements the functionality of the MMTEL specification in 3GPP TS 22.173 [37] for typical supplementary services such as call forwarding, conference calls, call-hold, suppression of the telephone number of the calling party and so on.

Like a simple SIP device described above, an IMS VoLTE device needs to register with the IMS system when it is switched on. In addition to the previously described SIP procedure, further actions are necessary in the IMS system. To be flexible, no IP configuration of the P-CSCF system is necessary in the device. Instead, the device receives the P-CSCF IP address from the network during the LTE, UMTS or GPRS attach procedure. In addition, the network informs the device during the attach process and also later on during routing and tracking area update procedures if the current radio network supports IMS services. Thus, it is possible, for example, to offer only IMS voice services while the mobile device is in the LTE network and to instruct a device to use circuit-switched services (cp. Chapter 1) while in the UMTS or GSM network of the same network operator.

Like in other parts of the network, the IMSI is used to identify a user. As the IMSI is stored on the SIM card, it is not necessary to manually configure it in the device.

To secure the transmission of signaling messages between a device and the P-CSCF, a security context is established during registration. This includes the use of an integrity checksum in all messages to ensure that they have not been modified accidentally or on purpose. Optionally, the signaling messages between the mobile device and the P-CSCF can also be encrypted.

Once registered, voice calls can then be initiated or received. Figures 4.31 and 4.32 demonstrate which messages are sent to establish a session between two IMS devices. The role of the MMTEL server is not shown in the example as in this basic scenario no messages are modified by the application server (e.g., there is no message modification to forward a call to the voice mail system).

In the first step, the originating device sends an SIP 'Invite' message to the SIP proxy, in this case to the P-CSCF. The P-CSCF in turn confirms the reception with an SIP '100 Trying' message and forwards the 'Invite' message to the S-CSCF. Here, the message is analyzed and forwarded to an I-CSCF whose task is to find the S-CSCF of the destination subscriber to which the message is then forwarded. The S-CSCF of the destination subscriber can either be located in the same network or in a different network in case the destination is a subscriber of a different network operator. In the latter case, Border Gateway Controllers (BGCs) are part of the transmission chain to properly separate the two IMS networks from each other. Once the message arrives at the S-CSCF of the destination subscriber, the P-CSCF responsible for the user's device is determined and the message is then forwarded. The P-CSCF then tries to

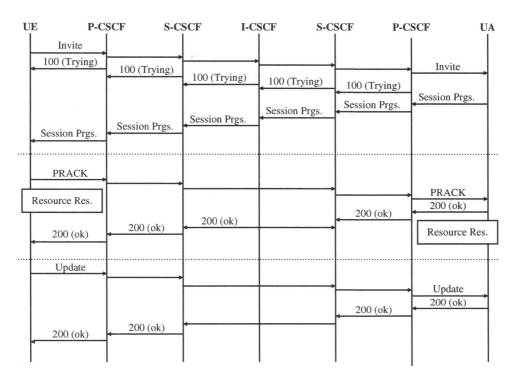

Figure 4.31   VoLTE Call Establishmetn Part 1. Source: Martin Sauter. Reproduced by permission of Martin Sauter

forward the 'Invite' message to the subscriber. If successful, the device returns an SIP 'Session Progress' message to the P-CSCF and from there via all other IMS servers back to the originating device.

Among other information, the SIP 'Invite' and 'Session Progress' messages contain the afore-mentioned SDP section in which all voice codecs are listed that are supported by the device. This is required to select a suitable speech codec and to establish a dedicated bearer with the suitable QoS settings on the air interface. As described before, it is the task of the P-CSCF to establish the dedicated bearer as it is the component in the IMS network that communicates with the underlying LTE network via the PCRF. Such interaction is possible as the P-CSCF not only forwards the SIP message but also parses it for the information it requires for bearer handling.

Once resources have been assigned for the speech channel, the destination device alerts the user and returns an SIP '180 Ringing' message to the originator, again via all IMS components involved in forwarding signaling messages. IP packets containing the speech data, however, are exchanged directly between the two devices. If both subscribers are in the same network, the speech packets are thus sent to the PDN-Gateway and are looped back into the network straight away instead of traversing the SGi interface to another network (cp. Figure 4.30). It should be noted at this point that in practice, speech packets might traverse a gateway in the network to implement a quick transfer of the voice call from LTE to a circuit-switched UMTS

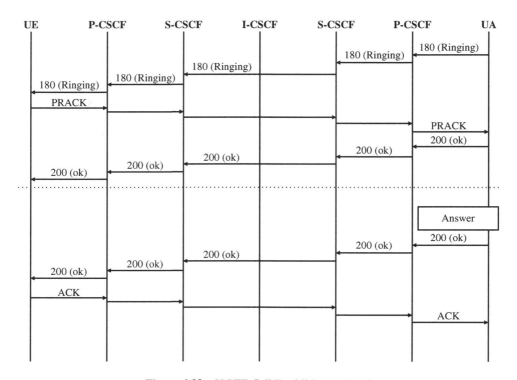

Figure 4.32 VoLTE Call Establishment Part 2

or GSM channel in case the user roams out of the LTE coverage area. This is part of the 3GPP Release 10 SR-VCC functionality described further below.

As the 3GPP IMS specifications contain too many implementation options, network operators decided to agree on a common set of options to use for their VoLTE system. This ensures that IMS speech services are compatible between networks and that software for devices can be developed that is usable in all IMS networks and between all mobile devices supporting the agreement. This lead to the GSMA IR.92 specification [38], which is also referred to as the VoLTE IMS profile. Like 3GPP specifications, this document is freely available on the Internet. On its own, it is not suitable as an introduction to IMS and VoLTE as it mainly contains references to details in relevant 3GPP specification documents. It is very well suited, however, as a basis to quickly discover which parts of which 3GPP documents are relevant for network operator-based VoLTE systems.

4.14.3 Single Radio Voice Call Continuity

As GSM and UMTS networks are likely to have a better geographical coverage than LTE for the foreseeable future in many countries, a mechanism is required to transfer a VoLTE call to UMTS or GSM. Handing over a call from UMTS to GSM is relatively simple because a circuit-switched connection in UMTS is transferred to a circuit-switched connection in GSM.

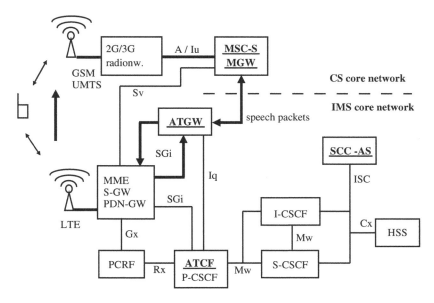

Figure 4.33   IMS and MSC components required for SR-VCC

However, as VoLTE is based on IMS and IP, it is required to hand over an ongoing IP-based voice call to a circuit-switched UMTS or GSM channel. As a device cannot be active in LTE and UMTS/GSM at the same time (Single Radio), a solution referred to as Single Radio Voice Call Continuity (SR-VCC) has been designed and improved in the standards over the years. SR-VCC is specified in 3GPP TS 23.237 [39]. The following description is based on the SR-VCC solution described in 3GPP Release 10. In practice, network operators can also implement earlier versions, which, however, work in a largely similar way.

Figure 4.33 shows which IMS components are involved in a VoLTE handover to UMTS or GSM. As before, the P-CSCF, I-CSCF and S-CSCF servers are responsible for establishing and maintaining the voice session. In addition to the MMTEL AS, a Service Centralization and Continuity Application Server (SCC-AS) is involved in the call establishment phase to collect all necessary information about a session to be prepared to hand over a voice call to the circuit-switched network as quickly as possible should it become necessary. The circuit-switched network is represented in Figure 4.33 by the MSC-Server (MSC-S) and the Media Gateway (MGW), which have been introduced in Chapter 1.

To be able to transfer a voice call as quickly as possible, voice data packets are no longer exchanged directly between the two mobile devices but are instead led over an Access Transfer Gateway (ATGW). The gateway is controlled by the Access Transfer Control Function (ATCF), which is part of the P-CSCF.

Figure 4.34 shows how a voice call is redirected in the network during an SR-VCC handover. In general, the connection is established as described before. The difference for SR-VCC is, however, that the speech data streams of both sides terminate at an ATGW for SR-VCC Release 10. The eNodeB base station is aware if the mobile device supports SR-VCC as this is signaled during the LTE attach procedure. It can thus initiate the handover procedure to a 3G or 2G

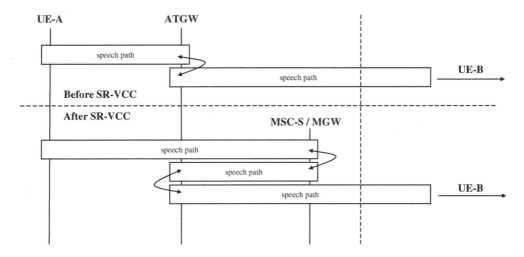

Figure 4.34   A speech connection before and after an SR-VCC handover

network when it becomes necessary by requesting a handover of an ongoing voice call from
the MME. The MME in turn sends a 'PS to CS Transfer Request' to the MSC-Server which
then reserves the required circuit-switched channels in the UMTS or GSM target cell and on
all required transport links. In addition, the MME instructs the ATCF to prepare the ATGW
for switching the voice data stream away from one of the subscribers toward the IP address
of the MGW of the MSC-Server. This is done with an SIP 'Invite' message. Once the ATCF
has responded with an SIP '200 OK' message and everything is prepared for a handover, the
MSC-Server then responds with an 'CS to PS Response' message to the MME, which then
triggers the handover by sending a 'Handover Command' message to the mobile device.

The mobile device then switches to the UMTS or GSM cell and continues the voice call over
the prepared circuit-switched channel. The circuit-switched data stream is converted back to
an IP data stream at the MSC's media gateway and is sent from there to the ATGW. Once the
ATGW receives the voice data stream from the circuit-switched network, it informs the ATCF
about the successful handover, which in turn informs the SCC-AS about the success of the
transfer. In the final step, the SCC-AS then sends an SIP 'Bye' message to the P-CSCF so that
it can remove the dedicated bearer for the speech data flow in the LTE network.

4.14.4   Internet-Based Alternatives

To a certain degree, Internet-based VoIP services, such as Skype, are an alternative to
operator-based voice services on PCs and mobile devices. With LTE, this trend is likely to
grow. The main disadvantage over a network operator-based voice solution is that ongoing
voice calls cannot be handed over to a circuit-switched bearer when the user leaves the
coverage area where fast IP connectivity via LTE, HSPA or EvDO is available. In addition,
Internet-based voice service cannot directly interact with the transport network and it is hence
difficult to prefer IP packets that contain voice, especially on the air interface in busy cells.

4.14.5   LTE Bearer Configurations for VoIP

Network operator-based voice services can directly interact with the transport network to ensure a low latency and constant bandwidth for a connection. This also helps the eNode-B to optimize transmission of VoIP packets over the air interface. With VoLTE, the following optimizations are activated during the call establishment phase.

In LTE, a dedicated bearer that is established alongside a default bearer is used to ensure QoS for a stream of similar IP packets. In HSPA, the concept is referred to as a secondary PDP context. Dedicated bearers or secondary PDP contexts are established when a service in the network requests a prioritization of IP packets belonging to a specific media stream between two IP addresses and TCP/UDP ports. In practice, the dedicated bearer then ensures the behaviors described in the following sections.

Unacknowledged Radio Bearers for Voice

For a voice stream, an Unacknowledged Mode Data Radio Bearer (UM DRB) is used. This refers to the configuration of the layer 3 RLC protocol on the air interface. On this layer, lost data is detected and repeated if it is configured in AM – the default RLC operating mode for user data in HSPA and LTE. For a voice data stream, however, lost voice packets should not be repeated as they would come too late to be useful. This is why a UM DRB is used. In addition to the unacknowledged mode bearer, other signaling bearers and the default bearer in AM are also active during a voice conversation.

The unacknowledged bearer for the voice stream is established by the voice service sending a request to the transport network during the establishment of the call to create a dedicated bearer for IP packets being exchanged between two particular IP addresses and two particular UDP ports. This stream is then mapped to a radio bearer for which no RLC error correction is used. All other IP packets, not matching the IP address and UDP port combination requested above are sent over an AM DRB without guarantees for latency and bandwidth.

It should be noted at this point that an extra dedicated bearer for the voice call does not require an additional IP address. Instead, only a single IP address is used as it is the combination of the IP addresses and UDP ports that distinguishes the packets that go through the UM bearer from those that use an AM bearer. For the application on top, that is, the VoLTE client application, all of this is transparent as the protocol stack below automatically decides which IP packet should be sent over which bearer.

Packet Loss and Guaranteed Bit Rate

To ensure that the packet loss in UM mode stays within reasonable limits, the radio transmission characteristics (power output, modulation, coding, …) for the UM bearer are configured to ensure that the packet loss rate does not exceed 1% – a value that the voice codec can still tolerate.

In addition, the UM DRB for voice is configured with a guaranteed bit rate and network resources are permanently allocated to the user during the call. One of the options to achieve this is to use semipersistent air interface scheduling as described in Section 4.5.1.2 that enables the mobile device to periodically send and receive data without requiring bandwidth assignments. This guarantees the bandwidth for the call and also saves the overhead of dynamic bandwidth assignments.

Header Compression and DRX

The main inefficiency of VoIP data streams is the overhead from the IP headers of each packet. To compensate, RoHC can be used between the base station and the mobile device.

Finally, reducing power consumption during a voice call during the times when no voice data is sent or received is also important. This can be achieved by using DRX. During the DRX period, the UE's transceiver is put into a sleep state. This is especially important for voice sessions as the bandwidth required is so small that the time between two IP packets containing voice data is very long. Keeping the receiver constantly switched on would waste a lot of energy in the mobile device.

To the voice service on top of the protocol stack, the voice optimization of the bearer is transparent. All that is required is that the voice service in the network requests appropriate QoS parameters to be used for a data stream via a network interface. In theory, this interface could also be used by the Internet-based voice services if offered to external services by network operators.

4.15 Backhaul Considerations

Owing to the high peak and average speeds of LTE, high-speed backhaul links are essential to ensure that the capabilities of the LTE air interface can be fully utilized. A three-sector eNode-B with a channel bandwidth of 20 MHz in each sector can easily achieve peak datarates that are three times 100 Mbit/s, that is, 300 Mbit/s in total. As eNode-Bs are usually colocated with UMTS and GSM base stations, the required combined backhaul bandwidth could hence be even higher. Today, three backhaul technologies are suitable for such high datarates.

Traditionally, copper-based twisted pair cables have been used to connect base station sites to the network. UMTS networks initially used 2 Mbit/s E-1 links and for some time, the aggregation of several links was sufficient for providing the necessary backhaul bandwidth. For LTE, this is not an option since peak datarates far surpass the capabilities of this backhaul technology. An alternative is very high speed DSL lines (VDSL) that can deliver datarates of the order of 100 Mbit/s. This might not be enough to cover the peak datarates required for a cell site but is a much better alternative compared to E-1 link bundling.

For higher datarates, copper-based cables have to be replaced with optical fibers. While the datarates that can be achieved over fibers match the requirements of a multiradio base station, it is costly to deploy, as in many cases new fiber cable deployments are required for buildings and often also alongside roads. Network operators that own both fixed-line and wireless networks can deploy and use a common fiber backhaul infrastructure to offer fixed-line VDSL and fiber connectivity to private and business customers and use the same network for wireless backhaul. This significantly improves the cost-effectiveness of the overall network deployment.

Wireless network operators that do not have fixed-line assets have two possibilities to connect their base stations to a fast backhaul link. The first option is to rent backhaul capacity from a fixed-line network operator. The second option is to use high-speed Ethernet-based microwave solutions that offer backhaul capabilities of several hundred megabits per second. The latest generation of microwave equipment is capable of speeds beyond one gigabit per second.

Once high-speed backhaul connectivity is available at a base station site, it can potentially be used by all equipment at that site. In many cases, this will be GSM, UMTS and LTE. As LTE is purely based on IP technology, the backhaul link should preferably offer native IP connectivity. UMTS base stations often require ATM connectivity, which can be simulated over IP. The current generation of UMTS digital modules in base stations is also capable of natively connecting Node-Bs over an IP connection to the RNC. GSM technology continues to be based on an E-1 timeslot-based architecture on the Abis backhaul link. Here, virtualization of E-1 connections can help to transparently tunnel the backhaul link over the installed IP connection. This way, all three base stations can be backhauled over a single link. In the future, GSM, UMTS and LTE multimode base stations might only contain a single digital backhaul module and thus the different traffic types can be transparently routed over a single IP connection.

4.16   LTE-Advanced (3GPP Release 10–12)

Most of the LTE functionality discussed in this chapter is specified in 3GPP Release 8. Release 9 only contains minor improvements for LTE, one of the reasons perhaps being that the industry was still working on the implementation and deployment of the new system when work started on Release 9 and therefore first had to evaluate the performance of the new system. Major new features to comply with the IMT-Advanced requirements of the ITU are included in 3GPP Release 10, 11 and 12 and are referred to as LTE-Advanced. The major goal of these enhancements is to further reduce cost, to further increase the maximum datarates per user and to improve throughput in cell edge and overlapping cell scenarios. It should be noted at this point that at the time of publication of this book, and for a number of years afterward, it is unlikely that LTE-Advanced functionality except for the carrier aggregation feature will be used in the field. The following sections list some of the most important new features of Release 10 and beyond. Details can be found in the 3GPP work plan [40], the original feasibility study for LTE-Advanced [41], Roessler and Kottkamp [42] and in Rumney [43].

4.16.1   Carrier Aggregation

A relatively simple way to further increase individual data transmission speeds is to increase the channel bandwidth. To remain backward compatible with 3GPP Release 8, the maximum carrier bandwidth of 20 MHz is not altered. Instead, carrier aggregation is used to combine the capacity of several individual carriers. The aggregated carriers can be adjacent or nonadjacent, they can be in a single band and also in different bands. An individual carrier is referred to in the standards as a component carrier (CC). One configuration, for example, is to combine carriers in LTE bands 7 (2600-MHz band) and 3 (1800-MHz band) to potentially achieve a total carrier bandwidth of 40 MHz in the downlink direction. Carriers can be aggregated asymmetrically in the downlink and the uplink directions. In the downlink direction, for example, carriers in two different bands can be aggregated to a combined 40-MHz channel, while in the uplink direction only a 20-MHz carrier in a single band is used.

For the future, further carrier aggregation configurations are envisaged that would result in even broader transmission channels. However, the use of even broader transmission channels is currently not possible in practice, as most network operators do not have enough spectrum to aggregate significant additional spectrum beyond 40 MHz.

In the United States and in some countries in Asia, Carrier Aggregation is used to combine two 10-MHz carriers in different frequency bands to reach the same aggregate speed as a single 20-MHz LTE Release 8 carrier used today in many European countries in the 1800-MHz and 2600-MHz bands, respectively. In the United States, for example, some network operators combine their 10-MHz spectrum in the 700-MHz band with an additional 10-MHz carrier in the 1700/2100-MHz band for a total aggregate bandwidth of 20 MHz. In other words, the United States requires LTE-Advanced Carrier Aggregation to reach the same downlink speed as a single 20-MHz carrier used in Europe. The first application of LTE-Advanced is thus not to increase data rates beyond the possibilities of LTE Release 8 if enough contiguous spectrum is available for a network operator but to compensate for a fractured frequency landscape. Therefore, inferring that some countries have more advanced LTE deployments based only on the use of LTE-Advanced Carrier Aggregation would be false.

4.16.2   *8 × 8 Downlink and 4 × 4 Uplink MIMO*

To further increase the datarates close to the center of the cell, LTE-Advanced introduces an 8×8 Single-User MIMO transmission mode. Compared to the 2×2 MIMO mode used by LTE in practice today and the resulting maximum transmission speed of 150 Mbit/s when a 20-MHz carrier is used, speeds of up to 600 Mbit/s could be reached. Together with the aggregation of two 20-MHz carriers, theoretical top speeds exceed 1 Gbit/s. In practice, however, it will be challenging to incorporate eight receive antennas in mobile devices. Similar challenges will be faced on the base station side as the number of antennas and the antenna sizes are further increased. This is challenging because of available space on top of the antenna masts and the additional stress on the mast due to additional wind forces.

In the uplink direction, current mobile devices only transmit a single data stream. The base stations, however, can use multiuser MIMO methods, as discussed earlier, to increase the overall bandwidth in the uplink direction of a cell by instructing several mobile devices to transmit simultaneously and then using MIMO techniques to separate the data streams. LTE-Advanced aims to increase the available datarates for a single user by introducing single-user MIMO methods with antenna configurations of up to 4×4. In an ideal situation, this results in a peak throughput of 300 Mbit/s in a 20-MHz carrier and 600 Mbit/s in a 40-MHz aggregated carrier. Again, practical considerations concerning the placement of four antennas in a small mobile device will limit the application of 4×4 MIMO in the uplink direction to larger mobile devices such as tablet computers, netbooks and notebooks.

To support Carrier Aggregation and the new MIMO schemes, three new device categories have been defined:

Category 6 devices can aggregate $20 + 20$ MHz carriers with a 2×2 MIMO or can support a 20 MHz configuration with 4 MIMO paths. In the Uplink direction, a 20-MHz channel with a single MIMO layer is supported or alternatively a 10-MHz channel with two MIMO paths.
Category 7 devices have the same capabilities in the downlink direction as that of category 6 devices. In the uplink, however, category 7 devices support a 20-MHz channel with 2 MIMO paths.
Category 8 devices support a $20 + 20$-MHz downlink carrier aggregation with an 8×8 MIMO in combination with a 4×4 MIMO in the uplink direction.

4.16.3   Relays

Small and inexpensive femtocells connected to a cheap backhaul link such as DSL are one way to increase throughput and to extend the coverage area of the network. Another complementary approach is relaying. Relay nodes, as standardized in 3GPP Release 10, act as standard LTE cells with their own physical cell-ID, broadcast channels, and so on. Unlike macrocells, however, which use a copper, fiber or microwave backhaul, relays use the LTE air interface to an LTE macrocell to transport the data via that cell to the core network. The relaying can take place on a carrier also used by a macro cell to serve mobile devices. Alternatively, a separate carrier channel that is exclusively reserved for the relay node can be used. With both options, areas can be covered without additional microwave equipment and without the need for a fixed-line backhaul connection.

4.16.4   HetNets, ICIC and eICIC

With LTE Release 8 and the enhancements discussed so far in this section, air interface parameters for peak data rates have been defined that far exceed what is technically possible in mobile devices today. Also, the theoretical peak data rates offered by these enhancements cannot be reached in most areas covered by a cell. In a significant part of the coverage area, the use of 64-QAM modulation is not possible due to an insufficient signal strength and interference from neighboring cells. The number of MIMO antennas is limited by the size of devices. Also, the numbers of sectors per cell size is limited to 3 or 4, in practice, due to the increase in the overlap areas between the sectors. Also, most network operators do not have more than 50–60 MHz of spectrum that they can use for LTE. As a consequence, other means have to be found to further improve the overall network and single-user throughput in the future to keep ahead of rising bandwidth demands.

The first approach, already put into practice today is to densify the macro network, that is, to install additional macro base stations to reduce the coverage area of a cell site. There are limitations to this approach as the number of suitable locations where macro base stations with large antennas can be installed is decreasing. Further densification therefore requires the use of much smaller base stations, referred to as 'small cells', which only cover an area with a diameter of a few dozen meters at most. Such small cells have a size similar to that of a typical Wi-Fi access point at home. In a waterproof and ruggedized casing, such cells can be installed outdoors or indoors in heavily frequented locations such as train stations, shopping malls and so on.

Although the price of such cells is significantly lower than that of a macro base station, there are a number of other factors that influence the use of small cells. First, electrical power needs to be available where the cell is located. Second, high-speed network connectivity must be available at the site, either over a standard twisted pair copper cabling or over an optical cable depending on the type of the small cell as will be described further below.

If small cells are used in a network, the homogeneous cell layout of macro cells is complemented by the fully overlapping network coverage provided by small cells. Therefore, such a network architecture is also referred to as a heterogeneous network or as an HetNet.

Another difficulty that has to be overcome in heterogeneous networks is the interference caused by the overlapping coverage of macro cells and small cells that would significantly reduce the capacity increase that small cells could provide. Therefore, the LTE specification offers a number of methods to reduce or to avoid inter-cell interference. These methods are referred to as an ICIC.

The first ICIC scheme was already defined in the first LTE specification, 3GPP Release 8, and is described in Section 4.11.2. The idea behind what is referred to as FFR is to use only some of the subcarriers at the cell edge by transmitting with a higher power on them than on other subcarriers. Neighboring cells would do the same but for subcarriers on different frequencies thus creating less interference. Naturally, a balance has to be found between the reduced interference on those subcarriers in cell edge scenarios and the reduced number of subcarriers in such areas that reduce the overall transmission speed for devices in such areas. In the center of such cells, all subcarriers can be used for data transmission and hence the scheme does not reduce peak data rates close to the cell. In effect, data rates close to the center of the cell may even benefit from FFR as well because of the reduced neighbor cell interference.

While the basic ICIC FFR scheme may be beneficial in a pure macro network environment, there are no benefits in a heterogeneous network environment where several small cells are located in the coverage area of a single macro cell. In such a scenario, the coverage area of the small cells fully overlaps with the coverage area of the macro cell and thus there is no benefit from reducing the power of some of the subcarriers. This is why in 3GPP Release 10, an additional ICIC scheme, referred to as eICIC was defined. With this scheme, the macro cell coordinates with the small cells as to which of its subframes it leaves empty in the time domain. The small cells will then use the empty macro cell subframes for their own data transmission and thus avoid interference from the macrocell in those subframes and reciprocally do not cause interference for the devices that are being served by the macro cell. Again, a balance has to be found between the gain of reduced interference and the loss of transmission capacity in the macro cell and the small cells operating in its coverage area because of the split of subframe resources. This can be done, for example, by adapting the number of blank subframes to the amount of data traffic being handled by the small cells which requires interaction between the macro cell and the small cells in its coverage area.

4.16.5 Coordinated Multipoint Operation

Release 8 ICIC and Release 10 eICIC are methods to enable neighboring cells to avoid interference by coordinating when and where each base station transmits in the downlink direction. In 3GPP Release 11, additional methods were defined to further improve the cooperation between different radio network elements. These methods are referred to as Coordinated Multipoint Operation (CoMP).

For a better understanding of these methods, it is perhaps worthwhile to take a quick look back at how UMTS reduces interference at the cell edge. Right from the start of the first UMTS networks, the soft handover mechanism has been used to transmit a voice data stream sent from several base stations simultaneously to a single user to improve reception conditions. In the uplink direction, several base stations can pick up the signal. This is possible in UMTS because there is a central controlling element in UMTS, the RNC that forwards a downlink data packet to several NodeBs for transmission and receives the same uplink data packet from several NodeBs in the uplink direction. Based on error correction information, it can then select which of the copies to forward to the core network. It is important to note, however, that the selection is based on packet level rather than on individual bits or layer 1 RF signals.

With the introduction of HSDPA for data transmission, the concept had to be partly abandoned because the RNC no longer controls when and how the data in the downlink direction is sent as this responsibility was moved to the NodeBs to improve the scheduling performance and to reduce the load on the RNC. However, in the uplink direction, the soft handover approach continues to be used for HSUPA (E-DCH). For circuit-switched voice calls, however, soft handover mechanisms in both downlink and uplink directions are still used today.

In LTE, soft handover mechanisms have never been defined as all scheduling decisions are based in the eNodeBs and because there is no central radio control instance as in UMTS. As a consequence, only hard handovers are possible up to 3GPP Release 10. With CoMP in 3GPP Release 11, a number of mechanisms with a similar effect as the UMTS soft handover are introduced in LTE. They are not called soft handover, however, because CoMP multipoint transmission methods are based on the distribution of ready-to-transmit RF signals rather than data packets that are modulated into an RF signal at several base stations.

3GPP TR 36.819 [44] describes four network deployment scenarios for different CoMP mechanisms. All mechanisms have in common that a central element is required for controlling RF transmissions in the downlink and in the uplink directions. This is a new concept in LTE where so far the eNodeBs are largely autonomous:

Scenario 1: Homogeneous Network – Intrasite CoMP: In this scenario, interference caused by the overlapping coverage of two sectors of a single base station is addressed. No optical cables are required in this scenario but the interference reduction effect is limited to the over-lapping areas of the sectors of a single cell and is not effective for the overlap areas between different cell sites.

Scenario 2: Homogeneous Network – RRHs: The first approach to reduce cell edge interfer-ence is to transmit a single RF signal to several RRHs. From a mobile device's point of view, only a single cell is visible. Owing to the high loss of modulated RF signals even over coaxial copper cables, the RF signals have to be sent over optical links which is a major challenge in practice. On the RRH side, the optical signals are converted to and from an electromagnetic signal.

Scenarios 3 and 4: Heterogenous Network Scenarios: A high-power signal from a macro cell is complemented by low power signals of small cells. These are either visible as individual cells for the mobile device, that is, small cells have their own physical cell id or would act as low-power RRHs in a transparent way from a mobile device's point of view.

In the downlink direction, the following methods have been defined in the CoMP work item: Firstly, with Joint Transmission (JT), the centralized scheduler can use several distributed RRHs to transmit a single signal to the mobile device, thus increasing the signal quality at the location of the mobile device. Secondly, a somewhat simpler scheme is Dynamic Point Selection (DPS), that is, the scheduler can quickly assign different antennas in its distribu-tion set for a mobile device depending on its changing location. And thirdly, a Coordinated Scheduling/Coordinated Beamforming (CS/CB) method has been defined to shape a trans-mission via several antennas in a way to concentrate the downlink signal energy in a beam to increase the received signal at the location of the mobile. This increases the signal strength there while reducing interference in other coverage locations.

In the uplink direction, two methods have been defined: Firstly, in Joint Reception (JR) mode, a signal can be received at several locations and forwarded to a central processing unit. Again, this requires optical cabling. And secondly a Coordinated Scheduling/Coordinated Beamform-ing approach has been defined to tune the signal energy in the direction of the recipient.

In situations with high uplink interference, it can be beneficial if a small cell receives the uplink signal while the downlink signal continues to be sent from the macrocell.

Whether CoMP features will be deployed in practice in the future depends on their effectiveness. The CoMP study in the TR mentioned above comes to the conclusion that datarates could be improved between 25% and 50% for mobiles at cell edges with neighboring cell interference.

4.16.6   Future LTE Uses: Machine Type Communication and Public Safety

The 3GPP Releases for LTE-Advanced also contain additions for two new usage scenarios. The first one is referred to as Machine-Type Communication (MTC) in 3GPP and deals with the special requirements of machine-to-machine communication and embedded devices whose numbers are expected to grow significantly over the coming years. From a network point of view, new methods have been specified to signal overload situations to delay tolerant devices so that they can hold back their communication requests for some time (e.g., 30 minutes). In addition, a new system broadcast message (SIB-14) has been defined that includes information concerning the new Extended Access Class Barring to reduce network load when many devices try to access the network simultaneously, for example, after a power outage.

At the other end of the spectrum, enhancements have been specified to allow the use of LTE for public safety purposes, for example, for police, fire departments and so on. In 3GPP Release 11, a new high-power UE power class has been defined for band 14 in the United States that is reserved for public safety purposes. High-power mobile devices that are installed in vehicles, for example, can have an output power of up to 33 dBm which equals 2 W. This has been done to reduce the number of required base stations for a potential public safety network. In addition to this, 3GPP Release 12 has started to specify direct device-to-device (D2D) communication which can also be useful in public safety networks in areas without network coverage. As for most other features described in this section, it is so far not clear when and how these features might be used in practice.

Questions

1. How many subcarriers are used for a 10-MHz FDD LTE channel?
2. What is the difference between an S1 and an X2 handover?
3. Describe the differences between the tasks for the MME and the tasks of the S-GW.
4. What is an RB?
5. How does a mobile device get access to the PUSCH?
6. What are the differences between ARQ and HARQ?
7. What is the difference between a default and a dedicated bearer?
8. What is the purpose of DRX in RRC Connected state?
9. How is mobility controlled in RRC Idle state?
10. What is the difference between a Cell Change Order and a Handover?
11. How can the LTE core network be interconnected with legacy core networks and why should this be done?
12. What is CS fallback?

13. What is the big disadvantage of Internet-based voice services compared to network operator-based voice services?
14. Describe different options for the backhaul connection of the eNode-B.

Answers to these questions can be found on the website to this book at http://www .wirelessmoves.com.

References

[1] The International Telecommunication Union, Framework and Overall Objectives of the Future Development of IMT-2000 Systems Beyond IMT-2000, ITU-R M.1645, 2003.

[2] 3GPP, Evolved Universal Terrestrial Radio Access (E-UTRA); User Equipment (UE) Radio Access Capabilities Release 8, TS 36.306.

[3] Sauter, M. (2009) Beyond 3G – Bringing Networks, Terminals and the Web Together, John Wiley & Sons Ltd, ISBN 978-0-470-75188-6.

[4] Next Generation Mobile Networks, Initial Terminal Device Definition, June 2nd, 2009.

[5] 3GPP, Evolved Universal Terrestrial Radio Access (E-UTRA); User Equipment (UE) Radio Transmission and Reception, version 9.2.0, TS 36.101.

[6] 3GPP, Evolved Universal Terrestrial Radio Access Network (E-UTRAN); S1 Data Transport, TS 36.414.

[7] 3GPP, Evolved Universal Terrestrial Radio Access Network (E-UTRAN); S1 Application Protocol (S1AP), TS 36.413.

[8] The Internet Engineering Task Force (IETF), Stream Control Transmission Protocol, RFC 4960, http:// tools.ietf.org/html/rfc4960.

[9] 3GPP, Evolved Universal Terrestrial Radio Access Network (E-UTRAN); X2 Application Protocol (X2AP), TS 36.423.

[10] M. Sauter, How File Sharing of Others Drains Your Battery, http://mobilesociety.typepad.com/mobile_life /2007/05/how_file_sharin.html, Accessed in 2010, May 2007.

[11] M. Sauter, IPv6 Crash Course – Part 4, http://mobilesociety.typepad.com/mobile_life/2009/12/ipv6-crash -course-part-4.html, Accessed in 2010, December 2009.

[12] M. Sauter, Why IPv6 Will Be Good for Mobile Battery Life, http://mobilesociety.typepad .com/mobile_life /2008/03/why-ipv6-will-b.html, Accessed in 2010, March 2008.

[13] P. Calhoun et al., Diameter Base Protocol, RFC 3588, September 2003.

[14] 3GPP, Evolved Universal Terrestrial Radio Access (E-UTRA); Physical Channels and Modulation, TS 36.211.

[15] 3GPP, Radio Resource Control (RRC); Protocol Specification, TS 36.331.

[16] 3GPP, Evolved Universal Terrestrial Radio Access (E-UTRA); Physical Layer Procedures, TS 36.213.

[17] 3GPP, Evolved Universal Terrestrial Radio Access (E-UTRA); Radio Link Control (RLC) Protocol Specification, TS 36.322.

[18] L.-E. Johnsson et al., The RObust Header Compression (ROHC) Framework, IETF RFC 4995 and 5795.

[19] G. Pelletier, RObust Header Compression Version 2 (ROHCv2): Profiles for RTP, UDP, IP, ESP and UDP-Lite, IETF RFC 5225.

[20] 3GPP, Evolved Universal Terrestrial Radio Access (E-UTRA); Medium Access Control (MAC) Protocol Specification, TS 36.321.

[21] 3GPP, General Packet Radio Service (GPRS) Enhancements for Evolved Universal Terrestrial Radio Access Network (E-UTRAN) Access, TS 23.401.

[22] 3GPP, Evolved Universal Terrestrial Radio Access Network (E-UTRAN); X2 Application Protocol (X2AP), TS 36.423.

[23] Agilent, Security in the LTE-SAE Network, www.home.agilent.com/upload/cmc_upload/All/Security_in_the _LTE-SAE_Network.PDF, Accessed in 2014, 2009.

[24] 3GPP, Evolved Universal Terrestrial Radio Access (E-UTRA); User Equipment (UE) Procedures in Idle Mode, TS 36.304.

[25] 3GPP, General Packet Radio Service (GPRS); Service Description; Stage 2, TS 23.060.

[26] 3GPP, Architecture Enhancements for Non-3GPP Accesses, TS 23.402.

[27] 3GPP, Self-configuring and Self-optimizing Network (SON) Use Cases and Solutions, (Release 9), TS 36.902.

[28] M. Sauter, Update: Self Organizing Networks, http://mobilesociety.typepad.com/mobile_ life/2010/01/update -lte-selforganizing-network.html, Accessed in 2010, January 2010.

[29] 3GPP, Circuit Switched (CS) Fallback in Evolved Packet System (EPS); Stage 2, TS 23.272.

[30] B. Klugg, HTC Thunderbolt Review: The First Verizon 4G LTE Smartphone, Anandtech, http://www.anandtech .com/show/4240/htc-thunderbolt-review-first-verizon-4g-lte-smartphone, 2011.

[31] B. Klugg, Why the iPhone 5 Lacks Support for Simultaneous Voice and LE or EVDO (SVLTE, SVDO), Anandtech, http://www.anandtech.com/show/6295/why-the-iphone-5-lacks-simultaneous-voice-and-lte-or -evdo-svlte-svdo-support-, 2012.

[32] J. Rosenberg *et al.*, SIP: Session Initiation Protocol, IETF RFC 3261

[33] G. Combs, Wireshark, http://www.wireshark.org.

[34] H. Schulzrinne, RTP: A Transport Protocol for Real-Time Applications, IETF RFC 3550.

[35] M. Handley, V. Jacobson, and C. Perkins, SDP: Session Description Protocol, IETF RFC 4566.

[36] V. Fajardo *et al.*, Diameter Base Protocol, IETF RFC 6733.

[37] 3GPP, IP Multimedia Core Network Subsystem (IMS) Multimedia Telephony Service and supplementary services; Stage 1, TS 22.173.

[38] The GSM Association, IR.92 IMS Profile for Voice and SMS, http://www.gsma.com/newsroom/wp-content /uploads/2013/04/IR.92-v7.0.pdf.

[39] 3GPP, IP Multimedia Subsystem (IMS) Service Continuity; Stage 2, TS 23.237.

[40] 3GPP, Release Work Plan, http://www.3gpp.org/ftp/Information/WORK_PLAN/Description_Releases/, Accessed in 2014, 2010.

[41] 3GPP, Feasibility Study for Further Advancements for E-UTRA (LTE-Advanced), Release 9, TR 36.912.

[42] A. Roessler, M. Kottkamp, LTE-Advanced (3GPP Rel. 11) Technology Introduction Whitpeper, Rohde und Schwarz, 2013.

[43] M. Rumney, 3GPP LTE Standards Update: Release 11, 12 and Beyond, Agilent, 2012.

[44] 3GPP, Coordinated multi-point operation for LTE physical layer aspects, 3GPP TR 36.819.

5

Wireless Local Area Network (WLAN)

In the mid-1990s, the first Wireless Local Area Network (WLAN) devices appeared on the market, but they did not get a lot of consumer attention. This changed rapidly at the beginning of this decade, when the hardware became affordable and WLAN quickly became the standard technology to interconnect computers, and later smartphones and tablets also wirelessly interconnected with each other and the Internet. This chapter takes a closer look at this system, which was standardized by the Institute of Electrical and Electronics Engineers (IEEE) in the 802.11 specification [1].

The first part of this chapter describes the fundamentals of the technology. Once the system became popular, a number of inherent security flaws were discovered and fixed. The chapter therefore also focuses on these issues and shows how WLAN can be used securely. And finally, this chapter also takes a look at the functionalities that were introduced in the subsequent versions of the specification.

5.1 Wireless LAN Overview

Wireless LAN received its name from the fact that it is primarily based on existing LAN standards. These standards were initially created by the IEEE for wired interconnection of computers and can be found in the 802.X standards (e.g., 802.3 [2]). In general, these standards are known as 'Ethernet' standards. The wireless variant, which is generally known as Wireless LAN, is specified in the 802.11 standard. As shown in Figure 5.1, its main application today is to transport Internet Protocol (IP) packets over layer 3 of the OSI model. Layer 2, the data link layer, has been adapted from the wired world with relatively few changes. To address the wireless nature of the network, a number of management operations have been defined, which are described in Section 5.2. Only layer 1, the physical layer, is a new development, as WLAN uses airwaves instead of cables to transport data frames.

From GSM to LTE-Advanced: An Introduction to Mobile Networks and Mobile Broadband,
Revised Second Edition. Martin Sauter.
© 2014 John Wiley & Sons, Ltd. Published 2014 by John Wiley & Sons, Ltd.

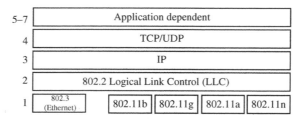

Figure 5.1 The WLAN protocol stack

5.2 Transmission Speeds and Standards

Since the creation of the 802.11 standard, various enhancements have followed. Therefore, a number of different physical layers, abbreviated as 'PHY', exist today in the standard documents. Each PHY has been defined in a different document and a letter has been put at the end of the initial 802.11 document name to identify the different PHYs (Table 5.1).

The breakthrough for WLAN was the emergence of the 802.11b standard that offered datarates from 1 to 11 Mbit/s. The maximum datarate that could be achieved in a real environment mainly depended on the distance between the sender and the receiver as well as on the number and kind of obstacles between them, such as walls or ceilings – in practice, around 5 Mbit/s could be achieved with this standard but only over short distances of a few meters.

To ensure connectivity over a larger distance, the number of bits used for redundancy was automatically adapted. This reduced the speed down to a few hundred kilobits per second under very bad conditions. Many vendors specify a maximum range of their WLAN adapters of up to 300 m. In practice, such a distance is achieved only outdoors where no obstacles absorb signal energy and only at very low speeds.

The 802.11b standard used the 2.4-GHz ISM (Industrial, Scientific and Medical) band, which can be used in most countries without a license. One of the most important conditions for the license-free use of this frequency band is the limitation of the maximum transmission power to 100 mW. It is also important to know that the ISM band is not technology restricted. Other wireless systems such as Bluetooth also use this frequency range.

The 802.11g standard specified a much more complicated PHY as compared to the 802.11b standard to achieve datarates of up to 54 Mbit/s. In practice, around 25 Mbit/s is reached on the

Table 5.1 Different PHY standards

Standard	Frequency band	Theoretical top speeds
802.11b [3]	2.4 GHz (2.401–2.483 GHz)	1–11 Mbit/s
802.11g [4]	2.4 GHz (2.401–2.483 GHz)	6–54 Mbit/s
802.11a [5]	5 GHz (5.150–5.350 GHz and 5.470–5.725 GHz)	6–54 Mbit/s
802.11n	2.4 GHz (as above)	6–600 Mbit/s
	5 GHz (as above)	
802.11ac	5 GHz (as above)	Up to 6.93 Gbit/s

application layer under good signal conditions. Even though standardization has significantly progressed, 11g devices are still in wide use. This variant of the standard also uses the 2.4-GHz ISM band and has been designed in a way to be backward compatible to older 802.11b systems. This ensures that 802.11b devices can communicate in new 802.11g networks and vice versa. More about the different PHYs can be found in Section 5.6.

In addition to the 2.4-GHz ISM band, another frequency range was opened for WLANs in the 5-GHz band. As with the 802.11g standard, datarates between 6 and 54 Mbit/s were specified. In practice, however, 802.11a devices never became very popular, as they had to be backward compatible to 802.11b and g, and the support of several frequency bands increased the overall hardware costs.

Owing to the rising datarates in local networks and of Internet connections via cable and ADSL, it was necessary to further increase the speed of Wi-Fi networks. After several years of standardization work, the companies involved finally agreed on a new air interface that is now specified in IEEE 802.11n. By doubling the channel bandwidth and by using numerous other improvements that are described in more detail later in this chapter, PHY data transfer speeds of up to 600 Mbit/s can be achieved. In practice, typical data transfer rates under favorable radio conditions are in the region of 70–150 Mbit/s. In addition, the specification supports both the 2.4-GHz and the 5-GHz bands. This has become necessary as the 2.4 GHz is widely used today, and in cities it is not uncommon to find many networks per channel. The 5-MHz band is still much less used today and hence allows higher datarates in favorable transmission conditions.

While most devices are 802.11n capable today, the IEEE has moved forward and has specified the next evolutionary Wi-Fi standard in 802.11ac. At the time of publication, first products were already available on the market implementing a subset of this specification. By using large bandwidths in the 5-GHz band, improved modulation and further methods to increase datarates that will be described later, a theoretical peak datarate of 6.9 Gbit/s has been specified. However, in practice, achievable datarates are much lower, but still it go significantly beyond what is possible with 802.11n devices.

Additional 802.11 standards, which are shown in the Table 5.2, specify a number of additional optional WLAN capabilities.

5.3   WLAN Configurations: From Ad Hoc to Wireless Bridging

All devices that use the same transmission channel to exchange data with each other form a basic service set (BSS). The definition of the BSS also includes the geographical area covered by the network. There are a number of different BSS operating modes.

5.3.1   Ad Hoc, BSS, ESS and Wireless Bridging

In ad hoc mode, also referred to as Independent Basic Service Set (IBSS), two or more wireless devices communicate with each other directly. Every station is equal in the system and data is exchanged directly between two devices. The ad hoc mode therefore works just like a standard wireline Ethernet, where all devices are equal and where data packets are exchanged directly between two devices. As all devices share the same transport medium (the airwaves), the packets are received by all stations that observe the channel. However, all stations except

Table 5.2 Additional 802.11 standard documents that describe optional functionality

Standard	Content
802.11e [6]	The most important new functionalities of this standard are methods to ensure a certain Quality of Service (QoS) for a device. Therefore, it is possible to ensure a minimum bandwidth and fast media access for real-time applications like Voice over Internet Protocol (VoIP) even during network congestion periods. Furthermore, this standard also specifies the direct link protocol (DLP), which enables two WLAN devices to exchange data directly with each other instead of communicating via an access point. DLP can effectively double the maximum data transfer speed between two devices
802.11f [7]	This standard specifies the exchange of information between access points to allow seamless client roaming between cooperating access points. It is used in practice to extend the range of a WLAN network. More about this topic can be found in Section 5.3.1
802.11h [8]	This extension adds power control and dynamic frequency selection for WLAN systems in the 5-GHz band. In Europe, only 802.11a devices that comply with the 802.11h extensions can be sold
802.11i [9]	This standard describes new authentication and encryption methods for WLAN. The most important part of 802.11i is 802.1x. More about this topic can be found in Section 5.7

the intended recipient discard the incoming packets because the destination address is not equal to their hardware address. All participants of an ad hoc network have to configure a number of parameters before they can join the network. The most important parameter is the service set identity (SSID), which serves as the network name. Furthermore, all users have to select the same frequency channel number (some implementations select a channel automatically) and ciphering key. While it is possible to use an ad hoc network without ciphering, it poses a great security risk and is therefore not advisable. Finally, an individual IP address has to be configured in every device, which the participants of the network have to agree on. Owing to the number of different parameters that have to be set manually, WLAN ad hoc networks are not very common.

One of the main applications of a WLAN network is the access to a local network and the Internet. For this purpose, the infrastructure BSS mode is much more suitable than the previously described ad hoc mode. In contrast to an ad hoc network, it uses an access point (AP), which takes a central role in the network as shown in Figure 5.2.

The AP can be used as a gateway between the wireless and the wireline networks for all devices of the BSS. Furthermore, devices in an infrastructure BSS do not communicate directly with each other. Instead, they always use the AP as a relay. If device A, for example, wants to send a data packet to device B, the packet is first sent to the AP. The AP analyzes the destination address of the packet and then forwards the packet to device B. In this way, it is possible to reach devices in the wireless and wireline networks without the knowledge of where the client device is. The second advantage of using the AP as a relay is that two wireless devices can communicate with each other over larger distances, with the AP in the middle. In this scenario, shown in Figure 5.2, the transmit power of each device is enough to reach the AP but not the other device because it is too far away. The AP, however, is close enough to

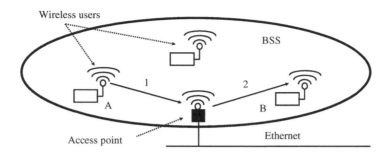

Figure 5.2 Infrastructure BSS

both devices and can thus forward the packet. The disadvantage of this method is that a packet that is transmitted between two wireless devices has to be transmitted twice over the air. Thus, the available bandwidth is cut in half. Owing to this reason, the 802.11e standard introduces the direct link protocol (DLP). With DLP, two wireless devices can communicate directly with each other while still being members of an infrastructure BSS. However, this functionality is declared as optional in the standard and not widely used today.

WLAN APs usually fulfill a number of additional tasks. Here are some examples:

- 100 Mbit/s or 1 Gbit/s ports for wireline Ethernet devices. Thus, the AP also acts as a layer 2 switch.
- At home, a WLAN AP is often used as an IP router to the Internet and can be connected via Ethernet to a DSL- or cable modem.
- To configure devices automatically, a Dynamic Host Configuration Protocol (DHCP) server [10] is also usually integrated into an AP. The DHCP server returns all necessary configuration information like the IP address for the device, the IP address of the DNS server and the IP address of the Internet gateway.

Furthermore, WLAN APs can also include a DSL or cable modem. This is quite convenient as fewer devices have to be connected to each other and only a single power supply is needed to connect the home network to the Internet. A block diagram of such a fully integrated AP is shown in Figure 5.3.

The transmission power of a WLAN AP is low and can thus only cover a small area. To increase the range of a network, several APs that cooperate with each other can be used. If a mobile user changes his position and the network card detects that a different AP has a better signal quality, it automatically registers with the new AP. Such a configuration is called an Extended Service Set (ESS) and is shown in Figure 5.4. When a device registers with another AP of the ESS, the new AP informs the previous AP of the change. This is usually done via a direct Ethernet connection between the APs of an ESS, and is referred to as the 'distribution system'. Subsequently, all packets arriving in the wired distribution system, for example, from the Internet, will be delivered to the wireless device via the new AP. As the old AP was informed of the location change, it ignores the incoming packets. The change in APs is transparent for the higher layers of the protocol stack on the client device. Therefore, the mobile device can keep its IP address and only a short interruption of the data transfer will occur.

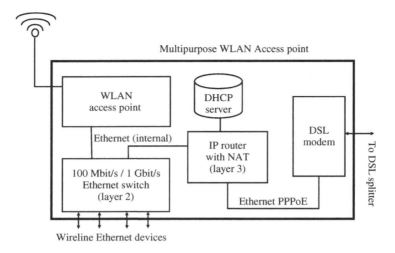

Figure 5.3   Access point, IP router and DSL modem in a single device

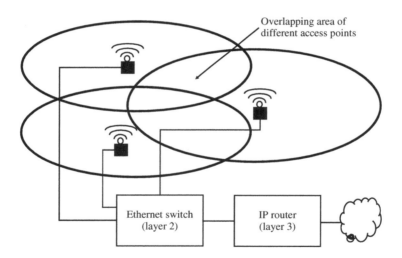

Figure 5.4   ESS with three access points

To allow a client device to transparently switch over to a new AP of an ESS, the following parameters have to match on all APs:

- All APs of an ESS have to be located in the same IP subnet. This implies that no IP routers can be used between the APs. Ethernet hubs, which switch packets on layer 2, can be used. In practice, this limits the maximum coverage area substantially because IP subnets are only suitable to cover a very limited area, like a building or several floors.

- All APs have to use the same BSS service ID, also called an 'SSID'. More about SSIDs can be found in Section 5.3.2.
- The APs have to transmit on different frequencies and should stick to a certain frequency repetition pattern as shown in Figure 5.5.
- Many APs use a proprietary protocol to exchange user information with each other if the client device switches to a new AP. Therefore, all APs of an ESS should be from the same manufacturer. To allow the use of APs of different manufacturers, the IEEE released the 802.11f standard (Recommended Practice for Multi-Vendor Access Point Interoperability) at the beginning of 2003. However, this standard is optional and by no means binding for manufacturers.
- The coverage area of the different APs should overlap to some extent so that client devices do not lose coverage in border areas. As different APs send on different frequencies, the overlapping poses no problem.

Another WLAN mode is wireless bridging, sometimes also referred to as a wireless distribution system. In this mode, the APs of an ESS can wirelessly forward packets they have received from client devices between each other. In practice, this mode is used if only one connection to the wired network exists but a single AP is unable to cover the desired area on its own. Usually, a wireless bridging AP also supports simultaneous BSS functionality. Therefore, only a single AP is required to offer service at a certain location to users and to backhaul the packets to the AP connected to the Internet.

5.3.2   SSID and Frequency Selection

When an AP is configured for the first time, there are two basic parameters that have to be set.
The first parameter is the basic SSID. The SSID is periodically broadcast over the air interface by the AP inside beacon frames, which are further discussed in Section 5.4. Note that the 802.11 standard uses the term 'frame' synonymously for 'packet' and this chapter also makes frequent use of it. The SSID identifies the AP and allows the operation of several APs at the same location for access to different networks. Such a configuration of independent APs should not be confused with an ESS, in which all APs work together and have the same SSID.

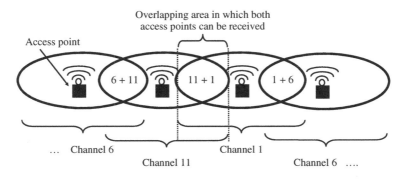

Figure 5.5   Overlapping coverage of access points forming an ESS

Usually, the SSID is a text string in a human readable form because during the configuration of the client device the user has to select an SSID if several are found. Many configuration programs on client devices also refer to the SSID as the 'network name'.

The second parameter is the frequency or channel number. It should be set carefully if several APs have to coexist in the same area. The ISM band in the 2.4 GHz range uses frequencies from 2.410 to 2.483 MHz. Depending on national regulations, this range is divided into a maximum of 11 (United States) to 13 (Europe) channels of 5 MHz each. As a WLAN channel requires a bandwidth of 25 MHz, different APs at close range should be separated by five ISM channels. As can be seen in Figure 5.5, three infrastructure BSS networks can be supported in the same area or a single ESS with overlapping areas of three APs. For infrastructure BSS networks, the overlapping is usually not desired but cannot be prevented if different companies or home users operate their APs close to each other. To be able to keep the three APs at least five channels apart from each other, channels 1, 6 and 11 should be used.

In practice, channels 12 and 13 are only allowed for use in Europe. Unfortunately, many WLAN drivers do not ask during software installation about the country in which the device is going to be used and block these channels by default. If it is unclear during the installation of a new AP as to which devices will be used in the network, channels 12 or 13 should not be selected, to enable all client devices to communicate with the AP.

802.11a and 11n systems use the spectrum in the 5-GHz range in Europe, between 5.170 and 5.350 GHz and between 5.470 and 5.725 GHz, for data transmission. In this 455 MHz bandwidth, 18 independent networks can be operated. This is quite significant, especially when compared to the three independent networks that can be operated in the 2.4-GHz band.

Figure 5.6   Client device configuration for a BSS or ESS

On a client device, the basic configuration for joining a BSS or ESS network is usually straight forward. To join a new network, the device automatically searches for active APs on all possible frequencies and presents the SSIDs it has discovered to the user as shown in Figure 5.6. The user can then select the desired SSID of the network to join. Selecting a frequency is not necessary, as the client device will always scan all frequencies for the configured SSID during power up. If more than one AP is found with the same SSID during the network search procedure, the client device assumes that they belong to the same ESS. If the user wants to join such a network, the device then selects the AP on the frequency on which the beacon frames are received with the highest signal strength. Further details about this process can be found in the Section 5.4.

In addition to selecting the SSID, activating encryption for the air interface is the second important step while setting up a BSS or an ESS. Fortunately, fewer and fewer APs are shipped with encryption disabled.

5.4 Management Operations

In a wired Ethernet, it is usually sufficient to connect the client device via cable to the nearest hub or switch to get access to the network. Physically connecting a wireless device to a WLAN network is of course not possible, as there is no cable. Also, a WLAN device has the ability to automatically roam between different APs of an ESS and is able to encrypt data packets on layer 2 of the protocol stack. As all of these WLAN operations have to be coordinated between the APs and the user devices, the 802.11 standard specifies a number of management operations and messages on layer 2, as well as additional Information Elements (IEs) in the Medium Access Control (MAC) header of data packets, which are not found in a wired Ethernet.

The AP has a central role in a BSS and is usually also used as a bridge to the wired Ethernet. Therefore, wireless clients always forward their packets to the AP, which then forwards them to the wireless or wired destination devices. To allow wireless clients to detect the presence of an AP, beacon frames are broadcast by the AP periodically. A typical value of the beacon frame interval is 100 milliseconds. As can be seen in Figure 5.7, beacon frames not only comprise the SSID of the AP but also inform the client devices about a number of other functionalities and options in a number of IEs. One of these IEs is the capability IE. Each bit of this 2-byte IE informs a client device about the availability of a certain feature. As can be seen in Figure 5.7, the capability IE informs the client device in the fifth bit, for example, that ciphering is not activated (privacy disabled). Other IEs in the beacon frame are used for parameters that require more than a single bit. Each type of IE has its own ID, which indicates to the client devices as to how to decode the data part of the IE. IE 0, for example, is used to carry the SSID, while IE 1 is used to carry information about the supported datarates. As IEs have different lengths, a length field is included in every IE header. By having an identifier and a length field at the beginning of each IE, a client device is able to skip over optional IEs that it does not recognize. Such IEs might be present in beacon frames of new APs that offer functionality that older client devices might not have implemented. This ensures backward compatibility to older devices.

During a network search, a client device has two ways to find available APs. One way is to passively scan all possible frequencies and just wait for the reception of a beacon frame. To speed up the search, a device can also send probe request frames to trigger an AP to send its system information in a probe response frame, without waiting for the beacon frame interval

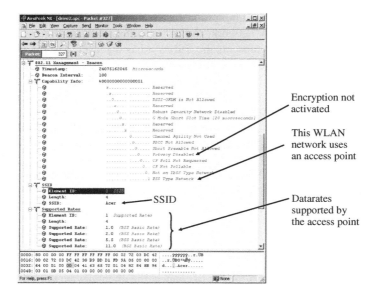

Figure 5.7   An extract of a beacon frame

to expire. Most client devices make use of both methods to scan the complete frequency range
as quickly as possible.

Once a client device has found a suitable AP, it has to perform an authentication procedure.
Two authentication options have been defined in the standard.

The first authentication option is called open system authentication. The name is quite mis-
leading as this option performs no authentication at all. The device simply sends an authentication frame with an authentication request to the AP, asking for open system authentication.
No further information is given to the AP. If the AP allows this 'authentication' method, it
returns a positive status code and the client device is 'authenticated'.

The second authentication option is called shared key authentication. This option uses a
shared key to authenticate client devices. During the authentication procedure, the AP challenges the client device with a randomly generated text. The client device then encrypts this
text with the shared key and returns the result to the AP. The AP performs the same operation
and compares the result with the answer from the client device. The results can match only if
both devices have used the same key to encrypt the message. If the AP is able to validate the
client's response, it finishes the procedure as shown in Figure 5.8 and the client is authenticated. Note that the use of the same key for all client devices can be a great security risk. This
is further discussed in Section 5.7.

Once authenticated successfully, the client device has to perform an association procedure
with the AP. The AP answers an association request message by returning an association
response message, which once more contains all necessary information about the wireless
network, for example, the capability IE. Furthermore, the AP assigns an association ID, which
is also included in the association response message. It is used later by the client device to
enter power-saving (PS) mode. Authentication and association with an AP are two separate
procedures. This allows a client device to quickly roam between different APs. Once a device

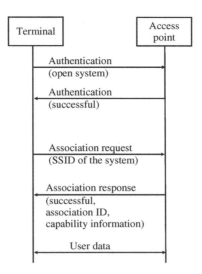

Figure 5.8 Authentication and association of a client device with an access point

is authenticated by all APs, it only has to perform an association procedure to roam from one AP to another.

Figure 5.8 shows the message flows of the authentication and association procedures. Acknowledgment (ACK) frames (see Section 5.5) are not shown for clarity.

Once the association with an AP has been performed successfully, user data packets can be exchanged. As a client device is informed via the capability IE if Wired Equivalent Privacy (WEP) encryption is activated for the network, it automatically starts ciphering all subsequent frames if the corresponding bit is set in the IE. As standard WEP encryption contains a number of severe security flaws, new algorithms and procedures have been standardized, which are available in all current products. More about this topic can be found in Section 5.7.

Authentication and encryption are independent of each other. Therefore, APs are usually configured to use the open system 'authentication' and only use the shared secret key for encryption of the data packets. Devices that do not know the shared secret key or use an invalid key can, therefore, authenticate and associate successfully with an AP but cannot exchange user data, as the encrypted packets cannot be decrypted by the receiver. Some AP manufacturers offer the option of specifically activating the shared authentication procedure. However, this does not increase the security of the system in any way. Usually, it just further complicates the initial configuration of a client device because the shared authentication procedure must be manually activated by the user.

If a client device resides in an ESS with several APs (see Figure 5.4), it can change to a different AP which is received with a better signal level at any time. The corresponding reassociation procedure is shown in Figure 5.9. To be able to find the APs of an ESS, the client device scans the frequency band for beacon frames of other APs when no data has to be transmitted. As all APs of the same ESS transmit beacon frames containing the same SSID, client devices can easily distinguish between APs belonging to the current ESS and APs of other networks. To change to a new AP, the client device changes to the send/transmit frequency of the new AP

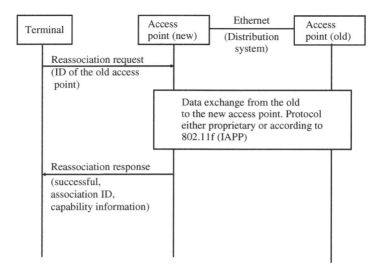

Figure 5.9   Reassociation (acknowledgment frames not shown)

and sends a reassociation request frame. This frame is similar to the association request frame and only contains an additional IE which contains the ID of the AP to which the client device was previously connected. The new AP then informs the previous AP via the wired Ethernet (distribution system) that the user has changed its association. The previous AP then acknowledges the operation and sends any buffered packets for the device to the new AP. Later, it deletes the hardware address and association ID from its list of served devices. In the future, all packets arriving for the client device via the wired distribution system will be ignored by the previous AP and will only be forwarded to the client device by the new AP. In the last step of the procedure, the new AP sends a reassociation response message to the client device.

At first, only the message exchange between the client device and the AP was standardized for the reassociation procedure. For a long time, however, no standard existed for the wired network between the two APs that are part of the procedure. Therefore, manufacturers developed their own proprietary messages to fill the gap. Therefore, it is preferable to use only APs from the same manufacturer to form an ESS to ensure a flawless roaming of the client devices. To tackle this shortcoming, the IEEE later standardized the procedure in the 802.11f Inter-Access Point Protocol (IAPP) recommendation. Implementation of the 802.11f standard, however, is optional.

The 802.11 standard also offers a PS mode to increase the operation time of battery-driven devices. If a device enters PS mode, the data transmission speed is decreased to some extent during certain situations. This is only a small disadvantage compared to the substantial reduction in power consumption that can be achieved.

The client device may enter a PS mode, for example, if its transmission buffer is empty and no data has been received from the AP for some time. To inform the AP that it will enter PS mode, the client device sends an empty frame to the AP with the PS bit set in the MAC header. When the AP receives such a frame, it will buffer all incoming frames for the client device for a certain time. During this time, the client device can power down the receiver. The time

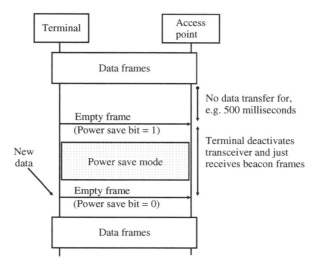

Figure 5.10   Activation and deactivation of the PS mode (acknowledgment frames not shown)

between reception of the last frame and activation of the PS mode is controlled by the client device. A typical idle time before the power save mode is activated is half a second.

If a client device wants to resume the data transfer, it simply activates its transceiver again and sends an empty frame containing a MAC header with the PS bit deactivated. Subsequently, the data transfer can resume immediately (See Figure 5.10).

For most applications used on mobile devices, like web browsing, data will only be delivered in rare cases once the PS mode has been activated. In order not to lose frames, they are buffered on the AP. Thus, a device in PS mode has to periodically activate its transceiver so that it can be notified of buffered frames by the AP. This is done via the Traffic Indication Map (TIM) IE, which the AP includes in every beacon frame. Each device has its own bit in the TIM, which indicates whether buffered frames are waiting. The client device identifies its bit in the TIM via its association identity (AID), which is assigned by the AP to the client device during the association procedure. Up to 2007 AIDs can be assigned by each AP. Therefore, the maximum size of the TIM IE is 2007 bits. To keep the beacon frames as small as possible, not all bits of the TIM are sent. The TIM, therefore, contains a length and offset indicator. This makes sense as in practice only few devices are in PS mode and therefore only a few bits are required.

As beacon frames are sent in regular intervals (e.g., every 100 milliseconds), the AP and client device agree on a listen interval during the association procedure after which the TIM has to be read. To negotiate the listen interval, the client device proposes an interval to the AP. If the AP accepts the proposed interval, it has to buffer any incoming frames for the device for this duration once the device activates the PS mode. It can be observed that a common listen interval is, for example, 3. The value implies that the client device has to check only every third beacon frame and can thus switch off its transceiver for 300 milliseconds at a time. When the client device exits PS mode temporarily to receive a beacon frame and the TIM bit for the device is not set, the transceiver is again switched off for 300 milliseconds before the procedure is repeated.

If the TIM bit is set, the client device does not go back to PS mode directly. Instead, a PS-poll frame is sent to the AP. The AP will send a single buffered frame to the client device for every PS-poll frame received. To inform the client device of further waiting frames, the 'more' bit in the MAC header of the frame is set. The client device then continues to send PS-poll frames as long as the 'more' bit is set in incoming frames.

Broadcast and multicast frames are buffered by the AP as well if at least one client device is currently in PS mode. Broadcast frames are not saved for every client device individually. Instead, the first bit of the TIM (AID = 0) is used as an indicator for the client devices in PS mode if broadcast data is buffered. These frames are then automatically sent after a beacon frame, which includes a Delivery Traffic Indication Map (DTIM) instead of an ordinary TIM. In order for client devices to be able to activate the receiver at the right time for buffered broadcast frames, a countdown timer inside the TIM announces the transmission of a DTIM.

5.5 The MAC Layer

The MAC protocol on layer 2 has similar tasks in a WLAN as in a fixed-line Ethernet:

- It controls access of the client devices to the air interface.
- A MAC header is put in front of every frame that contains, among other parameters, the (MAC) address of the sender (source) of the frame and the (MAC) address of the recipient (destination).

5.5.1 Air Interface Access Control

As the air interface is a very unreliable transmission medium, a recipient of a packet is required to send an ACK frame to inform the sender of the correct reception of the frame. This is a big difference from a wired Ethernet, where frames are not acknowledged. In all previous figures in this chapter, ACK frames were not shown for easier interpretation of the content. Figure 5.11 shows for the first time how frames are exchanged between a client device and an AP including the ACK frames. Each frame, regardless of whether it contains user data or management information (authentication, association, etc.,), has to be acknowledged with an ACK frame. The same or a different client device is allowed to send the next frame only when the ACK frame has been received. If no ACK frame is received within a certain time, the sender assumes that the frame was lost and thus resends the frame.

To ensure that the ACK frame can be sent before another device attempts to send a new data frame, the ACK frame is sent almost immediately after the data frame has been received. There is only a short delay between the two frames, the short interframe space (SIFS). All other devices have to delay their transmission by at least a distributed coordination function (DCF) interframe space (distributed coordination function interframe space, or DIFS for short).

Optionally, devices can also reserve the air interface prior to the transmission of a data frame. This might be useful in situations where devices can reach the AP but are too far away from each other to receive each other's frames. Under these circumstances, it can happen that two stations might attempt to send a frame to the AP at the same time. As the two frames will interfere with each other, the AP will not be able to receive either of the frames correctly. This scenario is also known as the 'hidden station problem'. To prevent such an overlap, a device

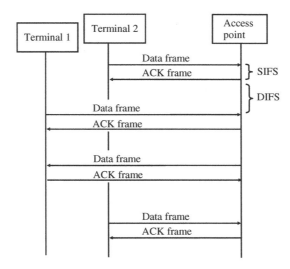

Figure 5.11   Acknowledgment for every frame and required interframe space periods

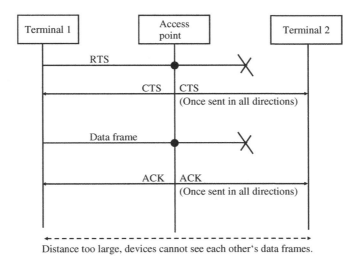

Figure 5.12   Reservation of the air interface via RTS/CTS frames

can reserve the air interface as shown in Figure 5.12 by sending a short RTS (Ready to Send) frame to the AP. The AP then answers with a CTS (Clear to Send) frame and the air interface is reserved. While the RTS frame might not be seen by all client devices in the network because of the large distance between them, the CTS frame can be seen by all devices because the AP is the central point of the network. Both RTS and CTS frames contain a so-called Network Allocation Vector (NAV) to inform other devices regarding the period of time during which the air interface is reserved. If a device uses an RTS/CTS sequence before sending, a frame can be configured in the driver settings dialog box of the network card. However, RTS/CTS sequences slow down the throughput of a device. Therefore, this mechanism should be used only if a very high network load is expected and the client devices are dispersed over a wider area.

As in a wired network, there is no central instance that controls the device which is allowed to send a frame at a certain time. Every device has to decide on its own as to when it can send a frame. To minimize the chance of a collision with frames of other devices, a coordination function is necessary. In WLAN networks, the DCF is used for this purpose. Distributing coordination to access the air interface is an approach completely different from that taken by all other systems described in this book. The other systems use a central logic that decides which user device is allowed to send at a certain time and for how long. The advantage of the DCF, however, is the easy implementation in all devices. The biggest disadvantage is the fact that no bandwidth can be reserved or guaranteed. This is mainly a problem for real-time applications like voice or video telephony if the network is already highly loaded with other traffic. As voice and video telephony over IP and over WLAN has become more and more popular, the IEEE has released the 802.11e standard for devices and applications that require a constant bandwidth and a deterministic medium access time. With this enhancement, devices can request a certain Quality of Service (QoS) from the AP to get precedence over transmissions from other devices. The enhancements also include a method to assign the air interface to a device for a specific time and thus guarantee a certain bandwidth and a maximum medium access time. 802.11e is backward compatible to the older 802.11b, g and a standards. Older devices that do not support the new standard can still be used in such a network without degrading the new QoS mechanism offered by the 802.11e standard.

Going back to the standard 802.11b DCF medium access scheme, DCF uses Carrier Sense Multiple Access/Collision Avoidance (CSMA/CA) to detect if another device is currently transmitting a frame. This method is quite similar to CSMA/Collision Detect (CD), which is used in fixed-line Ethernet, but it offers a number of additional functionalities to avoid collisions.

If a device wants to send a data packet and no activity is detected on the air interface, the packet can be sent without delay. If another device is already sending a data packet, the device has to wait until the data transfer has finished. Afterward, the device has to observe another delay time, the DIFS period, which has been described above. Then, the device yet again defers sending its packet for additional backoff time, which is generated by a random number generator. Therefore, it becomes very unlikely that several devices attempt to send data waiting in their output queue at the same time. The device with the smallest backup time will send its data first. All other devices will see the transmission, stop their backup timer and repeat the procedure once the transmission is over. In spite of this procedure if two devices still attempt to send packets at the same time, the transmissions will interfere with each other and thus no ACK frame will be sent. Both stations then have to retransmit their packets. If a collision occurs, the maximum possible backup time from which the random generator can choose is increased in the affected devices. This ensures that even in a high-load situation the number of collisions remains small.

The backoff time is divided into slots of 20 microseconds. For the first transmission attempts, the random generator will select 1 of the 31 possible slots in 802.11b and 11g devices. If the transmission fails, the window size is increased to 63 slots, then to 127 slots and so on. The maximum window size is 1023 slots, which equals 20 milliseconds. In the 802.11n standard, the first backoff window has been reduced to 15 slots, that is, 0.3 milliseconds.

In addition to the detection of an ongoing transmission and the use of a backoff time, each packet header contains an NAV field to inform the other devices of the time required to send the current frame and the following ACK frame. This additional feature is especially useful if the air interface is reserved via RTS and CTS frames as shown in Figure 5.12. Here, the first RTS frame contains the duration required to send the subsequent CTS frame, the actual data frame and the final ACK frame. The following CTS frame of the other device contains a slightly smaller NAV, which only contains the transmission duration for the subsequent data frame and the final ACK frame.

5.5.2   The MAC Header

The most important function of the MAC header is to address the devices in the local network. This is done by using 48-bit MAC addresses for the sender (source) and receiver (destination). The WLAN MAC addresses are identical to the MAC addresses that are used in a wired Ethernet. In a WLAN BSS, however, a frame is not directly sent from the sender to the receiver but is always sent to the AP first. Because of this, three MAC addresses are part of the MAC header as shown in Figure 5.13. The third MAC address is the AP address. When the AP receives a frame, it uses the destination address to decide if the receiver is a fixed or a wireless client and forwards the frame accordingly. Therefore, a client device does not need to know if the destination device is a wireless or a fixed Ethernet device.

Other important fields of the MAC header are the frame type and subtype. The frame type field informs the receiver if the current frame is a user data frame, a management frame (e.g., association request) or a control frame (e.g., ACK). Depending on the type of frame, the subtype field contains further information. For management frames, it indicates which management operation is contained in the frame (e.g., authentication, association, beacon frame, etc.).

The frame control flags are used to exchange additional management information between two devices. They are used, for example, to indicate to the destination if the user data is encrypted (WEP-enabled bit), if the device is about to change into PS mode (power management bit) or if the frame is intended for an AP ('to distribution system' bit).

If the frame contains user data, the Logical Link Control header (LLC header, layer 2) follows the MAC header. The most important job of the LLC header is to identify which protocol is used on layer 3.

5.6   The Physical Layer and MAC Extensions

On layer 1, the physical layer that is also referred to as the PHY, there are different modulation standards, as shown in Section 5.2, which are defined in the IEEE 802.11b, g, a, n and ac standards.

5.6.1   IEEE 802.11b – 11 Mbit/s

The breakthrough of WLAN in the consumer market was triggered by the introduction of devices compliant with the 802.11b standard, with a maximum speed of up to 11 Mbit/s. More

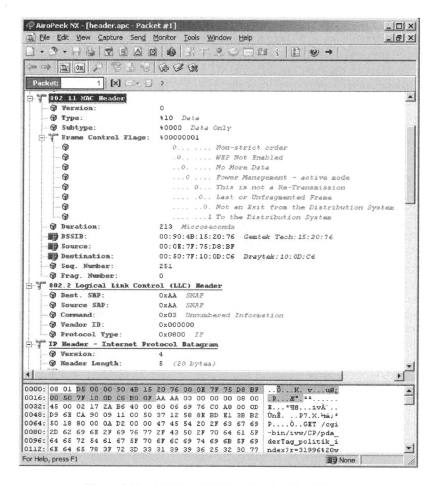

Figure 5.13   MAC and LLC header of a WLAN frame

recent PHYs described in the 802.11g, n and ac standards can achieve even higher speeds with the same bandwidth requirement of 22 MHz. The different PHYs are discussed later in detail. The following list shows some basic WLAN parameters and compares some of them to similar parameters of other systems:

- WLAN maximum transmission power is limited to 0.1 W. Mobile phone power, on the other hand, is limited to 1 – 2 W. Global System for Mobile Communications (GSM) and Universal Mobile Telecommunications System (UMTS) base stations have a typical power output of 20 W per sector and frequency.
- Each channel has a bandwidth of 22 MHz. Up to three APs can be used at close range in the ISM band without interfering with each other. GSM uses 0.2 MHz (200 kHz) per channel, and UMTS uses 5 MHz.

- Frame size is 4–4095 bytes. However, IP frames do not usually exceed 1500 bytes. This value is especially interesting for comparison with other technologies: A General Packet Radio Service (GPRS) packet, as shown in Section 2.3, consists of four bursts of 114 bits each and thus can only contain 456 bits. If coding Scheme 2 for error detection and correction is used, only 240 bits or 30 bytes remain for the actual packet. Therefore, an IP packet can be transmitted over a single WLAN frame, but it has to be split into several packets if it has to be transmitted over the air interface of a GPRS network.

- Transmission time of a large packet depends on the size of the packet and the transmission speed. If a large packet with a payload of 1500 bytes is transmitted with a speed of 1 Mbit/s, the transmission takes about 12 milliseconds. If reception conditions are good and the packet is sent with a transmission speed of 11 Mbit/s, the same transmission takes only 1.1 milliseconds. Note that the SIFS and the time it takes to send a short ACK frame as confirmation have to be added to these values to calculate the precise transmission time.

- Time between a data frame and an ACK frame (SIFS) is 10 microseconds or 0.01 milliseconds.

- If a transmission error occurs, a backoff procedure is performed as described in the previous Section. A backoff slot (of which 63 exist for the first retry) has a length of 20 microseconds or 0.02 milliseconds.

- At the beginning of the frame, a preamble is sent, which notifies all other devices that the transmission of a frame is about to start. The preamble is necessary to synchronize all listeners to the start of the frame. The preamble has a length of 144 microseconds or 0.144 milliseconds.

The preamble mentioned in the list above is part of the Physical Layer Convergence Procedure (PLCP) header, which is sent at the start of every frame. The PLCP header also contains information about the datarate used for the subsequent MAC frame. With the 802.11b standard, the MAC frame can be sent with a speed of 1, 2, 5.5 and 11 Mbit/s. This flexibility is necessary, as devices experiencing bad radio conditions can only send and transmit with a lower speed to compensate unfavorable radio conditions with a higher redundancy. In practice, the sender decides on its own which coding to use for a frame. Some devices also offer the possibility for the user to manually lock the speed to a fixed value (e.g. 5.5 Mbit/s). This helps in situations when the automatic speed selection algorithm cannot find a good value on its own. APs by some manufacturers record the speed at which individual stations send their frames and then use the same speed for subsequent packets to the clients. Beacon frames are usually sent at a speed of 1 or 2 Mbit/s. This allows even distant devices to detect the presence of an AP. However, this behavior is not mandatory and some APs transmit their beacon frames at a speed of 11 Mbit/s. This increases the overall speed of the network slightly, but has some disadvantages for distant devices.

For the coding of the actual user data in a frame, the direct sequence spread spectrum (DSSS) method is used for transmission speeds of 1 and 2 Mbit/s. Instead of transferring the bit itself, the DSSS algorithm converts the bit into 11 chips which are then transmitted over the air. Instead of sending a bit with the value of '1' the chip sequence '0,1,0,0,1,0,0,0,1,1,1' is transmitted. For a bit with the value of '0', the sequence is '1,0,1,1,0,1,1,1,0,0,0'. These sequences are also known as Barker code. As 11 values are transmitted instead of only one, redundancy increases substantially. Thus, a bit can be received correctly even if some of the chips cannot be decoded correctly at the receiver site.

UMTS also makes use of this technique, also known as spreading, to increase redundancy. However, there is an important difference. In a WLAN network, only a single station sends at a time (time division multiple access). UMTS additionally uses spreading to allow several devices to send at the same time (code division multiple access). This is possible in UMTS, as shown in Chapter 3, as orthogonal codes are used instead of fixed Barker sequences.

Once a bit has been converted into chips, the Barker chip sequence is sent over the air using Differential Binary Phase Shift Keying (DBPSK) with a transmission speed of 11 Mchips/s. The resulting bit rate is 1 Mbit/s. To transmit chips, DBPSK changes the phase of the signal for a '1' chip by 180 degrees. For a '0' chip, no phase of the carrier frequency remains unchanged.

To achieve a bit rate of 2 Mbit/s, DBPSK is replaced with Differential Quadrature Phase Shift Keying (DQPSK) modulation. Instead of one chip per transmission step, two chips are transmitted. The four (quadrature) possible values (00, 01, 10, or 11) of the two chips are encoded into 90 degree phase shifts of the carrier frequency per transmission step.

To further increase the data transmission speed without increasing the necessary bandwidth, the 802.11b standard introduces Complementary Code Keying (CCK), also known as high rate/direct sequence spread spectrum (HR/DSSS). Instead of coding a single bit into an 11-chip Barker sequence, CCK encodes the bits as follows.

For a datarate of 11 Mbit/s, all bits of a frame are arranged into blocks of eight bits as shown in Figure 5.14. The first two bits of a block are then transmitted using DQPSK and are encoded in phase shifts of 90 degrees.

The remaining six bits are used to generate an eight-chip symbol. As six bits are coded in an eight-bit symbol, the process adds some redundancy. The symbol is then split into four parts of two chips each, which are then modulated onto the carrier frequency as the phase changes.

As the chip rate remains the same as for 1 and 2 Mbit/s transmissions, CCK raises transmission speeds to 11 Mbit/s. A disadvantage compared to lower transmission speeds, however, is the fact that there is less redundancy in the resulting chip stream.

Different devices can send at different speeds depending on their reception conditions. Therefore, it has to be ensured that even devices that cannot receive high datarate frames correctly do not cause collisions because of their inability to detect an ongoing transmission of a high datarate frame. Therefore, it has to be ensured that at least the beginning of the frame is received correctly by all devices. Thus, the PLCP header of a frame is always sent

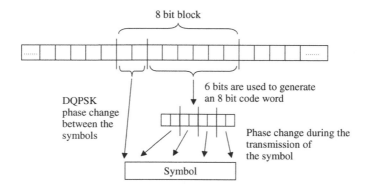

Figure 5.14   Complementary code keying for 11 Mbit/s transmissions

at a speed of 1 Mbit/s regardless of the speed and coding scheme of the rest of the frame. To inform other stations about the duration of the transmission, the PLCP header also contains information about the total duration of the transmission.

If the actual speed of an 11 Mbit/s WLAN network is compared to that of a 10 Mbit/s fixed Ethernet, it becomes apparent that is not quite as fast as its fixed-line counterpart. A 10 Mbit/s fixed-line Ethernet can reach a maximum speed of about 700–800 kB/s under ideal conditions. In an 11 Mbit/s WLAN, the maximum speed is about 300 kB/s between two wireless devices. This is due to the following properties that have already been described in this chapter:

- The PLCP header of a WLAN frame is always transmitted at 1 Mbit/s.
- Each frame has to be acknowledged with an ACK frame.
- In a wired network, a frame is directly sent from the source to the destination device. In a WLAN BSS, a frame is sent from the source to the AP, which then forwards the frame to the destination. The frame, therefore, traverses the air interface twice. In practice, this cuts the maximum transmission speed in half.

5.6.2   IEEE 802.11g with up to 54 Mbit/s

A first step to higher data transmission speeds was the 802.11g standard that introduced a new modulation scheme called Orthogonal Frequency Division Multiplexing (OFDM). This modulation scheme enables speeds up to 54 Mbit/s while using almost the same bandwidth as the 802.11b standard. Although today most 802.11b-only-capable devices have been replaced with newer equipment, 802.11g is still quite popular and the OFDM modulation scheme was reused with only few modifications in the subsequent PHYs that followed with 802.11n and 11ac.

The OFDM modulation scheme is fundamentally different from the modulation schemes used in the 802.11b standard. As shown in Figure 5.15, OFDM divides the bandwidth of a single 20-MHz channel into 52 subchannels.

The subchannels are 'orthogonal', as the amplitudes of the neighboring subchannels are exactly zero at the middle frequency of a subchannel. Therefore, they do not influence the amplitude of neighboring subchannels. OFDM does not transmit data by changing the phase of the carrier but by changing the amplitudes of the subchannels. Depending on the reception quality, a varying number of amplitude levels are used to encode a varying number of bits.

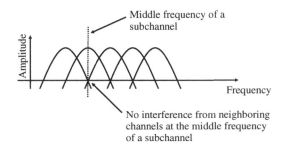

Figure 5.15   Simplified representation of OFDM subchannels

To demodulate the signal, the receiver performs a Fast Fourier Transformation (FFT) analysis for each transmission step. This method calculates the signal energy (amplitude) over the frequency band. The simplified result of an FFT analysis is shown in Figure 5.15. The x-axis represents the frequency band instead of the time as in most other graphs. The amplitude of each subchannel is shown on the y-axis.

Table 5.3 gives an overview of the datarates offered by the 802.11g standard. In practice, an algorithm dynamically selects the best settings depending on reception conditions.

Under ideal transmission conditions, the 64-quadrature amplitude modulation (64-QAM) can be used. Together with a 3/4 convolutional coder (three data bits are coded in four output bits) and a symbol speed of 250,000 symbols/s, a maximum speed of 54 Mbit/s is reached (216 bits per step × 250,000 symbols/s = 54 Mbit/s). It is to be noted that a similar convolutional coder for increasing redundancy is also used for GSM and UMTS (see Section 1.7.6).

802.11g client devices and APs are backward compatible to slower 802.11b devices. This means that 802.11g APs also support 802.11b client devices that can only communicate with a speed of up to 11 Mbit/s. 802.11g client devices can also communicate with older 802.11b APs. However, the maximum datarate is then, of course, limited to 11 Mbit/s. As slower 802.11b devices are not able to decode OFDM modulated frames, 802.11g devices in mixed configurations have to transmit a CTS packet to their selves for the reservation of the air interface prior to transmitting a frame. This ensures that 802.11b and g devices can be used simultaneously in a BSS. Furthermore, the PLCP header of a frame is sent at 1 Mbit/s for all devices to be able to receive the header correctly. While these procedures ensure interoperability, performance is reduced by about 20% because of the extra overhead of the CTS frames, which can only be sent at a maximum speed of 11 Mbit/s. Owing to these disadvantages, a 'G-only' mode can be activated in some APs to avoid this extra overhead.

Under ideal conditions, a maximum transfer rate of about 2500 kB/s can be observed in an 802.11g BSS. If two wireless devices communicate with each other, the maximum speed drops to about 1200 kB/s, as all frames are first sent to the AP, which then forwards the frames to the wireless destination device. As mentioned earlier, the 802.11e standard aims to overcome this problem by standardizing direct client-to-client communication in a BSS, provided the devices are sufficiently close to each other. Compared to the transfer speeds of the 802.11b standard, the 802.11g datarates are a dramatic improvement on older networks with a throughput of around 2000 kB/s, or 100 kB/s between two wireless client devices. There is still a big gap

Table 5.3 802.11g datarates

Speed (Mbit/s)	Modulation and coding	Coded bits per channel	Coded bits in 48 channels	Bits per step
6	BPSK, $R = 1/2$	1	48	24
9	BPSK, $R = 3/4$	1	48	36
12	QPSK, $R = 1/2$	2	96	48
18	QPSK, $R = 3/4$	2	96	72
24	16-QAM, $R = 1/2$	4	192	96
36	16-QAM, $R = 3/4$	4	192	144
48	64-QAM, $R = 2/3$	6	288	192
54	64-QAM, $R = 3/4$	6	288	216

to 100 Mbit/s wired Ethernet, in which maximum transfer speeds of over 7000 kB/s can be achieved.

5.6.3   IEEE 802.11a with up to 54 Mbit/s

The 802.11a standard is almost identical to the 802.11g standard. The main difference is the use of channels in the 5-GHz band, which makes it incompatible to 802.11b and g networks. Owing to the fact that a different frequency band is used, 802.11a devices do not have to be backward compatible. Therefore, PLCP headers can be sent with a speed of 6 Mbit/s instead of 1 Mbit/s. 802.11a networks are thus faster than mixed 802.11g/b networks and also have a slight advantage over 802.11g networks because they transmit the PLCP header at a higher speed. In practice, there are few remaining 802.11a networks today as the 5-GHz band is now used by 802.11n-compatible devices as described in the next section.

5.6.4   IEEE 802.11n with up to 600 Mbits/s

As has been shown in Section 5.6.2, data transfer rates of about 20–25 Mbit/s can be reached with 802.11g devices at the application layer. For many ADSL connections, this speed is still sufficient. Current ADSL2+, Very high-speed Digital Subscriber Line (VDSL) and cable modems, however, offer faster speeds and hence, 802.11g networks are no longer sufficient. With 802.11n, however, the speeds of these new technologies can be met. In addition to higher transmission speeds, another goal of 802.11n was the introduction of QoS mechanisms, so that applications such as Voice over Internet Protocol (VoIP) or video streaming can perform well even in loaded networks. Owing to the large number of companies that have been involved in the standardization work, the 802.11n standard is quite extensive and contains a high number of optional functionalities, of which most functionalities have not been implemented in practice. The following section describes the new functions of the High Throughput (HT) PHY and the MAC layer extensions that are defined as mandatory in the standard. In addition, options that are used in practice today are also described.

An easy way to increase throughput is to increase the channel bandwidth. In addition to the 20-MHz channels, it is now also possible to use a 40-MHz channel. In practice, this was already implemented by many vendors in a proprietary fashion in 802.11g products. As a consequence, devices from different manufacturers were often not interoperable at higher speeds. In practice, all 802.11n-capable PCs and Wi-Fi APs support 40-MHz channels in a standardized manner today. Some 802.11n-compatible smartphones, however, are still limited to 20-MHz channels.

In addition to using a wider channel bandwidth, the number of OFDM subchannels on a 20-MHz carrier has been increased from 52 to 56. The subchannel bandwidth of 312.5 kHz, however, remains the same. The additional subchannels occupy bandwidth that was not used up to this point, at the right and left sides of the carrier. The number of pilot subchannels remains unaltered at four. If two channels are combined to be a 40-MHz channel, 114 subchannels are used for data transmission and six subchannels are used as pilot channels, that is, for channel estimation. Table 5.4 compares the PHY carrier parameters of 802.11g to the 20-MHz and 40-MHz bandwidth options of 802.11n.

The initial 802.11 specification required the transmission of an ACK frame to confirm the correct reception of each data frame. This is important to ensure that frames are correctly

Table 5.4   Comparison of PHY carrier parameters of 802.11g vs. 11n

	20 MHz non-HT (as 802.11g)	20 MHz HT	40 MHz HT
Number of carriers	48	52	108 (2 × 54)
Number of pilot carriers	4	4	6
Total number of carriers	52	56	114 (2 × 57)
Unused carriers at the center	1	1	3

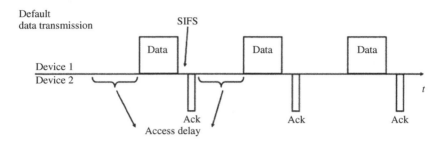

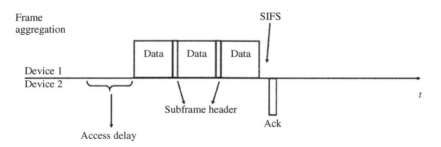

Figure 5.16   Default frame transmission compared to frame aggregation

received over the comparatively unreliable air interface and to be able to retransmit faulty data as quickly as possible. The disadvantage of this approach is that the air interface is not used efficiently. To reduce this overhead, the 802.11n standard has specified a method to aggregate frames on the MAC layer allowing frames to be transmitted together. This is much more efficient when transmitting large amounts of data as only a single ACK is required for the aggregated frames. The maximum aggregated frame size is 65,535 bytes. The disadvantage of this method is, however, that in case of a transmission error, the aggregated frame has to be completely retransmitted. Figure 5.16 compares the default data transmission to a transmission in which the frames are aggregated.

Another air interface parameter that was optimized is the OFDM guard interval (GI). It is required between two OFDM symbols to reduce the interference between consecutive symbols. In practice, a GI of 400 nanoseconds is sufficient in most transmission environments when compared to the 800 nanoseconds used previously. This significantly decreases the transmission time of an OFDM symbol from 4 to 3.6 microseconds, and hence, more symbols can be transmitted in a certain time frame.

To further increase transmission speeds, a new coding scheme was introduced with a reduced number of error detection and correction bits. In 802.11g, the least conservative coding scheme defined was 3/4, that is, three user data bits are encoded into 4 bits that are transferred over the air interface. Under very good signal conditions, 801.11n devices can now use a 5/6 coding rate, that is, 5 user data bits are encoded into 6 bits that are then transferred over the air interface.

All methods used simultaneously increase the maximum datarate by about 2.5 times compared to 802.11g. This results in a maximum speed on the air interface of 150 Mbit/s. As in previous Wi-Fi systems, application layer speeds are around half of this value owing to acknowledgement frames and other air interface properties.

As discussed earlier in this chapter, only three independent 20-MHz networks can be operated in the 2.4-GHz band. Especially in cities, many networks overlap each other. This significantly reduces the throughput of each network if a high amount of data is transferred on several networks that share the same channel. If an AP detects 20-MHz channels, the standard mandates that a network using a 40-MHz channel has to immediately switch to a 20-MHz channel and remain in this mode for at least 30 minutes after it has received the last frame of an AP of another network. A 40-MHz channel does not, therefore, result in a reliable and significant speed improvement in the 2.4-GHz band. In theory, the AP could change to another frequency and inform devices of the new channel number via a channel switch announcement message, but this is unlikely to improve the situation in the overcrowded 2.4-GHz band. In practice, some manufacturers have therefore decided to ignore the 20-MHz fallback requirement, and configuring a 40-MHz channel in the 2.4-GHz band results in a 40-MHz channel independent of whether there are other 20-MHz networks active or not. The 802.11n standard also applies to the 5-GHz band. Here, up to nine independent 40-MHz channels are available. As this frequency range would be much less used at the time of publication, it would be usually possible to find an unused channel. In practice, many APs and client devices support this band today so it has become a viable alternative. However, the downside of the 5-GHz band, is the shorter transmission range as higher frequency signals have more difficulties, such as permeating walls and other obstacles when compared to a 2.4-GHz signal. If an 802.11n device supports only the 2.4-GHz band or comes with additional 5 GHz, capabilities can usually only be noticed when networks in the 5-GHz band are not detected. Many of today's smartphones and entry-level notebooks, for example, support only 802.11n in the 2.4-GHz band. However, the number of 5-GHz-capable smartphones and notebooks, is rapidly increasing.

To further increase transmission speeds and network range, the 802.11n standard specifies a number of Multiple Input Multiple Output (MIMO) transmission schemes for 20-MHz and 40-MHz channels. Most devices today offer MIMO spatial multiplexing, which transmits several data streams over different transmission paths from the transmitter to the receiver over the same channel. This requires several antennas at both ends as each data stream originates from a separate antenna at the transmitter. Figure 5.17 shows how this is done in a simplified manner. In practice, the two data streams are usually not completely independent and hence a mathematical procedure is required on the receiver side to remove the effect of the two data streams interfering with each other on the way from the transmitter to the receiver.

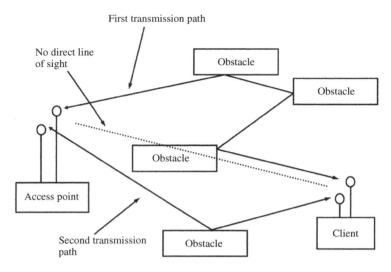

Figure 5.17   2 × 2 MIMO

The standard specifies that up to four MIMO channels and APs must support at least two independent transmissions chains. Other 802.11n devices, such as smartphones, are allowed to have only one transmission and reception chain. This makes sense because small mobile devices might not have enough space for a second antenna and focus less on the transmission speed enhancements of the 11n standard and more on power saving and the other additional functionalities offered. As devices can inform the AP of their capabilities during the association procedure, a 20-MHz channel and a single stream data transmission can be used for a smart-phone, while packets for another device can use the full 40-MHz bandwidth in combination with multistream MIMO transmission.

In practice, today, many devices use two or three MIMO channels, which can double or triple the theoretical peak data transmission rate as compared to those of single stream transmissions under ideal signal conditions. It should be noted at this point that some 802.11g APs also have two antennas. However, such devices make use of only reception (RX) diversity by detecting which antenna experiences better reception conditions.

Owing to many variables, such as the number of MIMO channels, long or short guard time, modulation, coding, etc., there are up to 77 possible combinations that result in different trans-mission speeds. Table 5.5 shows a number of examples.

Channel bundling and the shorter GI, for example, more than double the transmission speed because of the reduced time per symbol, the use of additional subcarriers and the reduced number of pilot channels. The effect of a shorter GI is most profound with a 40-MHz channel and two MIMO streams as the PHY transmission speed increases from 270 to 300 Mbit/s.

Together with the previously discussed enhancements, 2 × 2 MIMO (two transmitter antennas and two receiver antennas) can achieve a 5× improvement over 802.11g and a maximum transfer speed of 300 Mbit/s on the PHY. In a 4 × 4 MIMO system that uses four antennas on each side, the theoretical peak datarate is 600 Mbit/s. In practice, speeds of around 80–110 Mbit/s can be reached at the application layer under ideal conditions, that is, within

Table 5.5 Feature combinations and resulting transmission speeds

	20 MHz, no MIMO (Mbit/s)	20 MHz, two MIMO streams (Mbit/s)	40 MHz, two MIMO streams (Mbit/s)
802.11b	1, 2, 5.5, 11	–	–
802.11g	1, 2, 6, 9, 12, 18, 24, 36, 48, 54	–	–
802.11n, GI 800ns	6.5, 13, 19.5, 26, 39, 52, 58.5, 65	13, 26, 39, 52, 78, 104, 117, 130	27, 54, 81, 108, 162, 216, 243, 270
802.11n, GI 400ns	7.2, 14.4, 21.7, 28.9, 43.3, 57.8, 65, 72.2	14.4, 28.9, 43.3, 57.8, 86.7, 115.6, 130, 144.4	30, 60, 90, 120, 180, 240, 270, 300

distances of a few meters without walls between the transmitter and receiver, and by disabling backward compatibility (Greenfield mode). In addition, the AP needs to be equipped with gigabit Ethernet ports to be able to forward data at speeds exceeding 100 Mbit/s. Under less ideal conditions, devices automatically select a more robust modulation (16-QAM, QPSK or BPSK) and a more conservative coding such as 3/4, 2/3 or 1/2.

Another important property of 801.11n certified devices is the support of QoS mechanisms on the air interface as specified in 802.11e. With this extension, data packets of applications such as VoIP programs can be preferred. This way, voice telephony and other delay-sensitive applications can be used over the air interface even in heavily loaded networks as the packet delay is deterministic.

To announce the new capabilities introduced with 802.11n, a number of new parameters have been defined that are broadcast in beacon frames. The most important is the 'HT Capabilities' parameter (element ID 45) that describes which HT options are supported by the AP. The following list gives an overview of the options:

- indication of 40-MHz mode support;
- the number of supported MIMO streams modulation and coding schemes (MCS);
- support of the short guard time (400 nanoseconds);
- support of the optional MCS feedback mode. If supported, the receiver can inform the transmitter about current reception conditions. This helps the transmitter to adapt the modulation and coding rates accordingly;
- Space time block coding (STBC) diversity support (described in more detail below);
- Power Save Multipoll (PSMP) support, an enhanced PS mechanism;
- optional MIMO beamforming support (see below);
- support of optional dynamic antenna selection methods (see below).

The 'HT information' parameter (element ID 61) is the second new parameter contained in beacon frames. This parameter is used by the AP to inform clients as to which HT functionalities are currently used in the network and which must not be used, for example, to preserve backward compatibility. The parameter indicates the following:

- If 40-MHz transmissions may be used or if transmissions must be limited to the primary 20-MHz channel.

- The operation mode of the network: Greenfield, HT-mixed, non-member protection modes (to protect transmissions of clients that communicate with other APs that use the same channel).
- If there are devices in the network that do not support the Greenfield mode.
- Activation of overlapping BSS protection: If the AP detects beacon frames of other APs on the same channel that do not support HT extensions or operate in mixed mode, it can instruct clients with this bit to also activate mixed-mode support. Neighboring APs that detect this bit but do not detect non-HT-capable clients do not have to set the bit. This ensures that HT-capable networks can coexist with non-HT-capable networks on the same channel and limits the use of such measures to areas where it is necessary.
- If a secondary beacon is sent, the AP informs client devices if the beacon frame was sent on the primary 20-MHz channel of a 40-MHz channel or on the secondary 20-MHz channel.

The HT capability and HT information parameters are also included in association, reassociation and probe response frames. Client devices are additionally informed of all necessary parameters and the current configuration during initial communication and when reselecting to a different AP in the same network.

In addition to the HT parameters, 802.11n compatible APs also broadcast 802.11e QoS parameters in beacon frames as discussed in more detail in Section 5.8.

During the association procedure, each client device in return informs the AP of its HT capabilities. APs can adapt transmissions to individual client devices by using only the supported transmission options. An AP can therefore communicate over a 40-MHz channel with two MIMO streams and a short GI with one device while a frame for a device with fewer capabilities is sent over a 20-MHz channel with an 800 nanoseconds GI and without using MIMO.

Owing to the required backward compatibility for 802.11b, g, and a devices and the many optional extensions of 802.11n, a device can choose from many data transfer options before transmitting a frame. If a frame is sent to an 802.11b device, HR/DSSS modulation has to be used and an appropriate coding rate is selected depending on the current signal conditions. For 802.11g devices, OFDM modulation is used with fewer subchannels (non-HT format) compared to transmissions to 802.11n devices. Also, an 802.11g PLCP header has to be used. OFDM is also used for a transmission between two 802.11n devices. Compared to 802.11g, however, a shorter PLCP header is used which contains HT-specific information (HT Greenfield mode). If 802.11n and 802.11g devices are present in the network (HT-mixed mode) as shown in Figure 5.18, a backward compatible PLCP header is used. This header can also be decoded by 802.11g devices and includes a number of additional bytes. Also, fewer OFDM subchannels are used. If 802.11b devices are additionally present in the network, a CTS packet has to be sent preceding the data frame using HR/DSSS modulation. In addition, an 802.11n-compatible device has to be aware of the 802.11n functionalities supported by the receiver. This is required to control the OFDM modulation (e.g., using a short GI) and to be able to choose between a 20-MHz and a 40-MHz channel. Furthermore, the number of MIMO channels used and the coding rate depend on the capabilities of the receiver.

Even this already quite extensive list does not consider additional optional 802.11n functionalities that are described in more detail below.

For battery-driven devices, it is important that the Wi-Fi chip uses as little energy as possible while no data is being transferred. In general, the PS mode described earlier is used for this purpose. Some applications that periodically transmit data, such as VoIP, might prevent the use

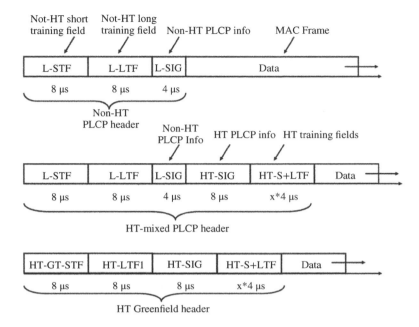

Figure 5.18   Different PLCP header variants

of this PS scheme, however, as the Wi-Fi chip has to monitor the channel for incoming data and therefore energy is wasted. For such situations, the optional 802.11n PSMP enhancement has been specified. With PSMP, a client can negotiate a transmission and reception pattern with the AP. If granted, the AP establishes a PSMP window and informs the client as to times at which data can be sent and received. The client then only activates its transceiver during the agreed window to receive data packets. Once the downlink window expires, an uplink opportunity window can be implicitly used without prior reservation of the channel. During all other times, the client's receiver can be fully deactivated to save power.

In PSMP mode, frames in both directions do not contain only user data but also ACK information for the received data frames. During a PSMP window, a device can transmit and receive several frames. If these are sent individually, an SIFS gap has to be inserted between the frames or, optionally, a shorter gap that is referred to as RIFS (Reduced Interframe Space) is attached. Furthermore, data frames can also be aggregated by using the frame aggregation extension described above to aggregate several frames into a single physical frame.

Figure 5.19 shows how a PSMP window can be used by several devices at the same time. For this purpose, a PSMP frame is sent at the beginning of each interval that contains information on which the device can transmit and receive data at which times during the PSMP window. According to the standard, a PSMP window should be inserted every 5–40 milliseconds, with a granularity of 5 milliseconds. For example, for VoIP applications, a good interval is 20 milliseconds, as speech codecs usually compress speech data over such a period and then transmit the result in a short data frame.

The PSMP windows and the transmission times for each client device are optimized for periodical transmissions with a constant bandwidth requirement. This also optimizes the use

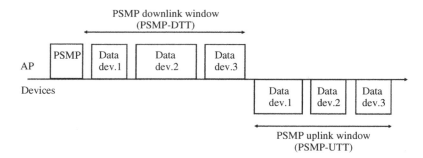

Figure 5.19   A Power Save Multipoll (PSMP) window in which several clients transmit and receive data

of the scarce air interface resources. In practice, however, a device might sometimes require additional bandwidth, for example, because of transmission errors or because of an additional frame that has to be sent. Such transmissions cannot occur during the normal transmission window as there is no space for unexpected transmissions. In such cases, the client device can request for additional uplink capacity via a flag in the MAC header of a frame. This flag is similar to the HSUPA 'happy' bit (Chapter 3). The AP can then schedule an additional PSMP window and announce this additional transmission opportunity in the PSMP management frame that precedes the next PSMP window. In case of an uplink transmission error, the AP returns a negative ACK to a client during the next PSMP downlink window and also inserts an additional PSMP window.

An additional PSMP functionality is the transmission of frames without requiring an ACK from the receiver. This is an interesting option for applications such as VoIP clients, as voice transmissions are delay sensitive and it might thus be preferable to ignore a faulty packet rather than to request a retransmission.

To make the use of MIMO as power efficient as possible, an additional PS functionality has been introduced with 802.11n for MIMO-capable devices. Even if no data is transferred, such devices must keep both receivers activated, as the AP can transmit a new frame at any time. To reduce power consumption for battery-driven devices, two optional MIMO SM PS modes have been specified. In static mode, a device transmits an 'SM Power Save Management Action Frame' message to the AP once it activates or deactivates the SM power save mode. Furthermore, a bit in the 'HT capabilities' parameter can be used by a device to indicate to an AP during the association procedure that it only wants to use a single stream. And finally, a dynamic SM power save mode is specified. While in this mode, the device deactivates all but one receiver and operates in single stream mode. Additional receivers are automatically activated once the AP transmits a frame to the device in single stream mode, for example, a RTS/CTS sequence. Further frames are then automatically sent over several MIMO streams without further signaling.

While MIMO spatial multiplexing increases the datarate, the range over which frames can be properly received remains the same as in single stream operations. The standard includes a number of methods to use the additional transmitter and receiver units (transceivers) to increase the range and the throughput of the network at certain distances and under less ideal transmission conditions.

One of these methods is MIMO beamforming, where the same data stream is sent over all transceiver chains. By intelligently combining transmission power and delayed transmissions, the signal energy can be directed in a desired direction. This way, the energy is not evenly distributed in all directions and a beamforming effect results. To direct the signal in the proper direction, feedback from the receiver is required. As a consequence, both the transmitting and the receiving devices need to support MIMO beamforming for the feature to be used.

Another method to increase the transmission range is STBC. If supported by the transmitter and receiver, a single data stream is sent over two separate 2×2 MIMO paths. With STBC, however, each data stream is coded differently and in such a way that the signals are orthogonal to each other. On the receiver side, this increases the signal-to-noise ratio, which helps to increase data transmission rates or the range or a combination of both as desired.

In case the transmitter does not support MIMO, a receiver has additional options to increase the signal quality. If the receiving device has several antennas, it can analyze which of its antennas experiences better reception conditions and then use the corresponding receiver chain for receiving the packet. In practice, even such a relatively simple method can yield significant improvements. If a device has only a single antenna and is stationary, it is sometimes helpful to move the antenna just a few centimeters to increase transmission speeds. Automatic antenna selection is not specific to 802.11n but was already used by 802.11g APs. A somewhat more complicated multiantenna scheme for single stream transmission is Maximum Ratio Combining (MRC). Here, the receiver analyzes the data stream it receives via each of its antennas and combines the signal energy of all receivers.

Which type of MIMO is used at a certain time depends on many factors. For devices that are close to the AP, MIMO spatial multiplexing is ideal to increase transmission speeds. The transceivers are then used to transmit several data streams in parallel. Under less ideal signal conditions, beamforming and STBC coding are a better choice to increase transmission speeds when these methods are supported by both the transmitter and the receiver. The achievable datarates are lower compared to MIMO spatial multiplexing as only a single data stream is used. Which of the MIMO schemes is used is a choice of the transmitting side and should be based on its knowledge of current signal conditions and the methods that are supported by the receiver. If an AP serves several client devices, the different methods can be used in parallel. The AP can then, for example, use MIMO spatial multiplexing with one client while a frame to another client is sent with STBC.

Another optional functionality that was specified in the 802.11n working group is MCS feedback. Without this enhancement, the receiver must either analyze the signal strength with which the last frame of a device was received or the MCS in that frame to decide which MCS it should use when it wants to transmit a packet of its own to the other device. Another method is to reduce the MCS when the number of transmission errors increases. In practice, this is not ideal and data transmissions are slower than necessary. With the MCS feedback functionality, a transmitter can now request details about the reception conditions from the receiver. When requested, a device returns its feedback implicitly in MAC headers of subsequent frames.

Owing to the many optional functionalities specified in 802.11n, it is difficult to judge a device's capabilities from the 802.11n product label on the box. In practice, APs support the 2.4-GHz and 5-GHz frequency bands today and a maximum datarate of 300 or 450 Mbit/s with two or three antennas, that is, two or three MIMO streams. On the devices side, this is not always the case as many smartphones and also entry-level notebooks still support only the 2.4-GHz band.

5.6.5   802.11ac Gigabit Wireless

In practice, new products usually support the basic 802.11n functionalities, such as MIMO, 40-MHz channels and the 5-GHz band. An exception is that there are still some smartphones that support only the 2.4-GHz band and some are limited to 20-MHz channels. Although the 802.11n standard contains many options beyond those used today to increase data transmission speeds, their technical implementation is difficult and they have remained unimplemented. As a consequence, standardization has moved on to specify features for improvements in speed that are easier to implement. The 802.11ac specification is the direct successor of 802.11n, and the first devices to implement the Very High Throughput (VHT) PHY appeared on the market in 2013. As with the previous versions of the standard, the first products implemented only a small subset of features to increase data transfer speeds beyond what is possible with 802.11n. In the years to come, it is expected that chipsets will also include the more advanced features 802.11ac has to offer.

The simplest way to further increase datarates is to extend the maximum channel bandwidth of 40 MHz specified in 802.11n to 80 MHz. This is the main feature implemented in early devices. The specification also includes the definition of 160-MHz channels, which will be supported by a second wave of 802.11ac products. A further option allows the use of two 80-MHz channels in different parts of the band that enable the use of both before and after sections of the spectrum that are reserved for other purposes. This is useful in some countries that reserve some parts of the 5-GHz band for applications such as weather radar. An overview of which parts of the 5-GHz band may be used in which parts of the world is given in [11]

In most parts of the world, the spectrum available in the 5-GHz band is around 400 MHz, which allows four to five non-overlapping 80-MHz networks. As with the previous PHYs, it is also possible to operate many networks on the same channel which then share the available bandwidth. When several networks are used on the same channel, devices of one network can sense the transmission of devices of the other network that are within range and can thus apply collision avoidance mechanisms as described at the beginning of this chapter. In practice, however, several networks on the same channel are usually not fully overlapping in a geographical sense. As a consequence, some devices in one network may not be able to sense transmissions of devices communicating with another AP and hence, the collision avoidance mechanism is not as effective as is the case when only one network is present on a single channel.

To improve the interworking between networks on the same channel, a number of features have been introduced with 802.11ac. As discussed before, a device senses the use of the air interface before it starts to transmit its own frame. If an ongoing transmission is detected, the device's own transmission is deferred. This is known as collision avoidance. In practice, this is achieved by two methods, namely, signal sensing and energy sensing. Signal sensing means that a device properly receives the beginning of a frame and is thus aware by decoding the header information of how long the channel will be busy. Energy sensing on the other hand blocks the use of the channel if a certain signal level is detected on the channel.

For backward compatibility reasons, 802.11ac divides the full channel bandwidth into 20-MHz chunks. An 80-MHz channel thus has four chunks. When the RTS/CTS scheme is used to reserve the channel, individual packets are sent simultaneously on all 20-MHz chunks of the channel, so 11ac networks on the same channel as 11n networks that support only 40-MHz channels and 11a networks that support only 20-MHz channels can properly detect the channel reservation (Clear Channel Assessment, CCA). This can also be used to reduce the channel bandwidth for a transmission if the CTS is, for example, received only on two 20-MHz channels instead of four due to interference from overlapping networks. This decision can be taken on a per-frame basis, which makes the system very flexible.

In practice, not all devices in a network may support 80 MHz transmissions. Therefore, a method has been specified to coordinate the use of several independent networks on the same channel to allow simultaneous transmissions in two networks. This is done by splitting the channel bandwidth into a primary channel and a non-primary channel. The channel bandwidth is half of the full channel bandwidth, for example, 40 MHz in an 80-MHz network. Two fully overlapping 80-MHz networks thus configure themselves in such a way that each network uses a different primary channel. If two 40-MHz devices belonging to two different 80-MHz networks want to transmit data simultaneously, they would each use the primary channel of their respective network that does not overlap with the primary channel of the other network. Each of the devices senses that the channel is free and no collision will occur. If a transmission on a non-primary channel is already ongoing, a device can either wait for another transmission opportunity or only transmit on the primary channel. Such dynamic bandwidth usage can significantly improve the overall spectrum usage in networks with devices that are unable to use the full channel bandwidth or in situations in which a 11ac network is used on the same spectrum as a network using a legacy PHY with a 20 or a 40-MHz channel.

It should be noted at this point that the 2.4-GHz band is too small for 80 or 160-MHz channels. As a consequence, 802.11ac is specified only for the 5-GHz band. In practice, 802.11ac-compatible chips obviously also support 802.11b, g and n in the 2.4-GHz band and 802.11a, n and ac in the 5-GHz band.

In addition to larger bands, a new modulation scheme has been defined for situations with exceptionally good channel conditions. Although 802.11n supports the transmission of up to 6 bits per transmission step with 64-QAM modulation, 802.11ac now allows 8 bits per transmission step, that is, 256-QAM modulation. Together with a code rate of 5/6, which is the number of overall bits to the number of user data bits plus error correction bits, it requires a 5 dB better receiver performance compared to that of 64-QAM [12]. This requires less noise and a stronger received signal which can partly be achieved with an improved receiver performance as chips get more sensitive and sophisticated over time. Adding 2 bits per transmission step increases the data transmission speed by about 30% compared to a 64-QAM transmission. Table 5.6 gives an overview of the MCS that have been specified for 802.11ac. MCS0 is the combination of very conservative modulation (BPSK) that transmits only 1 bit per step and a coding of 1/2, that is, the same number of data and error correction bits.

802.11ac has also retained the use of a short guard interval of 400 nanoseconds between OFDM symbols that was introduced in 802.11n. In practice, it can be observed that this feature is widely used in such networks which results in a performance increase of about 10% when compared to 802.11a and g.

Another way to increase theoretical maximum data transmission speed is to increase the number of MIMO streams. In 802.11ac up to 8×8 MIMO is supported, which goes significantly beyond the 4×4 MIMO mode of 802.11n. In practice, however, it is already difficult to realize the benefits of 4×4 MIMO, and the combination of 256-QAM and several MIMO streams requires an even more robust channel to a single subscriber than what is needed for 802.11n today. In practice, even 3×3 MIMO with 64-QAM, typically used in 802.11n networks today, provides only little gain over a 2×2 MIMO or even a single stream transmission in most practical environments. Additional MIMO paths could, however, become beneficial

Table 5.6 Modulation and coding schemes in
802.11ac

MCS	Modulation	Code rate
0	BPSK	1/2
1	QPSK	1/2
2	QPSK	3/4
3	16-QAM	1/2
4	16-QAM	3/4
5	64-QAM	2/3
6	64-QAM	3/4
7	64-QAM	5/6
8	256-QAM	3/4
9	256-QAM	5/6

when used in combination with beamforming to address several devices simultaneously as described in the next paragraph.

Beamforming is another option in the 802.11ac standard to concentrate signal energy in the direction of a client device. This requires that the AP becomes aware of the direction in which to concentrate the signal energy. For this purpose, the AP transmits channel sounding announcements, the so-called Null Data Packet (NDP) announcements. The name is derived from the channel sounding method, which is based on transmitting an empty frame whose OFDM symbols are analyzed by the mobile device for changes they have undergone during their transmission over the air. For this purpose, the APs sends an NDP announcement message to request beamforming-capable devices to respond. In a second step, NDP packets are sent by the AP and received by the client devices. These devices then analyze the OFDM symbols of the packets and calculate a response that describes how the OFDM symbols have been altered during transmission and return the result in a response packet. Based on this feedback, the AP can then calculate a steering matrix for each individual client device, which is then applied to data transmissions. The steering matrix describes how to distribute the signal energy across the available transmission chains and antennas and how to apply phase shifts to each chain. This way, the interference of the different wavefronts increases the signal in one direction while decreasing the overall signal in another direction. To account for changing signal conditions, this sounding procedure has to be repeated in the order of once per 100 milliseconds [12]. As the signal level increases in a desired direction, regulation requires that the overall power output of the AP is reduced by 3 dB during such a transmission to ensure that the overall transmission power limit, which is defined for an omnidirectional antenna, is not surpassed by the amplified directional signal. It remains to be seen by how much this transmit power restriction reduces the effect of beamforming.

Beamforming can be combined with MIMO transmission to direct several data streams to a single device and is referred to as Single-User (SU) beamforming. An even more sophisticated application of beamforming is to transmit one or more streams to several client devices simultaneously. This is referred to as Multi-User (MU) beamforming. Up to four devices can be serviced simultaneously, which could be especially useful when APs are used that support more MIMO streams than individual client devices. This could be the case, for example, if the AP supports four MIMO streams, whereas, especially, battery-powered devices support only one stream to reduce the computational overhead of MIMO reception to conserve battery power. By transmitting independent data streams to several devices, better use can be made of the transmission channel and thus a higher overall throughput can potentially be reached. The preamble of such a multi-user frame contains information to which client devices it is addressed to. Afterward, the multi-user frame stacks the individual data streams on top of each other and uses beamforming to direct the signal energy for each individual data stream in the desired direction. This way, the data for one device is not seen as noise from the point of view of another device, given that the two devices are well separated in space to make the streams independent of each other. As the amount of data as well as the modulation and coding for each device might be different, some parts of the multi-user frame are unused and padded. One question that arises is how the recipients of a multi-user frame can acknowledge reception as normal acknowledgement frames directly follow the transmission of a data packet. This is not possible for multi-user frames as only one acknowledgment frame can be sent at a time. Therefore, the acknowledgment frames have to be separated in time. This is achieved with the deferred block acknowledgement mechanism that is already known from 802.11n. This method was initially introduced to allow the acknowledgement of a transfer of several blocks after a certain time has elapsed to give the device some time to check if all transmissions were received successfully. In multi-user beamforming, only a single multi-user frame is sent so the delayed block ACK mechanism is used for a different purpose, that is, to separate responses from different devices in the time domain. All enhancements taken together increase the theoretical peak datarate to 6.93 Gbit/s. This would require a combination of a 160-MHz channel, eight MIMO streams, 256-QAM modulation, a minimal number of data streams and a short guard interval between packets. In practice, this is obviously difficult to realize. At the time of publication, pre-standard 802.11ac APs would support up to three MIMO streams over an 80-MHz channel, which results in a theoretical top speed of 1.3 Gbit/s. In practice, achievable speeds on the IP layer are far lower. Table 5.7 gives an overview of the typical speeds that can be reached in practice as reported in [13]. Although achievable datarates are significantly higher when compared to that of 802.11ac, they are currently nowhere near the theoretical maximum values.

It should be noted at this point that not all 802.11ac APs and client devices support the complete 5 GHz range. Many models support only the lowest 80 MHz part of the channel (channel

Table 5.7 Achievable 802.11ac datarates in practice

Date rate	Network setup
700 Mbit/s	3×3 MIMO, high end access point and PCI express Wi-Fi card in a PC with three external antennas, same room, very close range
300–380 Mbit/s	3×3 MIMO, standard access point and PCI express Wi-Fi card in a PC with three external antennas, close range
200 Mbit/s	2×2 MIMO, USB Wi-Fi stick with 2 antennas, close range
160–190 Mbit/s	3×3 MIMO, high end access point and PCI express Wi-Fi card in a PC with three external antennas, 20 m distance between client and access point with walls in between
100 Mbit/s	2×2 MIMO, USB Wi-Fi stick with two antennas, 20 m distance between client and access point with walls in between

numbers 36–48) as they do not support dynamic frequency selection (DFS). DFS is required in some countries, however, to automatically detect other users in the band (e.g., weather radar) and to automatically change to a different part of the channel.

5.7   Wireless LAN Security

WLAN security is a widely discussed topic as using a wireless network without encryption exposes users to many security risks.

In some cases, APs are still sold with encryption deactivated by default. If encryption is not configured by the owner of the network, any wireless device can access the network without prior authorization. This configuration is used in most public hotspots as it allows users to easily find and use the network. As the frames are not encrypted, however, it is easy to eavesdrop on their activities. Without protection on the network layer, it is left to the users to use virtual private network (VPN) tunnels and take other measures to protect themselves.

The use of such an open configuration for private home networks that use the wireless network to provide access to the Internet is even more critical. If encryption is not activated, neighbors can use the Internet connection without the knowledge of the owner of the Internet connection. Furthermore, it is possible to spy on the transmitted frames, for example to collect passwords, in the same way as it is possible in public Wi-Fi networks. As open APs also allow an eavesdropper to gain access to any PC that is connected to the wireless network, it potentially allows him to exploit weaknesses of operating systems which could then enable him to read, modify or destroy information.

5.7.1   Wired Equivalent Privacy (WEP)

To protect WLANs from unauthorized use and eavesdropping, WEP encryption is part of the 802.11b, g and a standards. Similar to GSM and UMTS, this encryption method is based on a stream-ciphering algorithm that encrypts a data stream with a ciphering sequence. The ciphering sequence is calculated for each frame by using a key and an initial vector (IV) (Figure 5.20). The IV changes for every frame to prevent easy reconstruction of the secret key by an attacker. In contrast to GSM or UMTS, however, WEP uses the same key for all users. While a single key is easy to manage, it creates a big problem especially if a WLAN is used by a company. As the same key has to be manually configured by all users in their devices, it is not possible to keep the key secret. In GSM or UMTS, the individual private key of each user is securely stored on the SIM card.

An even more serious problem is the fact that the first bytes of an encrypted frame always contain the same information for the LLC header. In combination with certain IVs, which are transmitted as clear text it is possible for an attacker to calculate the key. About 5–6 million frames are necessary to calculate the key with this approach. The length of the ciphering key plays only a secondary role. Tools that automate this process are freely available on the Internet. The attacker, therefore, simply has to bring his eavesdropping device into the range of a WLAN and wait until enough frames have been collected. The number of required frames sounds very high at first. However, the contrary is the case. If we assume for a moment that each of the 5 million frames contains about 300 bytes of user data, the key can be calculated by collecting 5,000,000 frames * 0.3 kB = 1.5 GB of information. Depending on the network load,

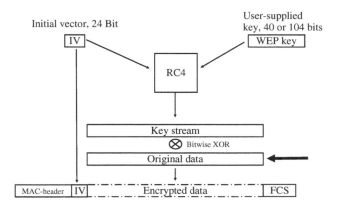

Figure 5.20   WEP encryption

the key can thus be generated in a time frame ranging from several weeks to only a few hours. It can be assumed that the larger time frame usually applies for home networks because of their low traffic rates. In company networks, however, many devices usually communicate with a server and thus create a high traffic volume. Therefore, the WEP encryption will only protect the network for a short time. Companies should therefore take additional security measures or use other encryption methods to secure their networks, as described in the following paragraph.

To increase the security of a WLAN network, many APs offer a number of additional security features. By activating the 'Hide SSID' option, an AP leaves the SSID fields of beacon frames empty. By doing this, the AP is only visible to users who know the SSID of the network and are able to manually configure the SSID in their device. The MAC address filter, another security feature, prevents devices from connecting to the network which have not previously been authorized by the administrator. These features, however, do not prevent an ambitious hacker with the tools described above to collect and automatically analyze frames to generate the ciphering key. The hacker can also easily retrieve the SSID that is no longer broadcast in beacon frames during the association procedure when the client devices send it. Statically defined MAC hardware addresses in the AP are also an easy measure to circumvent for a hacker, as a network trace reveals the MAC addresses of devices that are allowed to communicate with the network. This information can then be used to manually change the MAC address on the hacker's device to match the MAC address of a device that was previously used in the network.

5.7.2   WPA and WPA2 Personal Mode Authentication

Owing to the security problems presented above, the IEEE 802.11i working group created the 802.1x standard, which offers a solution to all security problems that have been found up to this point. As the ratification of the 802.11i was considerably delayed, the industry went forward on its own and created the Wireless Protected Access (WPA) standard. WPA contains all the important features of 802.11i and has been specified in such a way as to allow vendors to implement WPA on hardware that was originally designed for WEP encryption only.

The security issues of WEP are solved by WPA with an improved authentication scheme during connection establishment and a new encryption algorithm. As has been shown in

Figure 5.8, a client device performs a pseudo-authentication and an association procedure during the first contact with the network. With WPA, this is followed by another authentication procedure and a secure exchange of ciphering keys. The first authentication is therefore no longer necessary but has been kept for backward compatibility reasons. To inform client devices that a network requires WPA instead of WEP authentication and encryption, an additional parameter is included in beacon frames. This parameter also contains additional information about the algorithms to be used for the process. Early WPA devices only implemented Temporal Key Integrity Protocol (TKIP) for encryption. Current devices also support Advanced Encryption Standard (AES), which has become mandatory with the introduction of WPA2. Further details are discussed below.

Figure 5.21 shows the four additional steps that have been introduced by the WPA Preshared Key (PSK) authentication method with which client devices can authenticate themselves to the network and vice versa. The method is referred to as PSK authentication, as the same key is stored in the client devices and in the AP. During the process, the client device and the AP derive a common key pair for the ciphering of user data, each which is referred to as the session key.

In the first message, the AP sends a random number to the client device. On the client side, the random number is used in combination with the secret password (PSK) to generate a response. The password can have a length of 8–64 characters. The result is then sent to the AP in a response message together with another random number. In the next step, the AP compares the response with the expected value that it has calculated itself. These can only be identical if both sides have used the same password for the process. The client device is authenticated if the values match. The AP then creates a session key which it encrypts with the common password and sends it to the client device. The client device deciphers the session key with the common password and returns a confirmation to the AP that the message was received

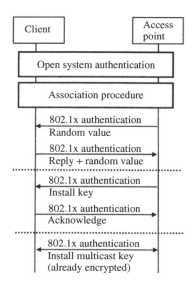

Figure 5.21   WPA-PSK authentication and ciphering key exchange

correctly. This message also implicitly activates ciphering in both directions. In the final step, the AP then informs the client device about the current key to decipher broadcast frames. While there is an individual session key for each client device, the key for broadcast messages is the same for all devices, as broadcast frames must be deciphered by all devices simultaneously.

The advantage of the use of individual session keys compared to the use of a password as an input to the encryption and decryption algorithms is that the session key can be changed during an ongoing connection. This prevents brute force attacks that try to obtain the key by trying out all different combinations or by analyzing a large amount of data collected over time. A typical value for the update of the session key is 1 hour.

5.7.3 WPA and WPA2 Enterprise Mode Authentication

In addition to the WPA-PSK authentication that uses a common key (Preshared Key) in the AP and all client devices, there is also an enterprise mode with an individual key for each device. The keys can be stored on a central authentication server as shown in Figure 5.22. This allows companies to have several APs to cover a larger area without the need to store the keys in each AP. In addition, individual passwords significantly increase overall security as network access can be granted and removed on a per user basis. The most popular protocols to communicate with an external authentication server are RADIUS (Remote Authentication Dial In User Service) and the Microsoft Authentication Service.

For WPA, a number of different authentication protocols have been specified to be compatible with as many external authentication servers as possible. These protocols are referred to as Extensible Authentication Protocols (EAPs). A popular authentication protocol is the Extensible Authentication Protocol Transport Layer Security (EAP-TLS) protocol that is described in RFC 5216. The protocol uses certificates that are stored on the client device and on the authentication server. An important part of the certificate system includes the public keys of the client device and the authentication server. These are used to generate the session keys that are exchanged between the client device and the network and are then used to encrypt the data traffic over the air interface.

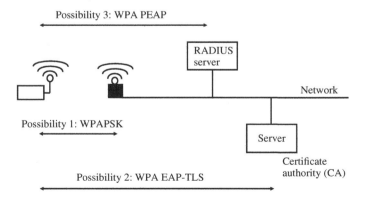

Figure 5.22 Three different options for WPA authentication

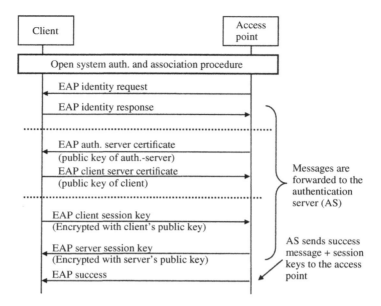

Figure 5.23   EAP-TLS authentication

After the session key has been encrypted by the sender with the public key of the receiver, it can be securely sent over the air interface and can only be decrypted on the receiver side with the private key, as shown in Figure 5.23. As the private keys are never exchanged between the two parties, it is not possible to obtain the session key by intercepting the message exchange during the authentication process. A disadvantage of certificates, however, is that the certificates have to be installed on the client device. This is more complicated compared to just assigning passwords, but much more secure. Not shown in Figure 5.23 is the exchange of session keys for broadcast frames, which is performed right after a successful authentication.

During the authentication phase, the AP only permits the exchange of data with the authentication server. Only after the authentication has been performed successfully and after the authentication server has informed the AP about the proper authentication will the AP grant full access to the network. At this point, the user data frames are already encrypted. Usually, the first user data packet is a DHCP request to receive an IP address from the network.

The EAP-TLS authentication procedure is very similar to TLS and Secure Socket Layer (SSL). These protocols are used by Secure Hypertext Transfer Protocol (HTTPS) for the authentication and the generation of session keys for secure connections between a web server and a web browser. The main difference between the EAP-TLS and HTTP TLS authentication procedures is that EAP-TLS performs a mutual authentication while HTTPS TLS is usually used only to authenticate the web server to the web browser. This is the reason why no certificate has to be installed in the web browser to establish an encrypted connection to a web server.

A potential alternative to EAP-TLS is Protected Extensible Authentication Protocol (PEAP). While this protocol is not used very often, it nevertheless has the advantage that a password is used on client devices instead of a certificate. This significantly simplifies the configuration of the client device.

5.7.4 EAP-SIM Authentication

Today, a growing number of GSM and UMTS devices are also equipped with a Wi-Fi module. This module can then be used to connect to the Internet at home, in the office or via public Wi-Fi hotspots. Mobile network operators that offer hotspot services are faced with the question of how they can authenticate their customers over Wi-Fi. A number of proprietary solutions are available on the market but all of them require some sort of interaction with the user. To simplify the process, the EAP-SIM protocol was specified in RFC 5216. Here, the authentication is performed with data contained on the SIM card and no user interaction is required.

EAP-SIM uses the same authentication method as was described in the Sections 5.7.2 and 5.7.3. Figure 5.24 shows the messages that are exchanged during the authentication process between a mobile device and the authentication server over an EAP-SIM compatible AP. After an open system authentication and an association procedure, the network initiates the EAP procedure by sending an EAP Identity Request message which the mobile device answers with an EAP Identity Response message. The identity that is returned in this message consists of the Identity Type Identifier, the IMSI that is read from the SIM card and a specific postfix of the mobile network operator.

Alternatively, the mobile device can also send a temporary identity to the network that has been assigned to it during a previous authentication procedure. This temporary identity is similar to the Temporary Mobile Subscriber Identity (TMSI) used in GSM and UMTS and hides the user's identity from potential eavesdroppers on the air interface.

In the next step, the network sends an EAP-SIM Start Request message. This message contains information on the supported EAP-SIM authentication algorithms. The mobile device selects one of them and answers with an EAP-SIM Start Response message. This message contains a random number, which is used later in the network together with the secret key Kc

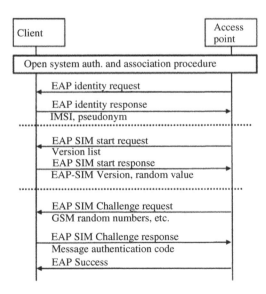

Figure 5.24 EAP-SIM authentication

for a number of calculations. As the secret GSM key Kc is stored in the network and on the SIM card, it is possible to use it as a basis to authenticate the device toward the network and vice versa.

At this point, the authentication server uses the subscriber's IMSI to contact the Home Location Register (HLR)/Authentication Center (AuC), as described in Chapter 1, to request a number of authentication triplets. The HLR/AuC responds with two or three triplets, which each contains a random number and a ciphering key Kc. These are used to generate the EAP-SIM session key and other parameters for the authentication process. These parameters are then encrypted and sent to the mobile device in a SIM Challenge Request message in addition to the two or three GSM random numbers, which are sent as clear text.

When the mobile device receives the GSM random numbers, it forwards them to the SIM card. The SIM card then uses them to generate the GSM signed responses (SRES) and the GSM ciphering keys (Kc), which are subsequently used to decipher the EAP-SIM parameters that have been received previously. If the encrypted SRES from the network is identical to the response received from the SIM card, the network is authenticated and a response can be returned. On the network side, the response message is in turn verified and if all values match, an EAP success message is returned to the mobile device. Subsequently, the mobile device is granted access to the network.

Figure 5.25 shows which protocols are used during the EAP-SIM authentication process. The mobile device is shown on the left and it sends its messages using the Extensible Authentication Protocol over Local Area Network (EAPOL) protocol. RADIUS is used for the communication between the AP and the authentication server. And finally, the authentication server uses the SS-7 signaling network and the Mobile Application Part (MAP) protocol to communicate with the HLR/AuC.

5.7.5   WPA and WPA2 Encryption

WPA introduces the TKIP to replace the weak WEP algorithms. With WEP, a 24-bit IV, the WEP key and the RC-4 algorithm were used to generate a ciphering sequence for each frame (cp. Figure 5.16). To improve security, TKIP uses a 48-bit IV, a master key and the RC-4 algorithm to create the ciphering sequence. This method is much more secure because of the longer IV and the periodic refresh of the master key, for example, once every hour.

The ciphering used with WPA does not fully meet the requirements of the 802.11i standard, but is nevertheless seen as sufficiently secure. The advantage of using RC-4, however, is that TKIP is compatible with the hardware that was originally designed for WEP.

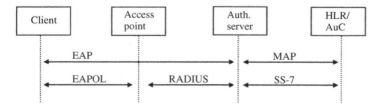

Figure 5.25   Protocols used in the EAP-SIM authentication process

To prevent attacks that exploit a weakness when previously received packets are replayed with a slight modification, the IV is increased by 1 in each frame. WPA-compatible devices ignore frames that use IVs that have already been used and are hence immune against such attacks.

For additional security, TKIP introduces a Message Integrity Code (MIC) that is included in each frame. The process of creation of MIC is referred to as 'Michael'. The difference to the cyclic redundancy check (CRC) checksum, which continues to be part of each frame, is as follows:

The CRC checksum is generated from the content of the frame with a public algorithm. The receiver can thus check if the frame has been altered, for example, by a transmission error. As the input parameters and the algorithms are known, an attacker could also modify the CRC. The CRC thus offers protection against transmission errors but not against modifications made by an attacker. The MIC calculation also uses a public algorithm and the content of the frame. Furthermore, a message integrity key is used as an input parameter, which was generated during the authentication process. This prevents a potential attacker from calculating the MIC and hence, the packet cannot be modified and replayed. To change the CRC checksum and the MIC, an attacker would have to overcome the RC-4 ciphering in combination with this additional WPA security measure.

If an error occurs during the transmission of the frame, both the MIC and the CRC will be invalid. The receiver can therefore distinguish between transmission errors and an attack on data integrity. The WPA standard requires that devices be disconnected from the network if they receive more than one frame per minute with a correct CRC but an invalid MIC. Subsequently, they have to wait for 1 minute before they reconnect. This effectively prevents attacks on user data integrity.

After the 802.11i standard has been finalized, the Wi-Fi alliance adapted WPA accordingly and the WPA2 specification now implements security as per the 802.11i standard while remaining backward compatible with WPA. This means that a WPA2-certified AP also supports 'WPA-only' devices.

In addition to the TKIP algorithm, which was introduced with WPA, WPA2 also supports the highly secure AES ciphering algorithm. As with WPA, there are two WPA2 flavors. If a device has been 'personal mode' certified, it supports authentication with an AP with the PSK procedure. For companies that often use more than one AP and who want to assign individual passwords, an AP should also support 'WPA2 enterprise mode'. In addition to PSK, such APs also support the 802.1x authentication framework and can therefore communicate with external authentication servers as described above.

5.7.6 Wi-Fi-Protected Setup (WPS)

The configuration of a device to join a wireless network that is protected by WPA/WPA2 is usually straightforward by typing in the password that has been configured in the AP. The Wi-Fi Alliance wanted to further simplify the process and created several methods that are referred to as Wi-Fi-Protected Setup (WPS). All APs and client devices have to implement WPS today to qualify for the Wi-Fi compatibility logos on devices and sales packaging. WPS is not a new encryption method but was designed to be a simple method to transfer the WPA/WPA2 key from the AP to a client device during the initial configuration of the client device so that the user does not have to type in a long password. WPS includes a number of methods of which today's APs usually support the Pushbutton method and the PIN method. The PIN method can usually be activated and deactivated in the AP and works as follows:

Step 1: A Diffie–Hellman key exchange procedure is performed to establish an encrypted channel for the information exchange that follows. This ensures that all data that is exchanged remains confidential and an eavesdropping attacker is unable to decode any of the values exchanged either during the authentication procedure that follows or later on in an offline attack. It is important to note that this key exchange is not for authentication purposes but just for establishing an encrypted channel over which sensitive data can be exchanged. Only once a bidirectional encrypted channel is established is authentication information exchanged. This approach can be compared to a password-protected website that uses the secure http (HTTPS) transfer protocol. HTTPS is used to provide an encrypted channel that cannot be decrypted by a third party, and the username and password provided by the user to a web page that was received over the encrypted channel serves as authentication.

Step 2: The AP and the client device generate a random number that is referred to as a 'Nonce'. Together with an eight-digit PIN they are used as the input to a hash function. The hash function generates a 256-bit result from the two values from which neither the PIN nor the random number can be deducted as the hash function is not reversible.

Step 3: The AP and the client device exchange their hash function results.

Step 4: Once each side has received the hash result of the other side, the nonce values are exchanged.

Step 5: Both devices now use the nonce value of the other device, add the PIN and execute the hash function over these parameters. If the result matches the hash values that have been transferred in step 1, both sides can be sure that the same PIN was used on both sides.

Step 6: After both sides have verified that the PIN was identical on both sides, the WPA/WPA2 password is transmitted. The string transmitted is the password the user would have typed in if WPS was not used.

Step 7: Once the client device has received the WPA/WPA2 password, a standard WPA/WPA2 connection establishment is performed.

The only weakness of the initially designed procedure is that it cannot protect against an active attack, that is, a man-in-the-middle attack in which an attacker is able to intercept frames from both devices, modify them and forward them to the destination. In practice, however, a number of weaknesses were unfortunately introduced during implementation that makes some devices very vulnerable to brute force attacks. If WPS is always active and the PIN always remains the same, it is possible to retrieve the WPA/WPA2 key with a brute force attack by trying out all possible PIN combinations. Such an attack is typically successful within only a few hours despite the 8-digit length of the PIN. This is because the PIN is validated in two parts of four digits. That means that an attacker only needs to perform 10.000 WPS attempts at most to get the first four digits. The second step can be performed even faster as one of the remaining four digits is used as a checksum to ensure the user has entered the PIN correctly. It is therefore deterministic. Some APs try to slow down attacks by only accepting a few WPS attempts per minute. This certainly slows down attacks but often not by a large degree. The only way to prevent such an attack is to use a PIN only once as was perhaps initially intended by those specifying the WPS authentication exchange. From a usability point of view, this is not very convenient as the PIN cannot be printed on the backside of a device. As a consequence, only few AP vendors have implemented a changing PIN. Therefore, some security experts recommend to disable WPS in an AP.

5.8 IEEE 802.11e and WMM – Quality of Service

Within a few years, Wi-Fi has revolutionized networking in offices and homes. Originally, these networks were mainly used for applications such as web browsing and access to files on a local server. Here, high bandwidths are required to transfer data quickly. Other aspects such as a guaranteed bandwidth and jitter were less important.

Today, applications such as VoIP and video streaming have additional requirements. Video streaming, for example, requires, in addition to a high bandwidth, a guaranteed bandwidth and maximum latency to ensure a smooth user experience. VoIP applications have similar requirements. While there is sufficient capacity on the network for all applications using it, such applications will function properly even without additional measures being taken. If, however, a multimedia transmission already requires a significant amount of the available bandwidth while other applications, potentially on other devices, start a file transfer or other bandwidth-intensive operations, it is likely that this transmission will interfere with the multimedia streaming. To prevent such issues, QoS measures were added with IEEE 802.11e. As with other extensions of the standards, there are some parts which must be supported by all devices while the support of others is optional.

To speed up the introduction of the QoS extensions, the Wi-Fi alliance has created the Wi-Fi Multimedia (WMM) specification, which is based on 802.11e. If an AP or mobile device is WMM certified, it contains all features that are declared as mandatory in the WMM specification and will be able to communicate with WMM devices of other vendors. To ensure that the QoS extensions are implemented in as many devices as possible, the Wi-Fi alliance requires that 802.11n devices also implement the WMM extensions. The following section now describes the 802.11e functionalities used by WMM. Subsequently, a number of additional features are described which are defined as optional.

The core of the QoS enhancements is an extension of the DCF that controls access to the air interface as described in Section 5.5.1. DCF requires that devices wait for a random time before starting their transmission to prevent collisions when several devices have data waiting in their transmission buffers simultaneously. This delay has been specified to be up to 31 slots of 20 microseconds in 802.11b and g. The value used by a device is determined by generating a random number between 1 and 31. In case the transmission fails, for example, because of a collision, the delay is increased to 63, 127, and so on up to a maximum of 1023 slots, which equals 20 milliseconds.

802.11e extends this channel allocation method with the Hybrid Coordination Function (HCF). HCF describes two channel access methods – Enhanced Distributed Channel Access (EDCA) and Hybrid Coordination Function Controlled Channel Access (HCCA). HCF is backward compatible to DCF, which means that both HCF and non-HCF-capable devices can be used in the network simultaneously. The following section describes EDCA, which is the basis for the WMM specification.

Instead of using the same window length for the random number generator, EDCA specifies four QoS classes with queues. Each QoS queue is then assigned a different window length before the air interface can be accessed. WMM defines queues for voice, for video, for background and for best effort transmissions. Each class with its queue has the following variable parameters:

- The minimum number of slots that a device has to wait for before it is allowed to transmit a frame (Arbitration Interframe Space Number, AIFSN).
- Shortest Contention Window (CWmin): The minimum number of slots that can be selected with the random number generator.
- Longest Contention Window (CWmax): Maximum number of slots from which the random number generator can select a value.
- Transmit Opportunity (TXOP): Maximum transmission time. The granularity of the parameter is 32 microseconds.
- Admission Control: Indicates if devices have to request for permission to use a transmission class. Further details are discussed below.

Figure 5.26 shows an example of how these values could be set in practice for the different priority classes. Speech frames, for example, have very stringent requirements concerning delay and jitter. Therefore, it is important in this QoS class that such frames can be sent after a short backoff time and are hence preferred over other frames. This is achieved by configuring a short waiting time (AIFSN), for example, two slots, and setting the contention window size to three slots. The maximum waiting time is hence only five slots. This way, voice frames will always be transmitted before other devices (which use best effort frames) have a chance

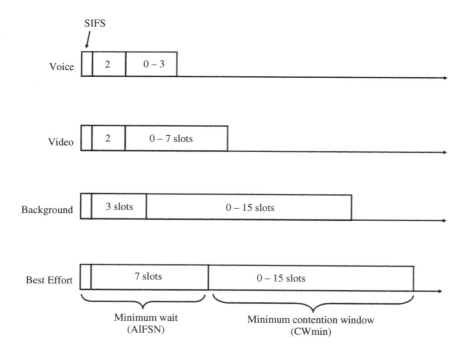

Figure 5.26   WMM priority classes with example values for CWmin, CWmax and TXOP

to access the air interface. This is the case in the scenario shown in Figure 5.26 – best effort frames have to wait at least for seven slots before their contention window begins. The values for CWmin, CWmax and TXOP are variable and many vendors allow the user to change these values via the user interface of the AP. If WMM is activated, the parameters are broadcast in the WMM parameter contained in beacon frames. In addition, these values are also included in association- and probe-response frames.

Furthermore, it is important for the implementation of QoS that applications have an easy method to use a certain QoS class for their data. In IP packets, for example, the Differentiated Services Code Point (DSCP) parameter of the IP header is used. If the application does not request a QoS to be used, the field is set to 'default'. Figure 5.27 shows an IP header of a voice packet, where the DSCP parameter is set to 'expedited forwarding'. The network driver of the wireless card then maps this field to a QoS class defined in 802.11e and hence, the data is preferred on the air interface as it is put into the 802.11e 'voice' service class queue.

In most cases, the prioritization of data frames on the air interface will be sufficient to ensure all QoS requirements. If, however, there are too many devices and applications in the network which transfer data with an elevated EDCA priority, collisions and hence, congestion can occur just as with the more simple DCF scheme. This means that network access times increase and datarates are reduced. This can only be prevented if devices register their QoS requirements such as datarate, frame size, and so on with the AP. The AP can then prevent other devices from using a certain QoS class once the current network load reaches the limit at which additional

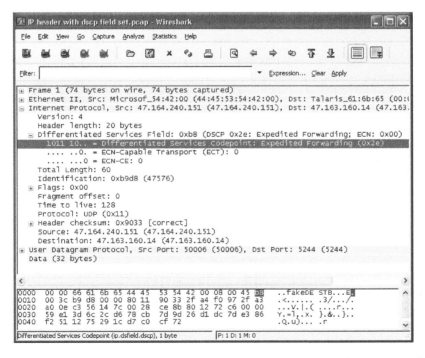

Figure 5.27 QoS field in an IP packet. Source: www.wireshark.org. Reproduced by permission of WireShark© 2010

streams can no longer be supported. These devices or applications can then choose to use a lower QoS class. For this purpose, the 802.11e standard specifies an optional admission control mechanism. Through the beacon frames, devices are informed if the AP requires access control for certain QoS classes. If a device does not support the admission control functionality, it must not use a QoS class for which admission control is used.

A device can register a new data stream by sending a Traffic Specification (TSPEC) in an Add Traffic Specification (ADDTS) management message to the AP. The AP then verifies if the network can support the additional traffic load and either grants or denies the request in a response message. The method by which the AP verifies that enough bandwidth is available for the application is not defined. In practice, parameters like the remaining bandwidth depend on the current reception conditions of all devices, and therefore it is not straightforward to conclude if an additional data stream can still be supported when the network already operates close to its maximum capacity.

In addition to QoS functionality, 802.11e also introduces enhancements to improve air interface usage efficiency. The most important functionality is packet bursting, which has already been introduced as a proprietary extension by many device vendors in 802.11g networks (Figure 5.28). With 802.11e, these methods are standardized and can therefore be used between clients and APs of different manufacturers. For packet bursting, several data frames need to be in the transmission buffer of a device. Instead of waiting for the default DCF backoff period after the ACK for the frame has been received, the next frame is sent after an SIFS period. In addition, the sender and receiver can agree on using a block ACK. Here, the sender bundles a number of frames and only expects an ACK once all frames have been sent. If the other device

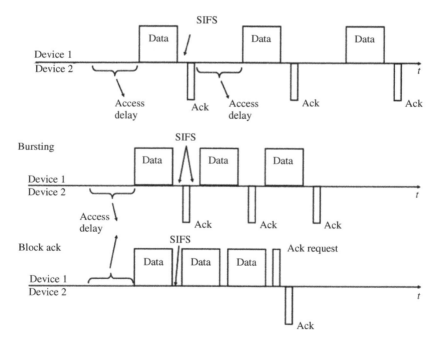

Figure 5.28 Packet bursting and block acknowledgments

has received the frames correctly, a single ACK frame is sufficient to confirm the reception. The ACK is sent either immediately (Immediate Block ACK) or somewhat later (Delayed Block ACK) to allow the receiver to check for errors in the received data.

The AP indicates the support for block ACK by including a capability information parameter in the beacon frames. Client devices indicate their block ACK capabilities to the AP during the association procedure. The 802.11n packet aggregation feature as described in Figure 5.16 can be used in combination with packet bursting and block ACK. Therefore, there are now several methods available in order to efficiently use the air interface when transferring large chunks of data compared to the initial specification.

In addition to the original PS mode and the PSMP extension that has been specified in 802.11n, an additional PS method referred to as Automated Power Save Delivery (APSD) was introduced with 802.11e. Again, the feature has a number of options. Using Unscheduled-Automated Power Save Delivery (U-APSD), which is optionally supported by WMM, the client device and AP negotiate that the client can enter a dormant state during which all incoming frames are buffered by the AP. During the negotiation phase, it is also specified as to which priority classes the algorithm is applied to and which frames continue to be delivered after wakeup with the normal PS mode. In addition, a Service Period (SP) is negotiated in which the device is active before it automatically enters the dormant state again.

The U-APSD negotiation sets no interval value after which the device has to return to the active state. Instead, a device transmits a trigger frame to the AP as soon as it is available again. Frames of QoS classes for which U-APSD has been activated will then automatically be delivered during the SP. The device automatically reenters the dormant state at the end of the SP. Frames belonging to QoS classes for which U-APSD has not been activated have to be retrieved with the standard PS methods by polling for each buffered frame. If an AP supports U-APSD, it is broadcast in beacon frames in the WMM parameter. From a client device's point of view, U-APSD operation can be negotiated during the association procedure via the QoS capability parameter or later by transmitting a TSPEC message. In addition, the 802.11e specification also contains a Scheduled-Automated Power Save Delivery (S-APSD) operation mode, which, however, is not used in WMM. Instead of trigger frames, a cyclic activity interval is used.

While most applications use a wireless network to establish a connection to a server on the Internet, there are a growing number of applications in homes and offices that require local connectivity, for example, video streaming or transferring large amounts of data between wireless devices in the network. The default method of exchanging data between two wireless devices in a network is to send the data to the AP, which then forwards it to the other device. In other words, the data is transmitted twice over the air interface and the overall throughput of the network is therefore cut in half. To improve the performance of data transfers between devices in the wireless network, 802.11e contains an extension referred to as the DLP. When two devices wish to communicate with each other directly, one of them sends a request to the AP. The AP in turn forwards the request to the other device. In case the other device is within the range of the first device and supports the DLP protocol, it returns an answer to the AP which returns it to the originator. Subsequently, the two devices can establish a direct connection and exchange data frames without involving the AP.

An alternative scheduling algorithm to the EDCA framework described above is HCCA. It is optional, however, and not part of the WMM specification. Unlike the distributed EDCA scheduling approach, HCCA is centrally scheduled and allows the AP to control the channel access of all devices in the network. This is done by periodically transmitting poll frames to each device, which then has the opportunity to transmit their data within a given time frame. As the AP can send the poll frames before other devices can access the air interface on their own because of the minimal backoff times they have to observe, it is ensured that only devices that have previously sent an ADDTS message to the network with a TSPEC message can transfer data. HCCA supports the previously mentioned QoS classes and can hence, in a similar way as EDCA, give precedence to frames with certain QoS requirements.

To get an idea of how the options described in this section are used in practice, a number of freely available network analysis tools can be used. A popular tool is Wireshark (http://www.wireshark.org). With the Linux operating systems, many WLAN adapters can be set into a mode in which they record Wi-Fi frames. Wireshark is also available for other operating systems, but a special Wi-Fi card is required for the recording. The website for the program contains a wide range of traces that can be downloaded, which can then be analyzed in offline mode. Another alternative is a Wi-Fi AP like the Linksys WRT54G, for which open source Linux-based software alternatives, such as OpenWRT and DD-WRT, are available. With programs such as Kismet that can be executed on the AP, data traffic can be recorded and then viewed offline using Wireshark. Further details can be found on the OpenWRT Wiki (http://www.openwrt.org).

5.9   Comparison of Wireless LAN and LTE

In the early days of Wi-Fi and UMTS, LTE's predecessor technology, the potential competition between the two wireless technologies was often debated. Today, both technologies have evolved and both have found their respective usage domains.

While presently Wi-Fi is mostly used at homes and in offices and also as a hotspot technology in public places such as airports and hotels, LTE has evolved to be a wide area high-speed connectivity technology. To demonstrate where LTE and Wi-Fi compete and cooperate today, the following sections describe the use of the two systems for Internet connectivity outside of homes and offices.

When comparing peak datarates of Wi-Fi and LTE outside of homes and offices, a significant difference can be observed at first. Although some hotspots still use the 802.11b standard today with peak datarates of 11 Mbit/s on the PHY, most have already been upgraded to 802.11g or even 802.11n and hence their theoretical peak datarate is specified as 300–450 Mbit/s. In the future, further peak datarate improvements can be made with 802.11ac and WMM QoS extensions can be used to prioritize data streams for applications such as VoIP. As shown in this chapter, peak datarates of 100 Mbit/s and more can be reached at the application layer. This compares to the current PHY datarates of LTE networks of 150 Mbit/s for a 20-MHz channel and achievable datarates in practice of several tens of megabits per second and even more under very favorable transmission conditions. This seems to suggest, at first, that Wi-Fi hotspots have a significant advantage over LTE. In public hotspots, however, the limiting factor is not the air interface speed but the bandwidth of the backhaul connection, usually a DSL line with a bandwidth of a few megabits per second. This bandwidth must then be shared by all users of the hotspot. The bandwidth offered by LTE cells must also be shared by all users in the area. A typical LTE base station that also includes UMTS cells, however, usually comprises of three sectors and hence the total capacity rises to several hundred megabits per second. To increase capacity, additional channels or frequency bands can be used. While a single user is still limited to the capacity of a single sector, this is usually more than enough for web browsing, video streaming, access to company intranets, and so on. As a consequence, users experience only little difference when using a Wi-Fi hotspot compared to when using a UMTS or LTEnetwork. The two types of networks also behave similarly for data transmissions in the uplink direction. Both offer speeds of a few hundred kilobits to several megabits (ADSL uplink bandwidth vs. HSUPA or LTE uplink).

It should not be forgotten, however, that a base station covers a much larger geographical area than a Wi-Fi hotspot. In addition, the base station also handles voice calls in its service area over UMTS or GSM, which reduces the capacity for data transmissions. In areas with a high data traffic density, base stations are, therefore, often equipped with additional UMTS carriers and LTE on different frequency bands as mentioned earlier, as well as significant backhaul capacity. Another option is to split the coverage area by installing additional base stations. Furthermore, small cells that cover only limited areas but which are significantly less expensive than macro base stations can also be used for the purpose. Over the years, Wi-Fi hotspots have been installed in many public places such as hotels, airports and train stations. In the future, it is likely that this trend will continue. Owing to their limited range, however, a suitable hotspot is not necessarily available at all locations desired. This can, for example, be an issue in hotels if the advertised Wi-Fi access is only available in certain areas.

In many countries, LTE networks cover significant areas and most cities with more than just a few thousand inhabitants are usually fully covered. UMTS can also be found in many countries, and GSM, GPRS and EDGE usually provide almost countrywide connectivity. With roaming agreements, it is furthermore assured that UMTS and GSM networks can be used in most countries around the world and it is only pricing that hinders a more global use while traveling. Wi-Fi hotspot operators also cooperate internationally, which allows, for example, a French customer to use a Wi-Fi hotspot with the credentials of a hotspot operator in his home country. Often, however, it cannot be assured in advance that this option is available at a certain location.

As LTE and UMTS are an enhancement of GSM and GPRS, international billing is usually in place through roaming agreements and is an integral part of the system. Some network operators have started to offer special data roaming tariffs in order to compete with hotspot offers, and roaming regulation in Europe, in general, has brought down data roaming prices to levels that are attractive to a wider audience. Wi-Fi hotspots do not have a standardized charging model and, owing to a missing standard, numerous variants exist, for example, prepaid top-up cards, online payment via credit cards or charging the use to a postpaid mobile phone contract. The latter variant is only possible, however, if the hotspot operator is also the operator of the customer's GSM/UMTS/LTE network or if a special international charging agreement exists. As a consequence, customers usually have to identify themselves manually and select a payment method before they get access to the network.

The technical realization of lawful interception for public Wi-Fi hotspots is not yet fully resolved. For all telecommunication networks including GSM, GPRS and UMTS, laws have been put in place by most countries that describe which interfaces network operators have to provide to the police and other security organizations. In many countries, there are no laws yet for small hotspot providers, and it remains to be seen as to how this will be dealt with in the future.

Wi-Fi has been designed for the coverage of small areas. The coverage area can be somewhat extended by using several APs that form an ESS as described at the beginning of this chapter. As all APs are part of the same IP subnet (cp. Figure 5.4), the range of a network is usually limited to a single building. For most Wi-Fi applications, this is sufficient and devices can seamlessly move between different APs within the same ESS. LTE and UMTS networks have been designed for countrywide coverage. As has been shown in Chapters 3 and 4, significant efforts have been taken to ensure that a connection is maintained while the user roams through the network, even at high speeds. This way, it is possible to communicate on the move and to maintain a continuous connection to the Internet.

The different cell ranges of the two systems are also a result of the different maximum transmission powers. Wi-Fi APs are limited to 0.1 W, and hence their maximum range is a few dozen meters. Inside buildings, the range of a single AP is even more limited because of the signal absorption by walls and other obstacles. LTE and UMTS base stations have a coverage range of a few hundred meters in urban areas and several kilometers in less densely populated areas. The transmission power per carrier is usually around 20 W. The coverage area of small LTE and UMTS cells is similar to that of Wi-Fi AP as their output power is similar.

As has been discussed in this chapter, the initial security measures of Wi-Fi were not sufficient and more secure authentication and encryption algorithms were only introduced later. Today, WPA and WPA2 authentication and encryption offer good security for home and office networks. Public hotspots, however, are usually not encrypted because of the complexity involved in the user obtaining and entering security credentials. It is uncertain if security measures will be introduced for public hotspots at some point. If this is not the case, it is easy for attackers to obtain usernames, passwords and cookies if the data transfer is not encrypted on a higher layer, for example, via secure HTTPS or secure e-mail exchange protocols such as secure Post Office Protocol (POP3S) and secure Simple Mail Transfer Protocol (SMTP). Hotspot users should therefore use additional security measures such as OpenVPN that is based on SSL or IPsec tunnels. These technologies unfortunately require advanced knowledge and, therefore, only few people use them in practice. In LTE and UMTS networks, security is part of the system concept and users do not have to deal with encryption settings. Encryption is automatically activated and secret keys that are stored on the SIM card of the user and in the AuC, which is part of the HLR in the network, are used for strong encryption.

The circuit-switched part of UMTS networks has been specifically designed for voice and video telephony, which are two of the main services of such networks. Wi-Fi networks are usually not used primarily for voice and video telephony, although there is a growing trend to migrate voice and video telephony from circuit-switched channels to voice and video over IP. VoIP and video applications such as Skype are increasingly used via notebooks and smartphones, and Wi-Fi networks are therefore being increasingly used for this kind of application as well.

Wi-Fi chips are integrated in virtually all smartphones on the market today and some have integrated VoIP capabilities that can be used over Wi-Fi. While their use is simple at home and in office networks, the manual authentication and payment process required in many public hotspots makes the use of such programs difficult on smartphones. Another issue with voice transmission is the lack of QoS measures on the air interface and on the backhaul link of Wi-Fi hotspots, which do not use the 802.11n standard.

In summary, Wi-Fi is a hotspot technology that gives users access to local devices and the Internet at a specific location and for a limited amount of time. Owing to the comparatively simple technology when compared with LTE and UMTS and owing to the license-free operation, installation is simple and operation of Wi-Fi networks are much cheaper than those of a

LTE and UMTS macro network. If a fast backhaul connection is available, Wi-Fi hotspots are an ideal way to connect users to the Internet. For connectivity in cars and trains and outside heavily frequented areas, where the operation of a Wi-Fi hotspot is economically not viable, the technology is not the right choice. Consequently, Wi-Fi access is sometimes also said to offer 'nomadic' Internet access, as the user has to remain in a certain area while using such a network.

LTE and UMTS, on the other hand, address the needs of mobile users that need to communicate while being stationary and while moving. With the introduction of HSPA and LTE, fast datarates enable users to communicate at multimegabit speeds at almost any location, that is, without looking for the next hotspot location. The complex technology that is required for mobility and voice calls makes LTE and UMTS networks much more expensive to deploy than Wi-Fi hotspots. The main usage scenarios for combined LTE and UMTS networks are, therefore, mobile voice and video telephony (over UMTS) and mobile Internet access (over UMTS and LTE), both for notebook connectivity and for applications on mobile devices such as web browsers, e-mail programs, instant messengers, and so on. UMTS and LTE are therefore sometimes also said to offer 'mobile' Internet access as connectivity is ensured even when users move at high speeds.

Questions

1. What are the differences between the 'ad hoc' and 'BSS' modes of a WLAN?
2. Which additional functionalities can often be found in WLAN APs?
3. What is an ESS?
4. What is an SSID and in which frames is it used?
5. What kinds of PS mechanisms exist in the WLAN standard?
6. Why are ACK frames used in a WLAN?
7. Why do 802.11g networks use the RTS/CTS mechanism?
8. Why are three MAC addresses required in BSS frames?
9. How can a receiving device detect the speed at which the payload part of a frame was sent?
10. What is the maximum transfer rate that can be reached in a data transfer between two 802.11g devices in a BSS?
11. What disadvantages does the DCF method have for telephony and video streaming applications?
12. Which security holes exist in the WEP procedures and how are they solved by WPA and WPA2 (802.1x)?

Answers to these questions can be found on the companion website for this book at http://www.wirelessmoves.com.

References

[1] IEEE, Part 11: Wireless LAN Medium Access Control (MAC) and Physical Layer (PHY) Specifications, ANSI/IEEE Std 802.11, 1999 Edition (R2003).
[2] IEEE, Part 3: Carrier Sense Multiple Access with Collision Detection (CSMA/CD) Access Method and Physical Layer Specifications, ANSI/IEEE Std 802.3, March 2002 Edition.

[3] IEEE, Part 11: Wireless LAN Medium Access Control (MAC) and Physical Layer (PHY) Specifications: High-Speed Physical Layer Extensions in the 2.4 GHz Band, ANSI/IEEE Std 802.11b, 1999 Edition (R2003).

[4] IEEE, Part 11: Wireless LAN Medium Access Control (MAC) and Physical Layer (PHY) Specifications – Amendment 4: Further Higher Data Rate Extensions in the 2.4 GHz Band, ANSI/IEEE Std 802.11g, 2003.

[5] IEEE, Part 11: Wireless LAN Medium Access Control (MAC) and Physical Layer (PHY) Specifications – High-Speed Physical Layer Extensions in the 5 GHz Band, ANSI/IEEE Std 802.11a, 1999.

[6] IEEE, Part 11: Wireless LAN Medium Access Control (MAC) and Physical Layer (PHY) Specifications – Amendment: Medium Access Control (MAC) Quality of Service Enhancements, IEEE Std P802.11e/D13, January 2005.

[7] IEEE, IEEE Trial-Use Recommended Practice for Multi-Vendor Access Point Interoperability via an Inter-Access Point Protocol Across Distribution Systems Supporting IEEE 802.11 Operation, IEEE Std 802.11F, 2003.

[8] IEEE, Part 11: Wireless LAN Medium Access Control (MAC) and Physical Layer (PHY) Specifications – Amendment 5: Spectrum and Transmit Power Management Extensions in the 5 GHz Band in Europe, IEEE Std 820.11h, 2003.

[9] IEEE, Part 11: Wireless LAN Medium Access Control (MAC) and Physical Layer (PHY) Specifications – Amendment 6: Medium Access Control (MAC) Security Enhancements, IEEE Std 802.11i, 2004.

[10] R. Droms, RFC 2131 – Dynamic Host Configuration Protocol, RFC 2131, March 1997.

[11] Wikipedia, List of WLAN Channels, http:/en.wikipedia.org/wiki/List_of_WLAN_channels, accessed December 2013.

[12] Gast, M. (2013) 802.11ac – A Survival Guide, O'Reilly, ISBN 978-1-449-34314-9.

[13] E. Ahlers, WLAN Wunschzettel, *C't Magazine*, Heise, 1, 2014.

6

Bluetooth

Although cables are ideal for exchanging data between stationary devices that are close together, there are significant disadvantages in a mobile environment. In practice, Bluetooth connectivity has become an alternative to cables for many close-range data exchange applications and is often used alongside the cellular radio technologies that were discussed in the previous chapters.

In the first part of this chapter, an introduction to the physical properties of Bluetooth and the protocol stack is given. Afterward, relevant Bluetooth profiles and how they are used in practice in a wide range of applications and scenarios are described.

6.1 Overview and Applications

Owing to the ongoing miniaturization and integration, more and more small electronic devices are used in everyday life. Bluetooth enables these devices to wirelessly communicate with each other without a direct line-of-sight connection. Although in the last decade, there was a wide range of applications of Bluetooth, it can be observed today that its use is now mostly focused on the following applications:

- Wireless connectivity from smartphones and notebooks to remote audio devices, such as headsets, hands-free telephony equipment, Bluetooth-enabled loudspeakers and in-car entertainment systems
- Exchange of files between smartphones and notebooks (e.g., pictures taken with a smartphone camera) and quick exchange of single-address book and calendar entries
- Connecting wireless keyboards and other input devices to notebooks and smartphones

Other applications such as, for example, sharing of the Internet connection from a smartphone to a notebook, calendar and address book synchronization and multi-player games between devices have migrated to other technologies such as Wi-Fi tethering and cloud-based services.

As there are a great number of different Bluetooth devices available from different vendors, reliable interoperability is of utmost importance for the success of Bluetooth and is a challenge

From GSM to LTE-Advanced: An Introduction to Mobile Networks and Mobile Broadband,
Revised Second Edition. Martin Sauter.
© 2014 John Wiley & Sons, Ltd. Published 2014 by John Wiley & Sons, Ltd.

to achieve in practice. New devices must therefore be approved by a Bluetooth Qualification Test Facility (BQTF) [1].

Table 6.1 lists the different Bluetooth protocol versions. In general, a new version is always downward compatible to all previous versions. This means that a Bluetooth 2.1 device is still able to communicate with a Bluetooth 4.0 device. Functionality, which has been introduced with a newer version of the standard, cannot of course be used with a device that supports only a previous version of the standard.

6.2 Physical Properties

Up to version 1.2 of the standard, the maximum datarate of a Bluetooth transmission channel is 780 kbit/s. All devices that communicate directly with each other have to share this datarate. The maximum datarate for a single user thus depends on the following factors:

- the number of devices that exchange data with each other at the same time;
- activity of the other devices.

The highest transmission speed can be achieved if only two devices communicate with each other and only one of them has a large amount of data to transmit. In this case, the highest datarate that can be achieved is 723 kbit/s. After removing the overhead, the resulting datarate is about 650 kbit/s. The bandwidth remaining for the other device to send data in the reverse direction is about 57 kbit/s. This scenario occurs quite often, for example, when transferring a file. In this case, one of the two devices sends the bulk of the data while the other device sends only small amounts of acknowledgement data. The left-hand side of Figure 6.1 shows the achievable speeds for this scenario.

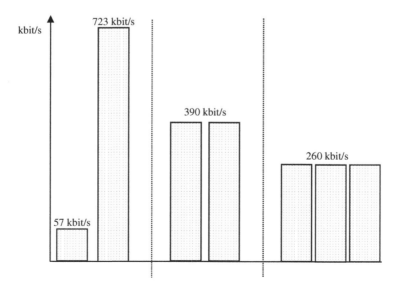

Figure 6.1 Three examples of achievable Bluetooth datarates depending according to the number of users and their activity

Table 6.1 Bluetooth versions

Version	Approved	Comment
1.0B	December 1999	First Bluetooth version, which was only used by a few first-generation devices
1.1	February 2001	This version corrected a number of errors and ambiguities of the previous version (errata list) and helped to increase the interoperability between devices of different vendors
1.2	November 2003	Introduction of the following new features:
		• faster discovery of nearby Bluetooth devices. Devices can now also be sorted based on signal strength, as described in Section 6.4.2
		• fast connection establishment, see Section 6.4.2
		• adaptive frequency hopping (AFH), see Section 6.4.2
		• improved speech transmission, for example, for headsets (enhanced-synchronous connection oriented (eSCO)) as described in Sections 6.4.1 and 6.6.3
		• improved error detection and flow control in the L2CAP protocol
		• new security functionality: Anonymous connection establishments
2.0	2004	Enhanced datarates extend the Bluetooth 1.2 specification with faster data transmission modes. Further details can be found in Sections 6.2 and 6.4.1. The complete standard can be found in [2]
2.1	2007	Security improvements and some functionality enhancements. The most important ones are
		• Secure simple pairing: Security improvements and simplification of the pairing process
		• Sniff subrating: Additional energy saving options for active connections with sporadic data exchange
		• Erroneous data reporting for eSCO packets
3.0 + HS	2009	Improvements concerning power management and introduction of the optional Bluetooth High-Speed (HS) mode. The HS mode uses Bluetooth for initial connection establishment and Wi-Fi for user data transmission. Most products sold today are Bluetooth 3.0 compatible but do not implement the optional HS mode.
4.0	2010	Integration of WiBree into Bluetooth as Low-Energy Option. Many devices support the low-energy option today but there are few applications that make use of the new capabilities.
4.1	2013	Introduces enhancements such as:
		• LTE co-existence in nearby bands
		• Auto re-connect capabilities when temporarily loosing signal
		• A device can act as a low-energy hub and peripheral simultaneously
		• L2CAP-dedicated channels for future IPv6 communication at the sensor level

If both ends of the connection need to send data as quickly as possible, the speed that can be achieved at each side is about 390 kbit/s. The middle section of Figure 6.1 shows this scenario.

If more than two devices want to communicate with each other simultaneously, the maximum datarate per device is further reduced. The right-hand side of Figure 6.1 depicts this scenario.

In 2004, the Bluetooth 2.0 + EDR (Enhanced Datarate) standard [2] was released. This enables datarates of up to 2178 kbit/s by using additional modulation techniques. This is discussed in more detail in Section 6.4.1.

To reach these transmission speeds, Bluetooth uses a channel in the 2.4-GHz ISM (Industrial, Scientific and Medical) band with a bandwidth of 1 MHz. Gaussian Frequency Shift Keying (GFSK) is used as modulation up to Bluetooth 1.2, while Differential Quadrature Phase Shift Keying (DQPSK) and eight-phase differential phase shift keying (8DPSK) are used for EDR packets. Compared to a 22-MHz channel required for wireless LAN, the bandwidth requirements of Bluetooth are quite modest.

For bidirectional data transmission, the channel is divided into timeslots of 625 microseconds. All devices that exchange data with each other thus use the same channel and are assigned timeslots at different times. This is the reason for the variable datarates shown in Figure 6.1. If a device has a large amount of data to send, up to five consecutive timeslots can be used before the channel is given to another device. If a device has only a small amount of data to send, only a single timeslot is used. This way, all devices that exchange data with each other at the same time can dynamically adapt their use of the channel based on their data buffer occupancy.

As Bluetooth has to share the 2.4-GHz ISM frequency band with other wireless technologies like Wireless Local Area Network (WLAN), the system does not use a fixed carrier frequency. Instead, the frequency is changed after each packet. A packet has a length of either one, three or five slots. This method is called frequency-hopping spread spectrum (FHSS). This way, it is possible to minimize interference with other users of the ISM band. If some interference is encountered during the transmission of a packet despite FHSS, the packet is automatically retransmitted. For single-slot packets (625 microseconds), the hopping frequency is thus 1600 Hz. If five-slot packets are used, the hopping frequency is 320 Hz.

A Bluetooth network, in which several devices communicate with each other, is called a piconet. In order for several Bluetooth piconets to coexist in the same area, each piconet uses its own hopping sequence. In the ISM band, 79 channels are available. Thus, it is possible for several WLAN networks and many Bluetooth piconets to coexist in the same area as shown in [3].

The interference created by WLAN and Bluetooth remains low and hardly noticeable as long as the load in both the WLAN and the Bluetooth piconet(s) is low. As has been shown in Chapter 4, a WLAN network only sends short beacon frames while no user data is transmitted. If a WLAN network, however, is highly loaded, it blocks a 25-MHz frequency band for most of the time. Therefore, almost a third of the available channels for Bluetooth are constantly busy. In this case, the mutual interference of the two systems is high, which leads to a high number of corrupted packets. To prevent this, Bluetooth 1.2 introduces a method called Adaptive Frequency Hopping (AFH). If all devices in a piconet are Bluetooth 1.2 compatible, the master device (see Section 6.3) performs a channel assessment to measure the interference encountered on each of the 79 channels. The link manager (see Section 6.4.3) uses this information to create a channel bitmap and marks each channel that is not to be used for the frequency-hopping sequence of the piconet. The channel bitmap is then sent to all devices of the piconet and thus, all members of the piconet are aware of how to adapt their hopping sequence. The standard does not specify a single method for channel assessment. Available choices are the Received Signal Strength Indication (RSSI) method or other methods that exclude a channel because of a high packet error rate. Bluetooth 1.2 also offers dual mode devices, which are equipped with both a WLAN and a Bluetooth chip, to inform the Bluetooth stack as to which channels are to be excluded from the hopping sequence. In practice, this is quite useful, as the device is aware as to which WLAN channel has been selected by the user, and it can then instruct the Bluetooth module to exclude 25 consecutive channels from the hopping sequence.

As Bluetooth has been designed for small, mobile and battery-driven devices, the standard defines three power classes. Devices like mobile phones usually implement power class 3 with a transmission power of up to 1 mW. Class 2 devices send with a transmission power of up to 2.5 mW. Class 1 devices use a transmission power of up to 100 mW. Only devices such as some Universal Serial Bus (USB) Bluetooth sticks for notebooks and PCs are usually equipped with a class 1 transmitter. This is because the energy consumption as compared to a class 3 transmitter is very high and should therefore only be used for devices where the energy consumption does not play a critical role. The distances that can be overcome with the different power classes are also quite different. While class 3 devices are usually designed to work reliably over a distance of 10 m or through a single wall, class 1 devices can achieve distances of over 100 m or penetrate several walls. The range of a piconet also depends on the reception qualities of the devices and the antenna design. In practice, newer Bluetooth devices have a much improved antenna and receiver design, which increases the size of a piconet without increasing the transmission power of the devices. All Bluetooth devices can communicate with each other, independently of the power class. As all connections are bidirectional, however, it is always the device with the lowest transmission power that limits the range of a piconet.

Security plays an important role in the Bluetooth specifications. Thus, strong authentication mechanisms are used to ensure that connections can be established only if they have been authorized by the users of the devices that want to communicate. Furthermore, encryption is also a mandatory part of the standard and must be implemented in every device. Ciphering keys can have a length of up to 128 bits and thus offer good protection against eavesdropping and hostile takeover of a connection.

6.3   Piconets and the Master/Slave Concept

As described previously all devices which communicate with each other for a certain time form a piconet. As shown in Figure 6.2, the frequency-hopping sequence of the channel is calculated from the hardware address of the first device that initiates a connection to another device and thus creates a new temporary piconet. Therefore, devices can communicate with each other in different piconets in the same area without disturbing each other.

A piconet consists of one master device that establishes the connection and up to seven slave devices. This seems to be a small number at first. However, as most Bluetooth applications only require point-to-point connections as described in Section 6.1, this limit is sufficient for most applications. Even if Bluetooth is used with a personal computer (PC) to connect with a keyboard and a mouse, there are still five more devices that can join the PC's piconet at any time.

Each device can be a master or a slave of a piconet. Per definition, the device that initiates a new piconet becomes the master device as described in the following scenario.

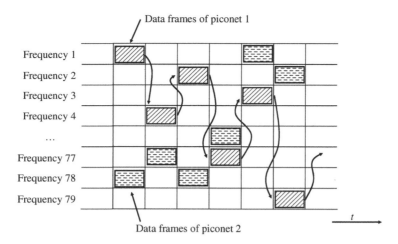

Figure 6.2   By using different hopping sequences, many piconets can coexist in the same area

Consider a user who has a Bluetooth-enabled mobile phone and headset. After initial pairing (see Section 6.5.1), the two devices can establish contact with each other at any time and thus form a piconet for the duration of a phone call. At the end of a phone call, the Bluetooth connection ends as well, and the piconet thus ceases to exist. In the case of an incoming call, the mobile phone establishes contact with the headset and thus becomes the master of the connection. In the reverse case, the user establishes an outgoing phone call by pressing a button on the headset and by using the voice-dialing feature of the mobile phone. In this case, it is the headset and not the mobile phone that establishes the connection and thus the headset becomes the master of the newly established piconet. If another person in the vicinity also uses a Bluetooth-enabled mobile phone and headset, the two piconets overlap. As each piconet uses a different hopping sequence, the two connections do not interfere with each other. Because of the initial pairing of the headset and the mobile phone, it is ensured that each headset finds its own mobile phone and thus always establishes a connection for a new phone call with the correct mobile phone.

The master of a piconet controls the order and the duration of slave data transfers over the piconet channel. To grant the channel to a slave device for a certain period of time, the master sends a data packet to the slave. The slave is identified by a three-bit address in the header of the data packet, which has been assigned to the device at connection establishment. The data packet of the master can have a length of one to five slots depending on the amount of data that has to be sent to the slave. If no data needs to be sent to the slave, an empty one-slot packet is used. Sending a packet to a slave device implicitly assigns the next slot to the slave, independently of the packet containing user data. The slave can then use the next one to five slots of the channel to return a packet. With Bluetooth 1.1, slaves answer on the following hopping frequency of hopping sequence. The Bluetooth 1.2 specification slightly changes this behavior and thus Bluetooth 1.2-compliant devices answer on the same frequency that the master has previously used. The slave sends an answer packet regardless of whether data is waiting in the buffer to be sent to the master. If no data is waiting in the slave's buffer, an empty packet is sent to acknowledge to the master that the device is still active and accessible.

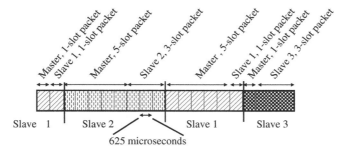

Figure 6.3 Data exchange between a master and three slave devices

After a maximum number of five slots, the right to use the channel expires and is automatically returned to the master even if there is still data waiting in the slave's output buffer to be sent. Afterward, the master device can decide whether the channel has to be granted to the same or a different slave device. If the master did not receive any user data from the slave and the master's output buffer for the particular slave is also empty, it can pause the data transmission for up to 800 slots to save power. As the duration of a slot is 625 microseconds, 800 slots equal a transmission pause of 0.5 seconds (see Figure 6.3).

As a slave cannot anticipate as to when a new packet of a master arrives, it is not able to establish a connection to additional devices. In some cases, it is therefore necessary that master and slave change their roles during the lifetime of the piconet. This is necessary, for example, if a smartphone has established a connection to a PC to synchronize data. As the smartphone is the initiator of the connection, it is the master of the piconet. While the connection is still established, the user wants to use the PC to access a picture file on another device and thus has to include this device in the piconet. This is only possible if the smartphone (master) and the PC (slave) change their roles in the piconet. This procedure is called a 'master–slave role switch'. After the role switch, the PC is the master of the piconet between itself and the smartphone. Now, the PC is able to establish contact with a third device while the connection to the smartphone remains in place. By contacting the third device and transferring the picture, however, the datarate between the PC and the smartphone is reduced.

6.4 The Bluetooth Protocol Stack

Figure 6.4 shows the different layers of the Bluetooth protocol stack and will be used in the following sections as a reference. The different Bluetooth protocol layers can only be loosely coupled to the seven-layer OSI model, as some Bluetooth layers perform the tasks of several OSI layers.

6.4.1 The Baseband Layer

The properties of the physical layer, that is, the radio transmission layer, have already been described. On the basis of the physical layer, the baseband layer performs the typical duties

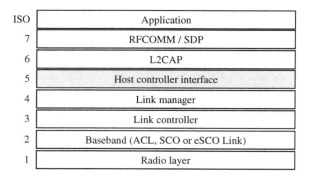

Figure 6.4   The Bluetooth protocol stack

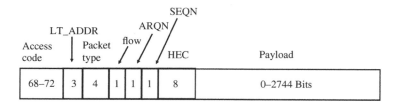

Figure 6.5   Composition of an ACL packet

of a layer 2 protocol, such as the framing of data packets. For the data transfer, three different packet types have been defined in the baseband layer.

For packet data transmission, Bluetooth uses Asynchronous Connectionless (ACL) packets. As shown in Figure 6.5, an ACL packet consists of a 68- to 72-bit access code, an 18-bit header and a payload (user data) field of variable size between 0 and 2744 bits.

Before the 18 header bits are transmitted, they are coded into 54 bits by a forward error correction (1/3 FEC) algorithm. This ensures that transmission errors can be corrected in most cases. Depending on the size of the payload field, an ACL packet requires one, three or five slots of 625 microseconds.

The access code at the beginning of the packet is used primarily for the identification of the piconet to which the current packet belongs. Thus, the access code is derived from the device address of the piconet master. The actual header of an ACL packet consists of a number of bits for the following purposes: The first three bits of the header are the logical transfer address (LT_ADDR) of the slave, which the master assigns during connection establishment. As three bits are used, up to seven slaves can be addressed.

After the LT_ADDR, the 4-bit packet-type field indicates the structure of the remaining part of the packet. Table 6.2 shows the different ACL packet types. Apart from the number of slots used for a packet, another difference is the use of FEC for the payload. If FEC is used, the receiver is able to correct transmission errors. The disadvantage of using FEC, though, is the reduction in the number of user data bits that can be carried in the payload field. If a 2/3 FEC is used, one error correction bit is added for two data bits. Instead of two bits, three bits will thus be transferred (2/3). Furthermore, ACL packets can be sent with a cyclic redundancy check

Table 6.2 ACL packet types

Packet type	Number of slots	Link type	Payload (bytes)	FEC	CRC
0100	1	DH1	0–270	No	Yes
1010	3	DM3	0–121	2/3	Yes
1011	3	DH3	0–183	No	Yes
1110	5	DM5	0–224	2/3	Yes
1111	5	DH5	0–339	No	Yes

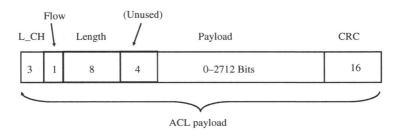

Figure 6.6 The ACL payload field including the ACL header and checksum

(CRC) checksum to detect transmission errors, which the receiver was unable to correct (see Figure 6.6).

To prevent a buffer overflow, a device can set the flow bit to indicate to the other end to stop data transmission for some time.

The ARQN bit informs the other end if the last packet has been received correctly. If the bit is not set, the packet has to be repeated.

The sequence bit (SEQN) is used to ensure that no packet is accidentally lost. This is done by toggling the bit in every packet. The following example shows how the bit is used in a scenario in which device-1 and device-2 exchange data packets: If device-2 receives two consecutive packets with the SEQN set in the same way, it indicates that device-1 was unable to receive the previous packet and has thus repeated its data packet. The repetition is necessary as it is not clear to device-1 if only the return packet is missing or if its own packet is also lost. Device-2 then repeats its packet that includes the acknowledgment for the packet of device-1 and ignores all incoming packets as long as no packet with a correct SEQN bit is received by device-1. Even if multiple packets are lost, all data is eventually delivered.

The last field in the header is the Header Error Check (HEC) field. It ensures that the packet is ignored if the receiver cannot calculate the checksum correctly.

The payload field follows the ACL header and is composed of the following fields. The first bits of the payload header field again contain some administrative information. The first field is called the logical channel (L_CH) field. It informs the receiver if the payload field contains user data (Logical Link Control and Adaptation Protocol (L2CAP)) packets, see Section 6.4.5) or Link Manager Protocol (LMP) signaling messages (see Section 6.4.3) for the administration of the piconet.

The flow bit is used to indicate to the L2CAP layer above that the receiver buffer is full. Finally, the payload header includes a length field before the actual payload part is transmitted. After the actual payload, an ACL packet ends with a 16-bit CRC checksum.

As no bandwidth is guaranteed for an ACL connection, this type of data transmission is not well suited to the transmission of bidirectional real-time data such as a voice conversation. For this kind of application, the baseband layer offers a second transmission mode that uses synchronous connection-oriented (SCO) packets. The difference to ACL packets is the fact that SCO packets are exchanged between a master and a slave device in fixed intervals. The interval is chosen in a way that results in a total bandwidth of exactly 64 kbit/s.

When an SCO connection between a master and a slave device is established, the slave device is allowed to send its SCO packets autonomously even if no SCO packet is received from the master. This can be done very easily as the timing for the exchange of SCO packets between two devices is fixed. Therefore, the slave does not depend on a grant from the master, and thus it is implicitly ensured that only this slave sends in the timeslot. This way, it is furthermore ensured that the slave device can send its packet containing voice data even if it has not received the voice packet of the master device.

The header of an SCO packet is equal to the header of an ACL packet with the exception that the flow, ARQN and SEQN fields are not used. The length of the payload field is always 30 bytes. Depending on the error correction mechanism used, this equals 10, 20 or 30 user data bytes. Table 6.3 gives an overview of the different SCO packet types.

The last line of the table shows a special packet type, which can contain both SCO and ACL data. This packet type can be used to send both voice data and signaling messages at the same time. As shown in Section 6.6.3, for the headset profile, an SCO connection between a mobile phone and a headset requires not only a speech channel but also a channel for signaling messages (e.g., to control the volume). The SCO voice data can then be embedded in the first 10 bytes of a 'DV' packet which are followed by up to 9 bytes for the ACL channel. The FEC and the checksum are only applied to the ACL part of the payload.

It has to be noted that it is not mandatory to use DV packets if voice and data have to be transmitted simultaneously between two devices. Another possibility is to use independent ACL packets in slots that are not used by the SCO connection. Finally, a third possibility to send both ACL and SCO information between two devices is to drop the SCO information of a slot and to send an ACL packet instead.

As no CRC and FEC are used for SCO packets, it is not possible to detect whether the user data in the payload field was received correctly. Thus, defective data is forwarded to higher layers if a transmission error occurs. This produces audible errors in the reproduced voice signal. Furthermore, the bandwidth limit of 64 kbit/s of SCO connections prevents the use of this transmission mechanism for other types of interactive applications such as audio

Table 6.3 SCO packet types

Packet type	Number of slots	Link type	Payload (bytes)	FEC	CRC
0101	1	HV1	10	1/3	No
0110	1	HV2	20	2/3	No
0111	1	HV3	30	None	No
1000	1	DV	10 (+ 0–9)	2/3	Yes

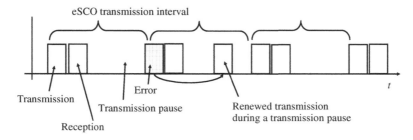

Figure 6.7   Retransmission of an eSCO packet caused by a transmission error

streaming in MP3 format that usually requires a higher datarate. Bluetooth 1.2 thus introduces a new packet type called eSCO, which improves the SCO mechanism as follows.

The datarate of an eSCO channel can be chosen during channel establishment. Therefore, a constant datarate of up to 288 kbit/s in full-duplex mode (in both directions simultaneously) can be achieved.

The eSCO packets use a checksum for the payload part of the packet. If a transmission error occurs, the packet can be retransmitted if there is still enough time before the next regular eSCO packet has to be transmitted. Figure 6.7 shows this scenario. Retransmitting a bad packet and still maintaining a certain bandwidth is possible, as an eSCO connection with a constant bandwidth of 64 kbit/s only uses a fraction of the total bandwidth available in the piconet. Thus, there is still some time to retransmit a bad packet in the transmission gap to the next packet. Despite transferring the packet several times, the datarate of the overall eSCO connection remains constant. If a packet cannot be transmitted by the time another regular packet has to be sent, it is simply discarded. Thus, it is ensured that the data stream is not slowed down and the constant bandwidth and delay times required for audio transmissions are maintained. Bluetooth 2.1 introduced an option to forward erroneous packets to higher layers with an error indication. This might be useful if a codec can correct small transmission errors by itself.

For some applications such as wireless printing or transmission of large pictures from a camera to a PC, the maximum transmission rate of Bluetooth up to version 1.2 is not sufficient. Thus, the Bluetooth standard was enhanced with a high-speed data transfer mode called Bluetooth 2.0 + EDRs. The core of EDR is the use of a new modulation technique for the payload part of an ACL or eSCO packet. While the header and the payload of the packet types described before are modulated using GFSK, the payload of an EDR ACL and eSCO packet is modulated using DQPSK or 8DPSK. These modulation techniques allow the encoding of several bits per transmission step. Thus, it is possible to increase the datarate while the total channel bandwidth of 1 MHz and the slot time of 625 microseconds remain constant. To be backward compatible, the header of the new packets is still encoded using standard GFSK modulation. Thus, the system becomes backward compatible as legacy devices can at least decode the header of an EDR packet and thus be aware that they are not the recipient of the packet. The same approach is used by WLAN (see Chapter 4) to ensure backward compatibility of 802.11n and 11g networks with older 802.11b devices. Furthermore, a coding scheme for the packet-type field was devised that enables non-EDR devices to recognize multislot EDR packets which are sent by a master to another slave device in order to be able to power down the receiver and thus save energy for the time the packet is sent. Table 6.4 gives an overview of all possible ACL packet

Table 6.4 ACL packet types

Type	Payload (bytes)	Uplink datarate (kbit/s)	Downlink datarate (kbit/s)
DM1	0–17	108.8	108.8
DH1	0–27	172.8	172.8
DM3	0–121	387.2	54.4
DH3	0–183	585.6	86.4
DM5	0–224	477.8	36.3
DH5	0–339	723.2	57.6
2-DH1	0–54	345.6	345.6
2-DH3	0–367	1174.4	172.8
2-DH5	0–679	1448.5	115.2
3-DH1	0–83	531.2	531.2
3-DH3	0–552	1766.4	265.6
3-DH5	0–1021	2178.1	177.1

types and the maximum datarate that can be achieved in an asynchronous connection. In this example, five-slot packets are used in one direction and only one-slot packet in the reverse direction. The first part of the table lists the basic ACL packet types which can be decoded by all Bluetooth devices. The second and third parts of the table contain an overview of the EDR ACL packet types. 2-DH1, 3 and 5 are modulated using DQPSK, while 3-DH1, 3 and 5 are modulated using 8DPSK. The numbers 1, 3 and 5 at the end of the packet type name describe the number of slots used by that packet type.

Owing to the number of EDR packet types, it is no longer possible to identify all packet types using the 4-bit packet-type field of the ACL header (see Figure 6.5). Since it was not possible to extend the field in order to stay backward compatible, the Bluetooth specifications had to go a different way. EDR is always activated during connection establishment. If master and slave recognize that they are EDR capable, the link managers of both devices (see Section 6.4.3) can activate the EDR functionality, which implicitly changes the allocation of the packet-type field bit combinations to point to 2-DHx and 3-DHx types instead of the standard packet types. While the DQPSK modulation is a mandatory feature of the Bluetooth 2.0 standard, 8DPSK has been declared an implementation option. Thus, it is not possible to derive the maximum possible speed of a device by merely looking at the Bluetooth 2.0 + EDR compliancy. Today, most Bluetooth 2.0 devices are capable of sending 3-DH5 packets.

Apart from ACL, SCO and eSCO packets for transferring user data, there are a number of additional packet types that are used for the establishment or the maintenance of a connection.

ID packets are sent by a device before the actual connection establishment to find other devices in the area. As the timing and the hopping sequence of the other device are not known at this time, the packet is very short and only contains the access code.

Frequency-hopping synchronization (FHS) packets are used for the establishment of connection during the inquiry and paging phases, which are further described below. An FHS packet contains the 48-bit device address of the sending device and timing information to enable a remote device to predict its hopping sequence and thus to allow connection establishment.

NULL packets are used for the acknowledgment of a received packet if no user data is waiting in the output buffer of a device that could be used in the acknowledgment packet. NULL packets do not have to be acknowledged and thus interrupt the mutual acknowledgment cycle if no further data is to be sent.

An additional packet type is the POLL packet. It is used to verify if a slave device is still available in the piconet after a prolonged time of inactivity due to lack of user data to be sent. Similar to the NULL packet it does not contain any user data.

6.4.2 The Link Controller

The link control layer is located on top of the baseband layer that was discussed previously. As the name suggests, this protocol layer is responsible for the establishment, maintenance and correct release of connections. To administrate connections, a state model is used on this layer. The following states are defined for a device that wants to establish a connection to a remote device.

If a device wants to scan the vicinity for other devices, the link controller is instructed by higher layer protocols to change into the inquiry state. In this state, the device starts to send two ID packets per slot on two different frequencies to request for listening devices with unknown frequency hopping patterns to reply to the inquiry.

If a device is set by the user to be detectable by other devices, it has to change into the inquiry scan state periodically and scan for ID packets on alternating frequencies. The frequency that a device listens to is changed every 1.28 seconds. To save power, or to be able to maintain already ongoing connections, it is not necessary to remain in the inquiry scan state continuously. The Bluetooth standard suggests a scan time of 11.25 milliseconds per 1.28-second interval. The combination of fast frequency change of the searching device on the one hand and a slow frequency change of the detectable device on the other hand results in a 90% probability that a device can be found within a scan period of 10 seconds.

To improve the time it takes to find devices, version 1.2 of the standard introduces the interlaced inquiry scan. Instead of only listening on one frequency per interval, the device has to search for ID packets on two frequencies. Furthermore, this version of the standard introduces the possibility to report the signal strength (RSSI) with which the ID fame was received to higher layers. Thus, it is possible to sort the list of detected devices by the signal strength and to present devices that are closer to the user at the top of the list. This is especially useful if many devices are in close proximity such as during an exhibition. In this environment, it can become quite difficult to send an electronic business card to a nearby device, as the result of the scan often reveals the presence of several dozen devices and it is necessary to scroll through a long list. If the list is ordered on the signal strength, however, it is very likely that the response of the device that should receive the electronic business card is received with a high signal level because of its closer proximity to the sender and is thus presented at the top of the list.

If a device receives an ID packet, it returns an FHS packet, which includes its address, frequency hopping and synchronization information.

After receiving an FHS packet, the searching device can continue its search. Alternatively, the inquiry procedure can also be terminated to establish an ACL connection with the detected device by performing a paging procedure.

To be detectable, master devices can also enter the inquiry scan state from time to time. Thus, it is possible to detect and connect to them even if they are already engaged in a connection with another device. It has to be noted, however, that some devices like mobile phones do not support this optional functionality.

If a user wants its device to remain invisible, it is possible to deactivate the inquiry scan functionality. Thus, a device can only initiate a paging procedure and thus a connection with the user's device, if it already knows the device's hardware address. It is useful to activate this setting once a user has paired all devices (see Section 6.5.1) that are frequently used together. In this way, the devices of the user remain invisible to the rest of the world but are still able to establish connections with each other. This drastically reduces the opportunity for malicious attacks on Bluetooth devices, which may try to take advantage of security holes of some Bluetooth implementations [4].

To establish an ACL connection by initiating a paging procedure, a device must be aware of the hardware address of the device to connect to from either a previous connection or as a result of an inquiry procedure. The paging procedure works in a similar way as the inquiry procedure, that is, ID packets are sent in a rapid sequence on different frequencies. Instead of a generic address, the hardware address of the target device is included. The target device in turn replies with an ID packet and thus enables the requesting device to return an FHS packet that contains its hopping sequence. Figure 6.8 shows how the paging procedure is performed and how the devices enter the connected state upon success.

The power consumption of a device that is not engaged in any connection and thus only performs inquiry and page scans in regular intervals is very low. Typically, the energy consumption in this state is less than 1 mW. As mobile phones have a battery capacity of typically 4000–5000 mWh, the Bluetooth functionality only has a small effect on the standby time of a mobile device.

After successful paging, both devices enter the connection-active state and data transfer can start over the established ACL connection.

During connection establishment, it can happen that the slave device is master of another connection at the same time. In such cases, the Bluetooth protocol stack enables the

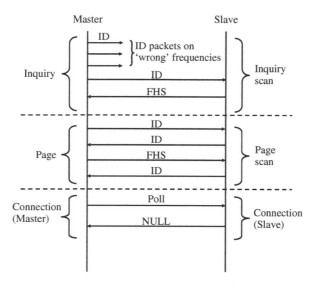

Figure 6.8 Establishment of connection between two Bluetooth devices

device to indicate during the connection establishment that a connection is possible only if a master–slave role change is performed after establishment of the connection. This is necessary, as it is not possible to be a master and a slave device at the same time. However, as a device needs to be a slave in order to be contacted, this feature allows a device to temporarily violate this rule to include another requesting device in its piconet.

During an active connection, the power consumption of a device mainly depends on its power class (see Section 6.2). Even while active, it is possible that for some time, no data is to be transferred. Especially for devices such as smartphones, it is very important to conserve power during these periods to maximize the operating time on a battery charge. The Bluetooth standard thus specifies three additional power-saving substates of the connected state.

The first substate is the connection-hold state. To change into this state, master and slave have to agree on the duration of the hold state. Afterward, the transceiver can be deactivated for the agreed time. At the end of the hold period, master and slave implicitly change back into the connection-active state.

For applications that only transmit data very infrequently, the connection-hold state is too inflexible. Thus, the connection-sniff state might be used instead, which offers the following alternative power-saving scheme. When activating the sniff state, master and slave agree on an interval and the time during the interval in which the slave has to listen for incoming packets. In practice, it can be observed that the sniff state is activated after a longer inactivity period (e.g., 15 seconds) and that an interval of several seconds (e.g., 2 seconds) is used. This reduces the power consumption of the complete Bluetooth chip to below 1 mW. If renewed activity is detected, some devices immediately leave the sniff state even though this is not required by the standard.

With Bluetooth 2.1, an additional sniff-subrating state was introduced to further reduce power consumption, especially for human interface devices (HIDs). With the new mechanism, devices in sniff state can agree on a further reduction of the sniff interval after a configurable timeout. Once the timeout expires, the connection enters the sniff-subrating state. The connection returns to the normal sniff state once a new packet is received and the timer is reset again.

The connection-park state can be used to even further reduce the power consumption of the device. In this state, the slave device returns its piconet address (LT_ADDR) to the master and only checks very infrequently if the master would like to communicate.

6.4.3 The Link Manager

The next layer in the protocol stack (see Figure 6.4) is the link manager layer. While the previously discussed link controller layer is responsible for sending and receiving data packets depending on the state of the connection with the remote device, the link manager's task is to establish and maintain connections. This includes the following operations:

- establishment of an ACL connection with a slave and assignment of a link address (LT_ADDR);
- release of connections;
- configuration of connections, for example, negotiation of the maximum number of timeslots that can be used for ACL or eSCO packets;

- activation of the EDR mode if both devices support this extension of the standard;
- conducting a master–slave role switch;
- performing a pairing operation as described in Section 6.5.1;
- activating and controlling authentication and ciphering procedures if requested by higher layers;
- control of AFH, which was introduced with the Bluetooth 1.2 standard;
- management (activation/deactivation) of power-save modes (hold, sniff and park);
- establishment of an SCO or eSCO connection and the negotiation of parameters like error correction mechanisms, datarates (eSCO only), and so on.

The link manager performs these operations either because of a request from a higher layer (see next section) or because of requests from the link manager of a remote device. Link managers of two Bluetooth devices communicate using the LMP over an ACL connection as shown in Figure 6.9. The link manager recognizes if an incoming ACL packet contains user data or an LMP message by looking at the L_CH field of the ACL header.

To establish a connection with higher layers of the protocol stack after a successful establishment of an ACL connection, the link manager of the initiating device (master) has to establish a connection with the link manager of the remote device (slave). This is done by sending an LMP_Host_Connection_Request message. Subsequently, optional configuration messages can be exchanged. The LMP connection establishment phase is completed by a mutual exchange of an LMP_Setup_Complete message. After this step, it is possible to transfer user data packets between the two devices. Furthermore, it is still possible at any time to exchange further LMP messages that are required for some of the operations that were described in the list at the beginning of this section.

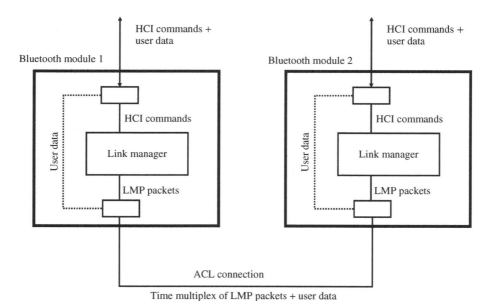

Figure 6.9   Communication between two link managers via the LMP

6.4.4   The HCI Interface

The next layer of the Bluetooth protocol stack is the Host Controller Interface (HCI). In most Bluetooth implementations, this interface is used as a physical interface between the Bluetooth chip and the host device. Exceptions are headsets, for example, which implement all Bluetooth protocol layers in a single chip because of their physical size and the limitation of using Bluetooth only for a single application, that is, voice transmission.

By using the HCI interface, the device (host) and the Bluetooth chip (controller) can exchange data and commands for the link manager with each other by using standardized message packets. Two physical interface types are specified for the HCI.

For devices like notebooks, the USB is used to connect to a Bluetooth chip. USB is the standard interface that is used in all PC architectures today to interconnect with printers, scanners, mouses and other peripheral devices. The Bluetooth standard references the USB specifications and defines how HCI commands and data packets are to be transmitted over this interface.

The second interface for the HCI is a serial connection, the universal asynchronous receiver and transmitter (UART). Apart from power levels, this interface is identical to the RS-232 interface used in the PC architecture. While an RS-232 interface is limited to a maximum speed of 115 kbit/s, some Bluetooth designs use the UART interface to transfer data with a speed of up to 1.5 Mbit/s. This is necessary, as the maximum Bluetooth datarate far exceeds the ordinary speed of an RS-232 interface used with other peripheral devices. The bandwidth that is used on the UART interface is left to the developers of the host device.

The following packet types can be sent over the HCI interface:

- Command packets, which the host sends to the link manager in the Bluetooth chip.
- Response packets, which the Bluetooth controller returns to the host. These packets are also called events, which are either generated as a response to a command or sent on their own, for example, to report that another Bluetooth device would like to establish a connection.
- User data packets to and from the Bluetooth chip.

On the UART interface, the different packet types are identified by a header, which is inserted at the beginning of each packet. The first byte is used to indicate the packet type to the receiver. If USB is used as a physical interface for the HCI, the different packet types are sent to different USB endpoints. The USB polling rate of 1 millisecond ensures that the user data and event packets, which are transmitted from the Bluetooth chip to the host, are detected with only minimal delay.

Today, most Linux distributions for PCs include Bluetooth support and contain a number of shell commands to trace the standardized HCI interface. The '*hcitool con*' command, for example, can be used to show the Bluetooth devices currently connected to the PC. The '*hcitool info <device address>*' command can be used to get further information about a connected device, while the '*hciconfig*' command executed with a number of different parameters gives further information about the capabilities of the Bluetooth chip in the PC. Perhaps, the most useful command is '*hcidump -X*', which allows tracing of all messages and data traversing the HCI interface between the PC's operating system (Linux) and the Bluetooth chip. For further analysis, '*hcidump -w dump-filename*' can be used to save all packets traversing the HCI interface into a file which can then be opened by packet trace software such as Wireshark for further analysis.

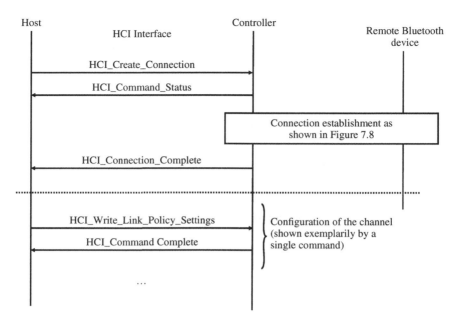

Figure 6.10 Establishment of a connection via the HCI command

Figure 6.10 shows how a Bluetooth module is instructed via the HCI interface to establish a connection with another Bluetooth device. This is done by sending an HCI_Create_Connection command, which includes all necessary information for the Bluetooth controller to establish the connection to the remote device. The most important parameter of the message is the device address of the remote Bluetooth device. The controller confirms the proper reception of the command by returning an HCI_Command_Status event message and then starts the search for the remote device. Figure 6.8 shows how this search is performed. If the Bluetooth device address is known, the inquiry phase can be skipped. If the controller was able to establish the connection, it returns an HCI_Connection_Complete event message to the host. The most important parameter of this message is the connection handle, which allows communication with several remote devices over the HCI interface at the same time. In the Bluetooth controller, the connection handle is directly mapped to the L_CH parameter of an ACL or SCO packet.

Furthermore, there are a number of additional HCI commands and events to control a connection and to configure the Bluetooth controller. A selection of these commands is presented in Table 6.5.

6.4.5 The L2CAP Layer

In the next step of the overall connection establishment, an L2CAP connection is established over the existing ACL link. The L2CAP protocol layer is located above the HCI layer and allows the multiplexing of several logical connections to a single device via a single ACL connection. Thus it is possible, for example, to open a second L_CH between a PC and a mobile phone to exchange an address book entry, while a Bluetooth dial-up connection is

Table 6.5 Selection of HCI commands

Command	Task
Setup_Synchronous_Connection	This command establishes an SCO or eSCO channel for voice applications (e.g., headset to mobile phone communication)
Accept_Connection_Request	The link manager informs the host device of incoming connections by sending a Connection_Request event message. If the host agrees to the connection request, it returns the Accept_Connection_Request command to the Bluetooth controller
Write_Link_Policy_Settings	This command can be used by the host devices to permit or restrict the hold, park and sniff states
Read_Remote_Supported_Features	If the host requires information about the supported Bluetooth functionality of a remote device, it can instruct the Bluetooth controller to request a feature list from the remote device by sending this message. Therefore, the host is informed about the kind of multislot packets that the remote device supports, the power-saving mechanisms that it supports, if AFH is supported, etc.
Disconnect	This message releases a connection
Write_Scan_Enable	With this command, the host can control the inquiry and page scan behavior of the Bluetooth controller. If both are deactivated, only outgoing connections can be established and the device is invisible to other Bluetooth devices in the area
Write_Inquiry_Scan_Activity	This command is used to transfer inquiry scan parameters to the Bluetooth controller, for example, the length of the inquiry scan window
Write_Local_Name	With this command, the host transfers a 'readable' device name to the Bluetooth module. The name is automatically given to remote Bluetooth devices searching for other Bluetooth devices. Thus, it is possible to assemble a list of device names instead of presenting a list of device addresses to the user as a result of a Bluetooth neighborhood search

already established, which connects the PC to the Internet via the mobile phone. If further ACL connections exist to other devices at the same time, L2CAP is also able to multiplex data to and from different devices. Such a scenario is shown in Figure 6.11. While a dial-up connection is established to slave 1, a file is transmitted over the same connection, and an MP-3 data stream is simultaneously received from slave 2.

An L2CAP connection is established from the host device by sending an L2CAP_ Connection_Request message to the Bluetooth controller. The most important parameter of the message is the protocol service multiplexer (PSM). This parameter decides as to which higher layer the user data packets are to be sent once the L2CAP layer is established. For most Bluetooth applications, PSM 0x0003 is used to establish a connection to the RFCOMM layer. This layer offers virtual serial connections to other devices for application layer programs and is described in more detail in Section 6.4.7. Furthermore, the L2CAP_Connection_Request message contains a connection identity (CID) which is used to identify all packets of a particular

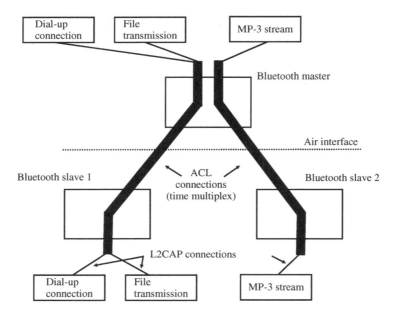

Figure 6.11 Multiplexing of several data streams

L2CAP connection. The CID is necessary, as the RFCOMM layer can be used by several applications at the same time, and thus the PSM is only unique during the connection establishment phase. If the remote device accepts the connection, it returns an L2CAP_Connection_Response message and also assigns a CID, which is used to identify the L2CAP packets in the reverse direction. Later, the connection is fully established and can be used by the application layer program. Optionally, it is now also possible to configure further parameters for the connection by sending an L2CAP_Configuration_Request command. Such parameters are, for example, the maximum number of retransmission attempts of a faulty packet and the maximum packet size that is supported by the device.

Another important task of the L2CAP layer is the segmentation of higher layer data packets. This is necessary if higher layer packets exceed the size of ACL packets. A five-slot ACL packet, for example, has a maximum size of 339 bytes. If packets are delivered from higher layers that exceed this size, they are split into smaller pieces and sent in several ACL packets. Thus, the header of an ACL packet contains the information if the packet includes the beginning on an L2CAP packet or if the ACL packet contains a subsequent segment. At the other end of the connection, the L2CAP layer can then use this information to reassemble the L2CAP packet from several ACL packets, which is then forwarded to the application layer.

6.4.6 The Service Discovery Protocol

Theoretically, it would be possible to begin the transfer of user data between two devices right after establishing an ACL and L2CAP connection. Bluetooth, however, can be used for many different applications, and many devices thus offer several different services to remote devices at the same time. A mobile phone, for example, offers services like wireless Internet connections (Dial-up Network, DUN), file transfers to and from the local file system, exchange of addresses and calendar entries, and so on. For a device to detect which services are offered by a remote device and how they can be accessed, each Bluetooth device contains a service database that can be queried by other devices. The service database is accessed via the L2CAP PSM 0x0001 and the protocol to exchange information with the database is called the Service Discovery Protocol (SDP). The database query can be skipped if a device already knows as to how a remote service can be accessed. As Bluetooth is very flexible, it offers services the option to change their connection parameters at runtime. One of these connection parameters is the RFCOMM channel number. More on this topic can be found in Section 6.4.7.

On the application layer, services are also called profiles. The headset service/headset profile ensures that a headset interoperates with all Bluetooth-enabled mobile phones that also support the headset profile. More about Bluetooth profiles can be found in Section 6.5.

Each Bluetooth service has its own universally unique identity (UUID) with which it can be identified in the SDP database. The dial-up server service, for example, has been assigned an UUID 0x1103. For the Bluetooth stack of a PC to be able to connect to this service on a remote device like a mobile phone, the SDP database is queried at connection establishment and the required settings for the service are retrieved. For the dial-up server service, the database returns information to the requesting device that the L2CAP and RFCOMM layers (see next section) have to be used for the service and also informs the requestor of the correct parameters to use.

The service database of a Bluetooth device furthermore offers a universal search functionality. This is required to enable a device to discover all services offered by a so far unknown device. The message sent to the database for a general search is called an SDP_Service_Search_Request. Instead of a specific UUID as in the example above, the UUID of the public browse group (0x1002) is used. The database then returns the UUIDs of all services it offers to other devices. The parameters of the individual services can then be retrieved from the database with SDP_Service_Search_Attribute_Request messages. For a service query, the database also returns the name for the requested service that can be set by the higher layers of the Bluetooth stack. Therefore, it is possible to have country- and Language-specific service names that are automatically assigned, for example, during the installation of the Bluetooth stack. The name, however, is just for presenting the service to the user. The Bluetooth stack itself always identifies a service by its UUID and never by using the service name (see Figure 6.12).

In practice, information that was initially retrieved from the service database of a remote device is usually stored on the application layer to access a remote device more quickly in subsequent communication sessions.

To finish the database request, the remote device releases the L2CAP connection by sending an L2CAP_Disconnection_Request message. If the device wants to establish a connection to one of the detected services right away, the ACL connection remains in place and another L2CAP_Connection_Request message is sent. This message, however, does not contain the PSM ID 0x0001 for the service database as before, but contains the PSM ID for the higher layer that needs to be contacted for the selected service. For most services this will be the RFCOMM layer, which offers a virtual serial connection. This service is accessed via

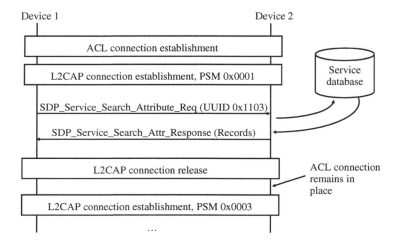

Figure 6.12   Establishment of connection to a service

PSM 0x0003. One of the few services that do not use the RFCOMM layer for data transfer is voice application (e.g., headset profile), which uses SCO connections for synchronous data transfer.

6.4.7   The RFCOMM Layer

As has been shown in Section 6.4.5, the L2CAP layer is used to multiplex several data streams over a single physical connection. The service database, for example, is a service that is accessed via the L2CAP PSM 0x0001. Other services can be accessed in a similar way by using other PSM IDs. In practice, some services also commonly use another layer, which is called RFCOMM and which is accessed with PSM 0x0003. RFCOMM offers a virtual serial interface to services and thus simplifies the data transfer.

How these serial interfaces are used depends on the higher layer service that makes use of the connection. The 'serial port' service, for example, uses the RFCOMM layer to offer a virtual serial interface to any non-Bluetooth application. From the application's point of view, there is no difference between a virtual serial interface and a separate, physical serial interface. Usually, the operating system assigns COM port 3,4,5,6, etc. to the Bluetooth serial interfaces. Which COM port numbers are used is decided during the installation of the Bluetooth stack on the device. Before Wi-Fi tethering became popular, these serial interfaces were then used during the installation of a new modem driver for the Windows DUN functionality. When an application such as the Windows DUN used the modem driver to establish a connection, the Bluetooth stack opened a connection to the remote device. This process could be performed automatically if the Bluetooth stack was previously assigned a certain COM port number to a specific remote device.

To simulate a complete serial interface, the RFCOMM layer simulates not only the transmit and receive lines but also the status lines like the request to send (RTS), Clear to Send (CTS), Data Terminal Ready (DTR), Data Set Ready (DSR), Data Carrier Detect (CD) and the Ring

Indicator (RI). In a physical implementation of a serial interface, these lines are handled by a UART chip. Thus, the Bluetooth serial port service simulates a complete UART chip. A real UART chip translates the commands of the application layer into signal changes on physical lines. The virtual Bluetooth UART chip on the other hand translates higher layer commands into RFCOMM packets, which are then forwarded to the L2CAP layer.

RFCOMM is also used by other services, like the file transfer service (Object Exchange (OBEX)) that is still in use today. By using different RFCOMM channel numbers, it is possible to select during connection establishment which of the services to communicate with. The channel number is part of the service description in the SDP database. For example, if a device asks the service database of a remote Bluetooth device for the parameters of the OBEX service, the remote device will reply that the service uses the L2CAP and RFCOMM layer to provide its service. Thus, the device will establish a L2CAP connection by using PSM 0x0003 to establish a connection to the RFCOMM layer (L2CAP to RFCOMM). Furthermore, the database entry contains the RFCOMM channel number so that the device can connect to the correct higher layer service. As the RFCOMM number can be assigned dynamically, the service database has to be queried during each new connection establishment to the service.

Figure 6.13 shows how different layers multiplex simultaneous data streams. While the HCI layer multiplexes the connections to several remote devices (connection handles), the L2CAP layer is responsible for multiplexing several data streams to different services per device (PSM and CID). This is used in practice to differentiate between requests to the service database (PSM 0x0001) and the RFCOMM layer (PSM 0x0003). Apart from the service database, many Bluetooth services use the RFCOMM layer and thus can only be distinguished because they use different RFCOMM channel numbers.

The RFCOMM channel number also allows the use of up to 30 RFCOMM services between two devices simultaneously. Thus, it is possible during a dial-up connection to establish a

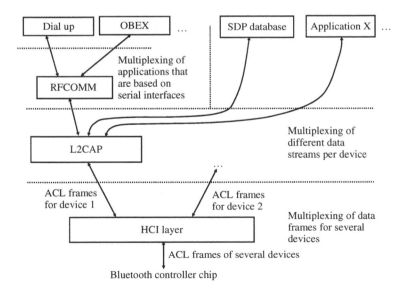

Figure 6.13   Multiplexing on different protocol layers

second connection to transfer files via the OBEX service. As both services use different RFCOMM channel numbers, the data packets of the two services can be time multiplexed and can thus be delivered to the right services at the receiving end.

6.4.8   Overview of Bluetooth Connection Establishment

Figure 6.14 gives an overview of how a Bluetooth connection is established through the different layers. To contact an application on a remote Bluetooth device, an ACL connection is initially established. Once the ACL link is configured, an L2CAP connection to the service database of the device is established by using the corresponding PSM number. Once the connection to the database is established, the record of the service to be used is retrieved. Then, the L2CAP connection is released while the ACL connection between the two devices remains in place. In the next step, contact to the application is established over the still existing ACL connection. This is done by establishing another L2CAP connection. Most services use the RFCOMM layer for further communication, which provides virtual serial interfaces. By using the RFCOMM channel number, the Bluetooth stack can finally connect the remote device to the actual service, for example, previously used modem service. How the two sides of the application communicate with each other depends on the application itself and is transparent for all layers described so far, including the RFCOMM layer. To ensure interoperability on the application layer between devices of different manufacturers, Bluetooth defines the so-called 'profiles', which are described in more detail in Section 6.6.

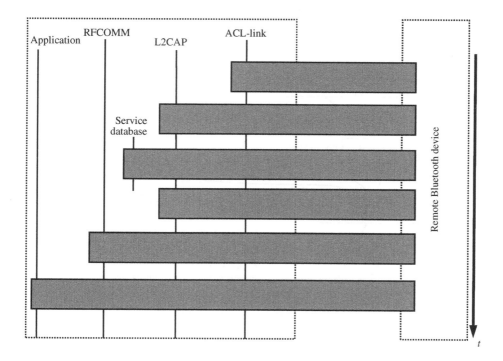

Figure 6.14   The different steps of a Bluetooth connection establishment

6.5   Bluetooth Security

As Bluetooth radio waves do not stop at the doorstep, the Bluetooth standard specifies a number of security functions. All methods are optional and do not have to be used during connection establishment or for an established connection. The standard has been defined thus: some services do not require security functionality. Which services are implemented without security is left to the discretion of the device manufacturer. A mobile phone manufacturer, for example, can decide to allow incoming file transfers without a prior authentication of the remote device. The incoming file can be held in a temporary location and the user can then decide to either save the file in a permanent location or discard it. For services like dial-up data, such an approach is not advisable. Here, authentication should occur during every connection establishment attempt to prevent unknown devices from establishing an Internet connection without the user's knowledge.

Bluetooth uses the SAFER + (Secure and Fast Encryption Routine) security algorithms, which have been developed by ETH Zurich and are publicly available. So far, no methods have been found that compromise the encryption itself. However, there have been reports on device-specific Bluetooth security problems as, for example, discussed in [4] and general weaknesses have been found concerning the initial key negotiation. If an attacker is able to record the initial pairing process that is described below, he can calculate the keys and later on decrypt the data. With version 2.1 of the Bluetooth standard, the new pairing mechanisms were hence introduced, which are described in Section 6.5.2.

6.5.1   Pairing up to Bluetooth 2.0

To automate security procedures during subsequent connection establishment attempts, a procedure called 'pairing' is usually performed during the first connection establishment between the two devices. From the user's point of view pairing means typing in the same PIN number on both devices. The PIN number is then used to generate a link key on both sides. The link key is then saved in the Bluetooth device database of both devices and can be used in the future for authentication and activation of ciphering. The different steps of the paring procedure are shown in Figure 6.15 and are performed as follows.

To invoke the pairing procedure, an LMP_IN_RAND message is sent by the initiating device over an established ACL connection to the remote device. The message contains a random number. The random number is used together with the PIN and the device address to generate an initialization key, which is called K_{init}. As the PIN is not exchanged between the two devices, a third device is not able to calculate K_{init} with an intercepted LMP_IN_RAND message.

By using K_{init}, which is identical in both devices, each side then creates a different part of a combination key. The combination key is based on K_{init}, the device address of one of the devices and an additional random number, which is not exchanged over the air interface. Then, the two combination key halves are XOR combined with K_{init} and are exchanged over the air interface by sending LMP_COMB_KEY messages. The XOR combination is necessary in order not to exchange the two combination key halves in clear text over the still unencrypted connection.

As K_{init} is known to both sides, the XOR combination can be reversed and thus the complete combination key is then available on both devices to form the final link key. The link key finally forms the basis for the authentication and ciphering of future connections between the two devices.

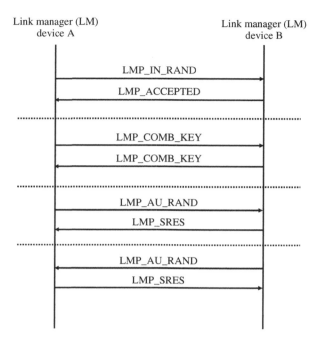

Figure 6.15   Pairing procedure between two Bluetooth devices

As the link key is saved in both devices, a pairing procedure and the input of a PIN by the user are only necessary during the first connection attempt. By saving the link key together with the device address of the remote device, the link key can be automatically retrieved from the database during the next connection establishment procedure. Authentication is then performed without requiring interaction with the user.

To verify that the link key was created correctly by both sides, a mutual authentication procedure is performed after the pairing. The way the authentication is performed is described in more detail in the next section. Figure 6.15 also shows how the complete pairing is performed by the link manager layers of the Bluetooth chips of the two devices. The only input needed from higher layers via the HCI interface is the PIN number to generate the keys.

6.5.2   Pairing with Bluetooth 2.1 (Secure Simple Pairing)

In 2005, Yaniv Shaked and Avishai Wool discovered a number of weaknesses that allow one to calculate the PIN and the link keys from the data exchanged during the pairing procedure. This might have been one of the reasons why Bluetooth 2.1 introduced the following new pairing mechanisms that are referred to as secure simple pairing:

The Numeric Comparison protocol: The major difference between this pairing mechanism and the classic protocol is that a public/private key exchange mechanism is used instead of a PIN. For this purpose, each device has a private and a public key. During the pairing process, each device sends its public key to the other end, which then encrypts a random number with the key and returns it to the originator. After both devices have received the encrypted random numbers, they use their private keys to decrypt the information which is then used to generate the link keys. The encryption and decryption only work one way, that is, the random number encrypted with the public key can only be decrypted with the private key. As the private keys are never transmitted over the air, an attacker cannot generate the link key from the intercepted message exchange. A similar authentication is also used by the EAP-TLS Wi-Fi authentication method (see Chapter 5) and during the first access to a web page via secure HTTP (HTTPS, SSL/TLS).

As the two devices do not yet know each other, however, an attacker could insert itself between the two devices and act as device B to device A and vice versa. This is also referred to as a Man-in-the-middle (MITM) attack. To prevent this, the Numeric Comparison protocol calculates a six-digit number after the generation of the link keys, which is then shown to the user on both devices. The user then has to confirm that the numbers are identical before the pairing process is finished. The method to calculate the six-digit number prevents MITM attacks as a device in the middle would alter the calculation and the numbers shown on the devices would not be the same. The Bluetooth Special Interest Group (SIG) states that by using this pairing mechanism, the chance of a successful MITM attack is below 1:1,000,000.

The Just Works protocol: This protocol is mostly identical to the Numeric Comparison protocol described above with the difference that no six-digit number is calculated and shown to the user. Hence, this pairing mechanism does not offer protection against MITM attacks but is still required, as some devices such as headsets do not have a display to show a generated number. Consequently, this pairing method should only be used if it is highly unlikely that no attacker is present during the pairing process. In case a MITM attack was successful during the pairing process, the attacker needs to be present during future communication sessions as otherwise the connection establishment process would fail.

The Passkey protocol: Here, a passkey (PIN) is used for authentication and, hence, this pairing option looks identical to the classic Bluetooth pairing method. Unlike in the classic pairing method, the PIN is not used as shown before, but instead private/public keys and random numbers are used during the pairing process. At the end of the pairing process, an acknowledgment for each bit of the PIN is generated, which is referred to as 'commitment' in the standard. The input parameters for the commitment algorithm on both sides are the public key, a different random number on each side and the current bit of the PIN. In the first step, both devices exchange the commitment for one bit. Subsequently, device A sends the random number used for the calculation so that device B can verify the commitment with a reverse algorithm. If the commitment is successfully verified, device B then sends its own random number to device A so that it can also verify the commitment. For the next bit, the procedure is performed in the reverse direction. An attacker in the middle cannot forge the commitments, as a bit of the PIN can only be reverse engineered from the commitment verification exchanges once the second random number has been sent. As the commitments are alternating, an attacker could only get one bit from each side before he would have to send a commitment. He is unable to do so, however, as he is not in the possession of the PIN.

The Out-of-Band protocol: Finally, Bluetooth 2.1 specifies a method to partly or fully perform authentication via a channel that is independent from the Bluetooth air interface. In practice, this method has been defined for use with Near Field Communication (NFC). During the authentication process, the devices have to be held very close to each other, a situation which prevents MITM attacks, as the attacker could potentially intercept the pairing process but would not be able to insert itself in the middle. The Bluetooth standard supports active NFC chips that can transmit and receive and also passive NFC chips that can only transmit when energy is induced via their antenna. This is necessary, as some devices such as headsets might not have space for an additional antenna. In such a case, the passive NFC chip could be put into the user manual of the device or on the packaging. During the pairing process, the Bluetooth device with an active NFC chip is then held close to the passive NFC chip. The passive NFC chip then transmits all necessary information to perform a secure pairing without user interaction.

The NFC method is also suitable for use when an action is to be performed when two devices are held close to each other. A practical example is when the user would like to print pictures stored on a mobile device. The user then holds the device that contains the pictures close to the printer. Both devices can then detect each other over the NFC interface and a connection is automatically established.

6.5.3 Authentication

Once the initial pairing of the two devices has been performed successfully, the link key can be used for mutual authentication during every connection request. Authentication is performed using a challenge/response procedure, which is similar to procedures of systems such as Global System for Mobile Communications (GSM), General Packet Radio Service (GPRS) and Universal Mobile Telecommunications System (UMTS). For the Bluetooth authentication procedure, three parameters are necessary:

- a random number;
- the Bluetooth address of the device initiating the authentication procedure;
- the 128-bit link key, which has been created during the pairing procedure.

Figure 6.16 shows how the initiating device (verifier) sends a random number to start the authentication procedure to the remote device (claimant). The link manager of the claimant then uses the BD_ADDR of the verifier device to request the link key for the connection from the host via the HCI interface.

With the random number, the BD_ADDR and the link key, the link manager of the claimant then calculates an answer, which is called the signed response[*] (SRES[*]), which is returned to the link manager of the verifier device. In the mean time, the verifier device has calculated its own SRES. The numbers can only be identical if the same link key was used to calculate the SRES on both sides. As the link key is never transmitted over the air interface, an intruder can thus never successfully perform this procedure.

6.5.4 Encryption

After successful authentication, both devices can activate or deactivate ciphering at any time. The key used for ciphering is not the link key that has been generated during the pairing process. Instead, a ciphering key is used, which is created on both sides during the activation of

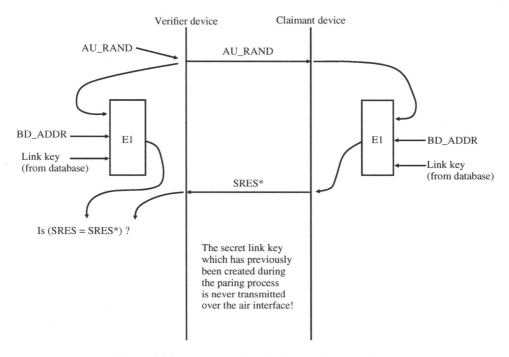

Figure 6.16   Authentication of a Bluetooth remote device

ciphering. The most important parameters for the calculation of the ciphering key are the link key of the connection and a random number, which is exchanged between the two devices when ciphering is activated. Since ciphering is reactivated for every connection, a new ciphering key is also calculated for each connection (see Figure 6.17).

The length of the ciphering key is usually 128 bits. Shorter keys can be used as well if Bluetooth chips are exported to countries for which export restrictions apply for strong encryption keys.

Together with the device address of the master and the lower 26 bits of the master's real-time clock, the ciphering key is used as input value for the SAFER + E0 algorithm, which produces a constant bit stream. As the current value of the master's real-time clock is known to the slave as well, both sides of the connection can generate the same bit stream. The bit stream is then modulo-2 combined with the clear text data stream. Encryption is applied to the complete ACL packet including the CRC checksum before the addition of optional FEC bits.

6.5.5 Authorization

Another important concept of the Bluetooth security architecture is the 'authorization service' for the configuration of the behavior of different services for different remote users. This additional step is required to open services to some but not all remote devices. Thus, it is possible,

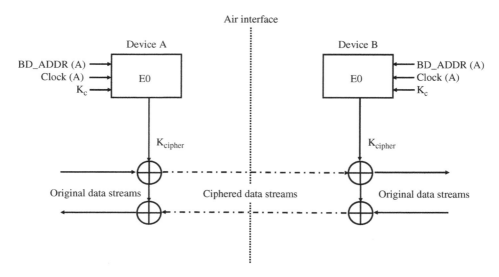

Figure 6.17   Bluetooth encryption by using a ciphering sequence

for example, to grant access rights to a remote user for a certain directory on the local PC to send or receive files. This is done by activating the OBEX service for the particular user and his Bluetooth device.

With the authorization service, it is possible to configure certain access rights for individual external devices for each service offered by the local device. It is left to the manufacturer of a Bluetooth device to decide how this functionality is used. Some mobile phone manufacturers, for example, allow all external devices, which have previously performed a pairing procedure successfully, to use the dial-up service. Other mobile phone manufacturers have added another security barrier and ask the user for permission before proceeding with the connection establishment to the service.

Bluetooth stacks for PCs usually offer very flexible authentication functionality for the service offered by the device. These include the following:

- A service may be used by an external device without prior authentication or authorization by the user.
- A service may be used by all authenticated devices without prior authorization by the user. This requires a one-time pairing.
- A service may be used once or for a certain duration after authentication and authorization by the user.
- A service may be used by a certain device after authentication and one-time authorization by the user.

Furthermore, some Bluetooth stacks offer the display of short notices on the screen if a service is accessed by a remote device. The notice is displayed for informational purposes only, as access is automatically granted.

6.5.6   Security Modes

The point at which ciphering and authorization is performed during the establishment of an authenticated connection depends on the implementation of the Bluetooth stack and the configuration of the user. The Bluetooth standard describes four possible configurations.

If security mode 1 is used for a service, no authentication is required and the connection is not encrypted. This mode is most suitable for the transmission of address book and calendar entries between two devices. In many cases, the devices used for this purpose have previously not been paired.

For security mode 2, the user decides if authentication, ciphering and authorization are necessary when a service is used. Many Bluetooth PC stacks allow individual configuration for each service. Security mode 1 therefore corresponds to using security mode 2 for a service without authentication and ciphering.

If a service uses security mode 3, authentication and ciphering of the connection are automatically ensured by the Bluetooth chip. Both procedures are performed during the first communication between the two link managers, that is, even before an L2CAP connection is established. For incoming communication requests, the Bluetooth controller thus has to ask the Bluetooth device database for the link key via the HCI interface. If no pairing has previously been performed with the remote device, the host cannot return a link key to the Bluetooth controller and thus the connection will fail. Security mode 3 is best suited for devices that only need to communicate with previously paired remote devices. Thus, this mode is not suitable for devices like mobile phones, which allow non-authenticated connections for the transfer of an electronic business card.

With version 2.1 of the Bluetooth specification, security mode 4 was introduced, which can be used with the Secure Simple Pairing mechanisms described above. This mode is similar to security mode 2 described above as a security category is selected on a per application basis:

- A secured link key is required, which requires that the initial pairing was performed with either the Numeric Comparison, Out-of-Band or Passkey protocol.
- A nonsecured link key is required, that is, the Just Works protocol was used during the pairing.
- No security is required at all.

6.6   Bluetooth Profiles

As shown at the beginning of this chapter, Bluetooth can be used for a great variety of applications. Most applications have a server and a client side. A client usually establishes the Bluetooth connection to the master and requests the transfer of some kind of data. Thus, the master and the client sides of a Bluetooth service are different. For example, for the transfer of a calendar entry from one device to another the client side establishes a connection to the server. The client then transfers the calendar entry as the sending component. The server, on the other hand, receives the calendar entry as the receiving component. To ensure that the client can communicate with servers implemented by different manufacturers, the standard defines a number of Bluetooth profiles. For each application (headset, calendar and address transfer, audio streaming, etc.), an individual Bluetooth profile has been defined, which describes how the server side and the client side communicate with each other. If both sides support the same profile, interoperability is ensured.

It is noteworthy that the client/server principle of the Bluetooth profile should not be confused with the master/slave concept of the lower Bluetooth protocol layers. The master/slave concept is used to control the piconet, that is, who is allowed to send and at which time, while the client/server principle describes a service and the user of a service. Whether the Bluetooth device, which is used as a server for a certain service, is the master or the slave in the piconet is thus irrelevant.

Table 6.6 gives an overview of a number of different Bluetooth profiles for a wide range of services. In practice, it can be observed, that the use of Bluetooth concentrates on a few profiles and some of them are described in more detail in the following sections.

Table 6.6 Bluetooth profiles for different applications

Profile name	Application
Headset profile	Profile for wireless headsets used with mobile phones. Voice quality transmissions only, not suitable for music
Hands-free profile	This profile is used to connect mobile phones with hands-free sets in cars
SIM access profile	Provides access for hands-free equipment in cars to the data stored on the SIM card of a mobile phone
Human interface device (HID) profile	Connects mouses, keyboards and joysticks to PCs, notebooks and smartphones
File transfer profile	This profile can be used to exchange files between two Bluetooth devices
Object push profile	Simple exchange of calendar entries, address book entries, etc.; used for ad hoc transfers
Advanced audio distribution profile	Profile for the transmission of high-quality audio, for example, music between an MP-3 player and a Bluetooth headset
Audio/video remote control profile	Profile to control audio/video devices remotely. This profile can be used, for example, with the advanced audio distribution profile to remotely control the audio player from the headset or an independent remote control
Dial-up networking (DUN) profile	Bluetooth connection between a modem or a mobile phone and a remote device like a PC or a notebook
FAX profile	Profile for FAX transmissions
Common ISDN access profile	Profile for interconnecting an ISDN adapter with a remote device like a PC or notebook
LAN access profile	IP connection between a smartphone, PC or notebook and a LAN and the Internet
Personal area network (PAN) profile	Same as the LAN access profile. However, the PAN profile does not simulate an Ethernet network card but instead uses Bluetooth protocols for this purpose
Synchronization profile	Synchronization of personal information manager (PIM) applications for calendar and address book entries, notes, etc.
Basic imaging profile	Transfer of pictures from and to digital cameras
Hard copy cable replacement profile	Cable replacement between printers and a remote device like a PC
Basic printing profile	Printing profile for mobile devices like mobile phones to enable them to print information without a printer driver

6.6.1   Basic Profiles: GAP, SDP and the Serial Profile

The Bluetooth standard specifies two profiles which are not visible on the application level. The Generic Access Profile (GAP) [2] defines as to how two devices can connect with each other in different situations and how to perform the connection establishment. The profile describes, among other things:

- the presentation of Bluetooth-specific parameters to the user like the device address (BD_ADDR) or the PIN;
- security aspects (security mode 1–3);
- idle mode behavior (e.g. inquiry, device discovery);
- connection establishment.

The GAP protocol thus ensures that the user interfaces for the configuration of the Bluetooth stack are similar on all devices. Furthermore, the GAP profile specifically defines which messages are sent during connection establishment, their order and which actions are taken when different options are discovered.

As shown in Section 6.4.6, each Bluetooth device has its own service database, in which each local service can store important data for the connection establishment to a remote device. The service discovery application profile [5] defines how the database is accessed and how it is structured for each profile.

The Serial Port Profile (SPP) [6] is also a basic profile on which many other profiles are based. As the name implies, this profile simulates a serial interface for any kind of application. The profile uses the RFCOMM layer, which already offers all necessary functionalities on a lower layer. If a device has implemented this profile, any higher layer application that is able to transfer data over a serial interface is able to communicate with remote Bluetooth devices. A special adaptation of the application to the Bluetooth protocol stack is not necessary because on the application layer the simulated Bluetooth serial interface behaves like a physically present serial interface. Figure 6.18 shows the protocol stack of the SPP.

Here is a practical example. The SPP can be used by a terminal program like Hyperterm to access a remote modem with a built-in Bluetooth interface. Before the Bluetooth connection can be used, the PC has to be paired with the modem. The Bluetooth configuration

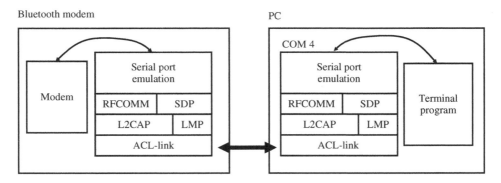

Figure 6.18   Protocol stack for the SPP

program is then used on the PC to assign a certain COM port number (e.g., COM 4) to the modem. The Bluetooth connection to the modem is automatically established whenever the terminal program is launched and the serial interface is accessed. All of this is transparent to the terminal program as it only sees the COM port which it treats as if it were a physically present interface.

6.6.2   Object Exchange Profiles: FTP, Object Push and Synchronize

To transfer structured objects such as files, business cards, calendar information, address book entries, etc., one of the several OBEX profiles is used as shown in Figure 6.19. An OBEX connection is established only between two devices for the duration of the transmission of one or several objects that are transmitted in sequence. OBEX services are based on the General Object Exchange profile, which is in turn based on the L2CAP and RFCOMM layers. Three specialized OBEX profiles then use General Object Exchange profile for specific services.

For the transfer of files and even complete directory structures, the File Transfer Profile (FTP) [7] has been developed. This should not be confused with the File Transfer Protocol of the Transmission Control Protocol (TCP)/IP world, which uses the same acronym.

The OBEX FTP is mostly used to transfer files between devices such as notebooks and smartphones. The files can be located at any position in the file system. The Generic Object Exchange Profile (GOEP) defines the following commands for this task, which are sent in a binary coding over an established RFCOMM connection: DISCONNECT, PUT, GET, SET-PATH and ABORT. Some PC Bluetooth stacks insert the directory tree of a remote Bluetooth device into the overall directory tree of the local device in a similar way as a remote file system on a local network. If the user clicks on the remote Bluetooth device in the directory tree, the general OBEX GET command is used to request the root directory of the remote Bluetooth device, which is then presented to the user in the local file manager. The user can then select one or several files for transfer to the local PC. For this purpose, the GOEP GET command is used. It is also possible to copy files or directories to the remote Bluetooth device. For this purpose, the general OBEX PUT command is used.

If the user changes to a subdirectory on the remote device, the OBEX SETPATH command is used in combination with another OBEX GET command to request the directory listing.

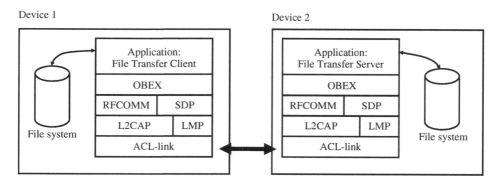

Figure 6.19   Protocol stack of the OBEX file transfer profile

```
<xml version="1.0">

<!DOCTYPE folder-listing SYSTEM „obex-folder listing.dtd">

<folder-listing-version="1.0">

    <folder name="Camera" modified="2004117T100840"
    user perm="RWD" group perm"W" />

    <folder name="other pics" modified="2004117T13321"
    user perm="RWD" group perm"W" />

</folder-listing>
```

Figure 6.20   XML-encoded directory structure

Figure 6.20 shows how the content of a directory is XML encoded in a human readable format and sent to the requesting device.

In the OBEX protocol layer, the CONNECT, DISCONNECT, PUT, GET, SETPATH and ABORT commands and the corresponding answers are processed as packets. The first byte of a packet identifies the command. The command field is followed by 2-byte length field and the parameters of the command. A parameter can be a directory name, a directory listing or the contents of a requested file. The standard uses the term 'header' for a parameter, which is somewhat confusing. To be able to recognize the type of a parameter, each parameter contains a type information in the first byte. The type of a parameter can, for example, be 'filename' or 'body' (the content of the file).

The maximum size of a packet is 64 kB. To transfer bigger files, that is, 'header' of type 'body', the file is automatically split into several packets by the OBEX layer.

Although the FTP profile is not commonly used anymore, a somewhat simpler application of the GOEP, that is, the object push profile [8] (Figure 6.21), has remained quite popular. This profile is used if the user wants to transmit a single calendar entry, address book entry or a single file via Bluetooth to another device. The profile works in the same way as the FTP, as it also uses general OBEX commands like PUT and GET. The object push profile, however, does not support directory operations or the deletion of files. This simplification accelerates

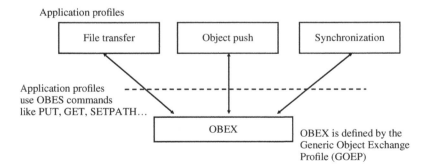

Figure 6.21   The FTP, object push and synchronization profiles are based on GOEP

the process for single objects, as only a few decisions have to be made by the user before the object is transmitted.

Many devices allow an incoming object push transfer without prior authentication and ciphering. The object is then stored in a buffer and only inserted into the calendar, the address book or copied to the file system once the user has authorized the transfer.

For the transmission of calendar and address book entries, the Bluetooth standard requires the use of vCalendar and the vCard format, which are standardized in [9–13]. This is a precondition to exchange address book and calendar entries between any program and any end-user device. For other objects such as pictures, the file name extension (e.g., .gif, .jpg, etc.) can be used by the receiver to make a decision on how to treat the received object.

Even though the profile is called 'object push', it also defines an optional business card pull functionality which can be used to send a predefined business card to a remote device upon its request. The business card exchange feature extends the functionality to automatically send the business card stored in the retrieving device during a request for a business card to the remote device.

The third profile, which is based on GOEP, is the synchronization profile [14]. Like the file transfer profile described before, it is not used much in practice anymore but shall be described for completeness. The synchronization profile allows automated synchronization of objects like calendar, address book entries, notes, and so on. Again, general OBEX commands like GET and PUT are used. While the object push profile can only transfer a single-address book entry to a remote device, the synchronization profile describes how to synchronize all records of a database. During the first synchronization attempt, all entries of the database on both devices are exchanged with each other. During all subsequent synchronizations, only objects that have changed since the last synchronization session are updated on both sides. This is achieved by recording every change of an object in a journal. To allow the exchange database records of products of different vendors, the synchronization profiles also use the standardized vCard and vCalendar formats.

The Bluetooth standard does not itself define how the synchronization is performed, but uses the synchronization system defined in the Infrared Mobile Communications (IrMC) standard [15] of the Infrared Data Association.

6.6.3   Headset, Hands-Free and SIM Access Profile

Wireless headsets for mobile phones were the first Bluetooth devices on the market. To establish a voice channel between the mobile phone and the headset, the headset profile [16] is used. This profile is special, as it is one of the few profiles which use SCO packets (see Section 6.4.1) for a connection. The SCO connection has a bandwidth of 64 kbit/s and carries the bidirectional audio stream between the headset and the mobile phone. If both devices are Bluetooth 1.2 compatible, eSCO packets are used for the voice path to add error correction and AFH. These features, which have been introduced with Bluetooth 1.2, particularly increase the speech quality if the error rate on the Bluetooth link increases because of an increased distance between the two devices or if there are obstacles in the transmission path which decrease the channel quality. If one of the two devices is not yet compatible with Bluetooth 1.2, the link manager layer automatically ensures that only SCO packets are sent and that AFH remains deactivated.

To use a headset with a mobile phone, the two devices have to be paired initially. Subsequently, the mobile phone tries to establish a connection to the Bluetooth headset for every incoming call. For the signaling between the headset and the mobile phone, referred to as

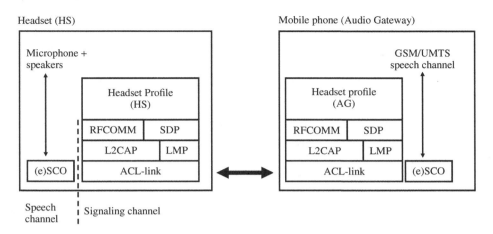

Figure 6.22   The headset profile protocol stack

the Audio Gateway (AG) in the Bluetooth headset standard, an ACL connection is used. The signaling connection uses the L2CAP and RFCOMM layers for communication as shown in Figure 6.22.

To exchange commands and the corresponding responses between the AG and the headset, the AT command language is used, which was initially designed for the communication between a data terminal and a modem. The headset profile not only reuses some of the well-known AT commands, but also defines a number of extra commands to account for the special nature of the application. Figure 6.23 shows how the AG establishes a signal channel

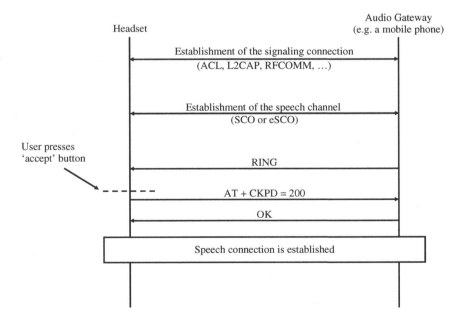

Figure 6.23   Establishment of the signaling and the speech channels

based on an ACL connection to send an unsolicited 'RING' response to the headset. The headset then informs the user about the incoming call by generating a 'ringing tone'. The user can then answer the call by pressing a key on the headset. When the user presses the accept button, the headset sends the following command to the AG to open the speech path: 'at + ckpd = 200'. The mobile phone then accepts the call and starts acting as an AG between the mobile network and the headset.

To conduct an outgoing call, the headset is also able to establish a new connection to the AG. Together with the speech dialing function, which is usually part of the mobile phone, it is possible to initiate outgoing calls via the headset without any interaction with the mobile phone.

Owing to the small size of the headset, only few functionalities of the remote device can be controlled via the headset. Thus, the only additional functionality of the headset profile is to control the volume of an ongoing conversation. This is done via +vgm AT commands to control the volume of the microphone, and +vgs commands to control the volume of the loudspeaker. In this way, it is also possible to control the volume settings of the mobile phone via the headset.

The headset can also be paired with a PC in the case where the Bluetooth stack on the PC supports the headset profile and where it has implemented the AG role. Therefore, a headset can be used with voice over IP software for telephone calls via the PC. Furthermore, it is possible to redirect the inputs and outputs of the PC's soundcard to the headset to stream music, MP3 files, and so on. to the headset. This application is not very useful, however, as the (e)SCO channel is limited to 64 kbit/s and has been optimized for the transmission of mono audio signals only. Furthermore, the frequency band is limited to 300–3400 Hz. A much more suitable profile for this task is the advanced audio distributed profile, which is described in more detail in Section 6.6.4.

Closely related to the headset profile is the hands-free profile [17], which addresses the special needs of hands-free devices in cars that cannot be fulfilled by the headset profile. The most important feature of this profile is to replace the wired connection between the hands-free car kit and the mobile phone. By using this profile, the mobile phone need not be installed in the car and can thus remain in the pocket or bag of the user. Despite the similar purpose of a headset and a hands-free set, an additional profile was necessary, as a hands-free set today typically offers much more possibilities to interact with the mobile phone than a headset.

The basic mode of operation of the hands-free profile is identical to that of the headset profile. Commands and replies are exchanged between the hands-free unit and the mobile phone (AG) via AT commands. Furthermore, the headset profile also uses SCO or eSCO connections for the voice path. In addition to the functionality of the headset profile, the hands-free profile offers the following functionalities:

- The transmission of the caller's number to the hands-free kit (CLIP).
- The hands-free set can reject incoming calls.
- The hands-free set can send a phone number to the AG, which the user has typed in via the keypad of the hands-free set.
- Call hold and multiparty calls.
- Transmission of status information such as remaining battery capacity and mobile network reception conditions of the mobile phone.

- Transmission of a roaming indicator to allow the hands-free set to indicate to the user that the phone is registered in a foreign network.
- Deactivation of the optional echo canceller of a mobile phone if the hands-free kit uses an integrated echo canceller.

Another possibility to use a headset and a hands-free car kit is the SIM access profile [18]. Contrary to the headset and hands-free profile, the mobile phone is not used as an AG, that is, as a bridge to the mobile phone network, but only offers access to its SIM card to an external device. Figure 6.24 shows this scenario. The external device, which will in most cases be a hands-free car kit, contains its own GSM/UMTS mobile phone except for the SIM card. When the hands-free kit is activated at the beginning of a trip, it establishes a Bluetooth connection to the mobile phone with which it has previously been paired. Activating the SIM access server in the mobile phone deactivates the mobile phone's radio module. This is necessary as the radio module in the hands-free kit is used for communication with the mobile network. Another big advantage of this method is the fact that the hands-free kit is usually connected to the power system of the car and an external antenna, which is not possible with the headset and the hands-free profile.

Figure 6.24 also shows the protocol stack that is used by the SIM access profile. On the basis of an L2CAP connection, the RFCOMM layer is used for a serial transmission between the hands-free kit (SIM access client) and the mobile phone (SIM access server). Apart from SIM access profile commands for activating, deactivating and resetting the SIM card, the Bluetooth connection is also used to send SIM card commands and responses. The commands and responses are sent as Application Protocol Data Units (APDUs) (see Section 1.10 and Figures 1.50 and 1.51). Instead of exchanging APDUs between the radio part of the hands-free kit and the SIM card in the mobile phone via an electrical interface, the APDUs are exchanged via the Bluetooth channel. For the higher layers of the software of the hands-free kit, it is completely transparent that the SIM card is not embedded in the device but queried via a Bluetooth connection.

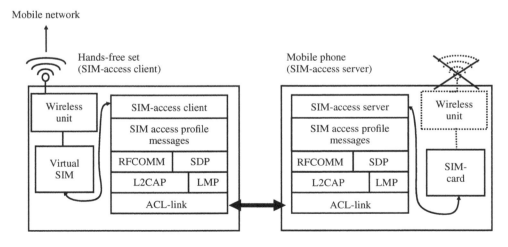

Figure 6.24   Structure of the SIM access profile

By using APDUs, it is possible not only to read and write files on the SIM card but also to invoke the GSM or UMTS security mechanism embedded in the SIM card. This is done by sending an authentication command to the SIM card, including a random number (RAND) as described in Section 1.6.4. Furthermore, the SIM application toolkit can be used over the Bluetooth connection as these messages are also embedded into APDUs as described in Section 1.10.

6.6.4   High-Quality Audio Streaming

Both the headset and the hands-free profile have been designed to carry telephony grade (mono) voice channels with a limited bandwidth. For high-quality audio streaming, a much higher quality is required. Therefore, the Advanced Audio Distribution Profile (A2DP) [19] has been designed to carry audio data streams with bandwidths ranging from 127 to 345 kbit/s depending on the audio stream type. As these datarates cannot be achieved by using SCO links, ACL links were selected to carry the audio stream. Some headsets support the A2DP profile as well as the standard headset profile and can be used for both audio streaming and voice telephony.

Figure 6.25 shows the protocol stack used by the A2DP profile. The profile is based on GAP, which allows remote devices to query the supported features of the profile in the SDP database. Above the L2CAP layer, the Audio Video Distribution Transfer Protocol (AVDTP) [20] is used to carry the audio data stream. As the protocol name implies, it can be used to carry both audio and video streams, and can thus be considered to be a generic transfer protocol for multimedia streams. The A2DP profile simply uses the protocol to transfer audio streams. Apart from the actual data stream, the protocol is also used to exchange control information between the two devices that are required for codec negotiation and to configure parameters like the bandwidth to be used for the stream. Higher layer control functionalities like switching to the next music track or pausing the transmission from a remote device are not part of AVDTP

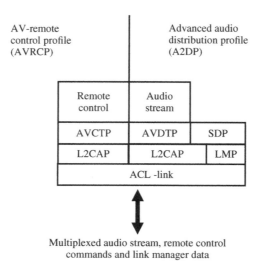

Figure 6.25   The protocol stack used for A2DP and remote control

and are handled by the Audio/Video Control Transport Protocol (AVCTP), which is described further below.

The standard allows devices to handle several Bluetooth applications simultaneously and to communicate with several remote devices at the same time. If this is supported by a device, it is, for example, possible to transfer a file between a notebook and a device while transmitting audio using the A2DP profile to another device. It should be noted, however, that the A2DP session requires a significant percentage of the overall capacity of the piconet, so that file transfer speed might be lower. If all devices support the Bluetooth version 2.0 + EDR standard, this is less of a problem as the total bandwidth of EDR piconets is about 2 Mbit/s. Remember that Bluetooth version 1.2 only supports 723 kbit/s for standard devices, of which about 345 kbit/s is used for the highest quality audio codec.

The A2DP profile specifies two roles for a connection: The audio source is typically an MP-3 player, a multimedia mobile phone or a microphone. The audio sink role is typically implemented in a headset or a Bluetooth-enabled loudspeaker set.

To ensure that A2DP-compliant devices share at least a single common codec for audio transmissions, the profile contains the description of a proprietary audio stream format, called sub-band codec (SBC), which is mandatory for implementation in all A2DP-compliant devices. A short description of this codec can be found below. Furthermore, the standard defines how audio streams encoded with MPEG 1−2 audio, MPEG-2,4 AAC and ATRAC shall be transported via the AVDTP. The implementation of these codecs is optional. The standard also offers the possibility to transport other codecs over AVDTP. To ensure interoperability, it is defined that a device supporting additional codecs must always be able to recode the audio stream into SBC if the remote device does not support the codec.

On a high level, the SBC codec works as follows: At the input, the SBC coder expects a PCM-coded audio signal at a certain sampling frequency. For high audio quality, the standard suggests using either 44.1 or 48 kHz. The codec then separates the frequency range of the input signal into several frequency slices, which are also referred to as sub-bands. The standard suggests splitting the signal into either four or eight sub-bands, each dealing with a certain frequency range of the input signal. Subsequently, a scaling factor is calculated for each sub-band, which gives an indication of the loudness of the signal in the sub-band. The scaling factors are then compared with each other to encode more important sub-bands with a higher number of bits. The recommendation for the number of bits to be used for this purpose ranges from 19 for middle-quality mono audio channels to up to 55 for high-quality joint stereo channels. The results of the different sub-bands are then compressed with a variable compression factor. Using the lowest compression factor to achieve the highest audio quality finally results in a bit stream of about 345 kbit/s.

To transfer user commands from the audio sink device (e.g., headset) like volume control, next/previous track, pause, and so on back to the audio source device (e.g., MP-3 player), the Audio/Video Remote Control Profile (AVRCP) [21] is used. The profile uses the AVCTP [22] as shown in Figure 6.26 to send the commands from controller devices and to receive responses from target devices. To achieve interoperability between controller and target devices, the remote control profile specifies the following target device categories:

- Category 1: Player/recorder.
- Category 2: Monitor/amplifier.
- Category 3: Tuner.
- Category 4: Menu.

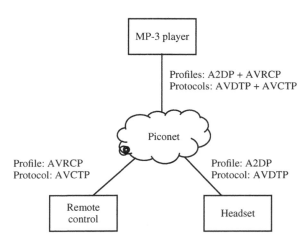

Figure 6.26 Simultaneous audio streaming and control connections to different devices

Depending on the device category, the standard then defines a number of control commands (operation IDs) and indicates for each device category if the support is mandatory or optional. Here are some examples of standardized control commands: 'select', 'up', 'right', 'root menu', 'setup menu', 'channel up', 'channel down', 'volume up', 'volume down', 'play', 'stop', 'pause', 'eject', 'forward' and 'backward'. Vendor-specific control commands can be added to the list of commands which, however, reduce the interoperability between devices and should therefore only be added with care.

It has to be noted that there is no interaction between the audio streaming session that uses the A2DP profile and a control session that uses the remote control profile. Thus, it is possible to form a piconet where an MP3 player streams audio to a headset while it receives commands (e.g., volume control commands) from a third device such as a remote control.

6.6.5   The Human Interface Device (HID) Profile

An application that has become more popular in recent years is to connect input devices such as keyboards and mice to devices such as notebooks and tablets. Although most wireless mice use a proprietary radio protocol and USB receiver, wireless connectivity of keyboards used in combination with tablets is based on Bluetooth technology as no proprietary receiver can be connected to such a device. The profile used for this application is the HID profile.

The HID profile establishes two L2CAP connections. The first connection is used for a control channel on which data is transferred synchronously, that is, in a request and response manner. The second L2CAP connection is required for the HID interrupt channel that is used for carrying asynchronous information, for example, notifications when the user has pressed or released a key. As the HID is a generic profile, information stored in the SDP database informs the host device what kind of input or output messages the device supports. Input messages can,

```
Frame 176: 19 bytes on wire (152 bits), 19 bytes captured (152 bits)
    Encapsulation type: Bluetooth H4 with linux header (99)
    [...]
    [Protocols in frame: hci_h4:bthci_acl:btl2cap:bthid]
    Point-to-Point Direction: Received (1)
Bluetooth HCI H4
    [Direction: Rcvd (0x01)]
    HCI Packet Type: ACL Data (0x02)
Bluetooth HCI ACL Packet
    .... 0000 0010 0011 = Connection Handle: 0x0023
    ..10 .... .... .... = PB Flag: First Automatically Flushable Packet
(2)
    00.. .... .... .... = BC Flag: Point-To-Point (0)
    Data Total Length: 14
Bluetooth L2CAP Protocol
    Length: 10
    CID: Dynamically Allocated Channel (0x0041)
    [PSM: HID-Interrupt (0x0013)]
Bluetooth HID Profile
    1010 .... = Transaction Type: DATA (0x0a)
    .... 00.. = Parameter reserved: 0x00
    .... ..01 = Report Type: Input (0x01)
    Protocol Code: Keyboard (0x01)
    0... .... = Modifier: RIGHT GUI: False
    .0.. .... = Modifier: RIGHT ALT: False
    ..0. .... = Modifier: RIGHT SHIFT: False
    [...]
    Reserved: 0x00
    Keycode 1: a (0x04)
    Keycode 2: <ACTION KEY UP> (0x00)
    [...]
0000   02 23 20 0e 00 0a 00 41 00 a1 01 00 00 04 00 00
0010   00 00 00
```

Figure 6.27   HID input message sent from a keyboard

for example, be keyboard notifications or mouse movements. Output messages can be sent by the host device, for example, to a force feedback joystick.

As HID devices are usually battery driven, power consumption has to be as low as possible. On the Bluetooth side, host and HID device therefore enter the Bluetooth Sniff mode after the establishment of the L2CAP control and interrupt channels. A typical sniff rate observed in practice is 40 milliseconds. Sniff subrating can be used to further reduce power consumption between keyboard activity input messages or between mouse movement notifications.

Figure 6.27 shows an abbreviated HID input message that was sent from a keyboard to a notebook. As can be seen in the figure, the message size is only 19 bytes and thus very small despite the ACL, L2CAP and HID protocols stacked on each other. Furthermore, the message shows that PSM 13 was used to establish the HID interrupt channel. The payload of the message is only a single byte (0x04h), which represents the lowercase 'a' character that the user has pressed on the keyboard.

Questions

1. What are the maximum speeds that can be achieved by Bluetooth and on what do they depend?
2. What is FHSS and which enhanced functionalities are available with Bluetooth 1.2 in this regard?
3. What is the difference between inquiry and paging?
4. What kinds of power-saving mechanisms exist for Bluetooth devices?
5. What are the tasks of the link manager?
6. How can several data streams for different applications be transferred simultaneously by the L2CAP protocol?
7. What are the tasks of the service discovery database?
8. How can several services use the RFCOMM layer simultaneously?
9. What is the difference between authentication and authorization?
10. Why are such a high number of different Bluetooth profiles required?
11. Which profiles can be used to quickly transfer files and objects between two Bluetooth devices?
12. What are the differences between the hands-free profile and the SIM access profile?
13. What are the advantages of integrating both wireless LAN and Bluetooth in a single device?

Answers to these questions can be found on the companion website for this book at http://www.wirelessmoves.com.

References

[1] Bluetooth Qualification Program, https://www.bluetooth.org/en-us/test-qualification/qualification-overview.
[2] Bluetooth Special Interest Group, Bluetooth Specification Version 2.0 + EDR [vol. 0], http://www.bluetooth.org, November 2004.
[3] J.-H. Jo and N. Jyand, Performance Evaluation of Multiple IEEE 802.11b WLAN Stations in the Presence of Bluetooth Radio Interference, *IEEE International Conference on Communications*, Anchorange, USA, vol. 26, pp. 1163–1168, May 2003.
[4] A. Laurie and B. Laurie, Serious Flaws in Bluetooth Security Lead to Disclosure of Personal Data, https://events.ccc.de/congress/2004/fahrplan/event/66.en.html, 2003.
[5] Bluetooth Special Interest Group, Bluetooth Specification Version 1.1 Part K:2 – Service Discovery Application Profile, http://www.bluetooth.org, February 2001.
[6] Bluetooth Special Interest Group, Bluetooth Specification Version 1.1 Part K:5 – Serial Port Profile, http://www.bluetooth.org, February 2001.
[7] Bluetooth Special Interest Group, Bluetooth Specification Version 1.1 Part K:12 – File Transfer Profile, http://www.bluetooth.org, February 2001.
[8] Bluetooth Special Interest Group, Bluetooth Specification Version 1.1 Part K:11 – Object Push Profile, http://www.bluetooth.org, February 2001.
[9] T. Howes, M. Smith, and F. Dawson, MIME Content Type for Directory Information, RFC 2425, September 1998.
[10] F. Dawson, T. Howes, and M. Smith, vCard MIME Directory Profile, RFC 2426, September 1998.
[11] F. Dawson and D. Stenerson, Internet Calendaring and Scheduling Core Object Specification (iCalendar), RFC 2445, September 1998.
[12] S. Silverberg, S. Mansour, F. Dawson, and R. Hopson, iCalendar Transport Independent Interoperability Protocol (iTIP) Scheduling Events, BusyTime, To-dos and Journal Entries, RFC 2446, September 1998.
[13] F. Dawson, S. Silverberg, and S. Mansour, iCalendar Message-based Interoperability Protocol (iMIP), RFC 2447, November 1998.

[14] Bluetooth Special Interest Group, Bluetooth Specification Version 1.1 Part K:13 – Synchronization Profile, http://www.bluetooth.org, February 2001.

[15] Infrared Data Association, Specifications for Ir Mobile Communications (IrMC) V1.1, http://www.irda.org, March 1999.

[16] Bluetooth Special Interest Group, Bluetooth Specification Version 1.1 Part K:6 – Headset Profile, http://www.bluetooth.org, February 2001.

[17] Bluetooth Special Interest Group, Hands-Free Profile, Version 1.0, http://www.bluetooth.org, April 2003.

[18] Bluetooth Special Interest Group, SIM Access Profile Interoperability Specification, Revision V10r00, http://www.bluetooth.org, May 2005.

[19] Bluetooth Special Interest Group, Advanced Audio Distribution Profile Specification, Version 1.0, http://www.bluetooth.org, May 2003.

[20] Bluetooth Special Interest Group, Audio/Video Distribution Transfer Protocol Specification, Version 1.0, http://www.bluetooth.org, May 2003.

[21] Bluetooth Special Interest Group, Audio/Video Remote Control Profile, Version 1.0, http://www.bluetooth.org, May 2003.

[22] Bluetooth Special Interest Group, Audio/Video Control Transport Protocol Specification, Version 1.0, http://www.bluetooth.org, May 2003.

Index

From GSM to LTE-Advanced: An Introduction to Mobile Networks and Mobile Broadband Revised,
Second Edition. Martin Sauter.
© 2014 John Wiley & Sons, Ltd. Published 2014 by John Wiley & Sons, Ltd.

Lightning Source UK Ltd.
Milton Keynes UK
UKOW07n1533141015

260535UK00002B/2/P